Basic Concepts of Mathematics
for
Elementary Teachers

Basic Concepts of Mathematics for Elementary Teachers

ROY DUBISCH

Professor of Mathematics
University of Washington

ADDISON-WESLEY PUBLISHING COMPANY

Reading, Massachusetts

Menlo Park, California · London · Amsterdam · Don Mills, Ontario · Sydney

This book is in the
ADDISON-WESLEY SERIES IN MATHEMATICS

Copyright © 1977 by Addison-Wesley Publishing Company, Inc. Philippines Copyright 1977 by Addison-Wesley Publishing Company, Inc.

All rights reserved. No part of this publication may be reproduced, stored in a retrieval system, or transmitted, in any form or by any means, electronic, mechanical, photocopying, recording, or otherwise, without the prior written permission of the publisher. Printed in the United States of America. Published simultaneously in Canada. Library of Congress Catalog Card No. 76–1742.

ISBN 0-201-01167-0
CDEFGHIJKL-MA-89876543210

Preface

The mathematics presented here is intended to provide you with the background for successful teaching of mathematics in the primary school (grades K–6). As you can see from the numerous illustrations from primary-school mathematics texts, the material presented is definitely closely related to the topics studied by children.

Although the emphasis is on mathematical concepts underlying the primary-school mathematics program, some consideration is also given to methods of presenting mathematics to children, so as to make clear how the presentation here relates to the primary-school classroom. Detailed discussion, however, is left to a class on methods.

Contemporary school mathematics programs place considerable emphasis on the "discovery" approach, where children are encouraged to discover mathematical relationships and to develop concepts for themselves. Since you are already familiar with much of the mathematics that children study, such a discovery approach to elementary mathematics cannot be followed here in the way it is done in primary school. (For example, you already know that $2 + 3 = 3 + 2$, $5 + 8 = 8 + 5$, etc., and so can't discover this for yourself at this time as a young child can!) However, various discovery-type exercises at a higher level are included here to illustrate this approach to mathematics.

Another point of emphasis in contemporary school mathematics programs is the use of games and puzzles. Here, too, most of the simple games and puzzles of interest to young children are hardly of interest to adults. Nevertheless, some of these are presented here along with some more sophisticated games and puzzles to challenge you and to help you see the possibilities in this approach to mathematics.

Still another point of emphasis is on the use of manipulative aids such as stick bundles, abaci, geoboards, and blocks of various kinds. Some of these are

vi   Preface

illustrated here, and if you have access to a math lab, either in this course or in a methods course, you may explore these aids in detail.

Certainly one of the central aims of this text, in addition to providing you with the necessary technical background for successful teaching of primary-school mathematics, is to convince you that primary-school mathematics can be a creative, "fun" subject and not necessarily a dull drill-type subject. Your own attitude toward mathematics is bound to affect your pupils' attitude. And children will inevitably learn any subject better if they and their teacher enjoy it.

Mathematics at any level is mastered by doing and not just by observing what someone else has done. Thus the exercises form an integral part of the text and should be done with attention to meaning and not just to get an answer. In the early chapters of the text, in particular, some of the exercises dealing with definitions and rules are in programed format with answers given directly below the exercises. It is, of course, very important that you cover up the answers as you consider these exercises and that, if you have an incorrect response, you seek to find out *why* your response is incorrect before going on to the next exercise. Generally, the exercises in programed format deal with definitions, basic properties, and problems that can be done mentally; the other sets of exercises involve problems where written work is needed.

Several articles from journals are included to extend remarks made in the text and to stimulate your interest in reading such articles now and in the future. Additional articles are listed in the References for Further Reading, given at the end of each chapter.

It is my hope that you will not only learn a good deal of mathematics by reading this book but that you will also learn to enjoy mathematics (if you do not already!) and then use this knowledge to teach mathematics to children so they in turn will both understand and enjoy it.

Seattle, Washington   R. D.
October 1976

Contents

Chapter 1 Some Basic Concepts

1.1	Introduction	1
1.2	Sets and set language	1
1.3	One-to-one correspondence of sets	5
1.4	Infinite sets	9
1.5	Concept of a whole number	11
1.6	Geometry	14
	Chapter Test	18

Chapter 2 Basic Ideas of Addition and Subtraction of Whole Numbers

2.1	Union of sets	21
2.2	Some remarks on the language of logic	24
2.3	Definition of the addition of whole numbers	24
2.4	Intersection of sets	29
2.5	Subtraction and set complementation	31
2.6	Subtraction as related to addition	33
2.7	Number lines	34
2.8	The use of frames	36
2.9	Number activities	39
2.10	Discovery	42
	Chapter Test	44
	Reading: The addition table: Experiences in practice–discovery	48

Chapter 3 Naming Whole Numbers

3.1	Early beginnings	55
3.2	Concept of a base	55
3.3	Base five number language	56

viii   **Contents**

3.4 Numerals in base ten   58
3.5 Picturing base ten numerals for children   58
3.6 Numerals in base five   61
3.7 Changing from one base to another   62
  Chapter Test   67
  Reading: Early Mayan mathematics   69

Chapter 4   Addition and Subtraction Algorithms for Whole Numbers

4.1 Introduction   73
4.2 Addition with stick bundles and abacus   76
4.3 The addition algorithm in symbols   78
4.4 Subtraction with stick bundles and abacus   80
4.5 Subtraction algorithms in symbols   82
4.6 Addition and subtraction in other bases   86
  Chapter Test   92
  Reading: A new algorithm for subtraction?   98

Chapter 5   Basic Concepts of Multiplication and Division of Whole Numbers

5.1 Primary-school beginnings   103
5.2 Multiplication in terms of Cartesian products of sets   104
5.3 Division as partition and as successive subtraction   108
5.4 Division as the inverse of multiplication   110
5.5 Properties of multiplication and division   112
5.6 Number activities involving multiplication and division   117
  Chapter Test   119
  Reading: Zero, the troublemaker   122

Chapter 6   Multiplication and Division Algorithms for Whole Numbers

6.1 Introduction   127
6.2 Multiplication algorithms   128
6.3 Division algorithms   132
6.4 Multiplication and division in other bases   134
6.5 The G.C.D. and the L.C.M.   136
  Chapter Test   141
  Reading: A simplified presentation for finding
    the L.C.M. and the G.C.F.   144

Summary Test for Chapters 1 through 6   146

Chapter 7   The Integers

7.1 Introduction   149
7.2 Addition of integers   152
7.3 Subtraction of integers   155
7.4 Multiplication of integers   159
7.5 Division of integers   161

Contents ix

	7.6 A formal approach to inequalities for integers	162
	7.7 Absolute value	166
	Chapter Test	167
	Reading: Grisly grids	170

Chapter 8 More on Geometry

	8.1 Introduction	175
	8.2 Definitions in geometry	176
	8.3 Congruence and similarity	178
	8.4 Tessellations	182
	8.5 The geoboard	185
	8.6 Geometry of space	189
	8.7 Discovery in geometry	191
	Chapter Test	201
	Reading: Geometric activities for early childhood education	206

Chapter 9 Rational and Irrational Numbers

	9.1 What are rational and irrational numbers?	215
	9.2 Equivalent fractions	218
	9.3 Comments on our definitions	220
	9.4 Terminology	221
	9.5 Decimal fractions	223
	9.6 Fractions in bases other than ten	231
	9.7 Negative rational numbers	233
	9.8 Inequalities and absolute value	234
	9.9 A formal approach to rational numbers	238
	9.10 Percent	239
	Chapter Test	241
	Reading: The equation method of teaching percentage	245

Chapter 10 Computations with Rational Numbers

	10.1 Introduction	24'
	10.2 Addition of rational numbers	24'
	10.3 Subtraction of rational numbers	25:
	10.4 Multiplication of rational numbers	25:
	10.5 Division of rational numbers	26(
	10.6 Computations with rational numbers written in bases other than ten	26(
	10.7 Ratio and proportion	26'
	10.8 Properties of inequalities	27(
	10.9 The field properties	27(
	Chapter Test	27:
	Reading: Addition of unlike fractions	27∠

x   Contents

Chapter 11   **Measurement**

11.1 Introduction   279
11.2 Length   281
11.3 Area   287
11.4 Volume   294
11.5 Angles   296
11.6 Other measurements   297
    Chapter Test   299
    Reading: Grids, tiles, and area   302

Chapter 12   **Recapitulation**

Sets   310
Number and number systems   310
Structure and properties   310
Numeration   311
Sentences   311
Operations   311
Geometry   311
Reasoning and proof   311
Problem solving   312
Measurement   312

Summary Test for Chapters 7 through 12   313

Chapter 13   **Binary Operations**

13.1 Examples and definitions   317
13.2 Properties of binary operations   319
13.3 Clock arithmetic   320
    Chapter Test   325
    Reading: An application of modular number systems   327

Chapter 14   **Functions**

14.1 The "Guess My Rule" game   329
14.2 Function "machines"   330
14.3 Definition of a function   333
14.4 Functions as mappings   336
14.5 Functions as sets of ordered pairs   338
14.6 Graphs of functions   340
    Chapter Test   346
    Reading: Let's consider the function!   349

Chapter 15   **Transformation Geometry**

15.1 Moving figures in a plane   355
15.2 Translations   356
15.3 Rotations   358
15.4 Reflections   362

Contents   xi

15.5	Combining movements	366
15.6	Symmetries of figures in a plane	368
	Chapter Test	370
	Reading: Informal geometry through symmetry	375

Chapter 16 **Probability**

16.1	Introduction	381
16.2	Experimental probability	381
16.3	The sample space of an experiment	383
16.4	Events	385
16.5	Definition of probability	387
16.6	Some properties of $P(E)$	390
16.7	A counting principle	392
	Chapter Test	395

Chapter 17 **Statistics**

17.1	Introduction	399
17.2	Organizing statistical data	399
17.3	Graphical representation of statistical data	402
17.4	Measures of central tendency	408
	Chapter Test	412
	Reading: Graphs in the primary grades	416

Chapter 18 **Number Theory**

18.1	Introduction	421
18.2	Tests for divisibility	422
18.3	Casting out nines	424
18.4	Some unsolved problems in number theory	426
18.5	Fibonacci numbers	431
	Reading: Modular arithmetic	438

List of Symbols	443
Answers to Summary Tests and Selected Exercises	445
Index	457

Chapter 1

Some
Basic Concepts

1.1 INTRODUCTION

All children arrive in school with *some* mathematical background. Many can count up to twenty or more, and they all have some notion of size and of concepts such as flatness and roundness. Many of their ideas, however, lack precision or are simply verbalizations without real understanding. For example, as the Swiss psychologist Jean Piaget has found in his research, many young children (from 4 to 6) are really quite vague about such a basic concept as equinumerous. That this is the case is shown by an experiment Piaget and many others have performed in which the experimenter provides himself and the child with an equal number of blocks and the child verifies that he and the experimenter do indeed have an equal number. When, however, either the child's or the experimenter's blocks are spaced farther apart, it is quite possible that the child may think that the number of blocks increases or decreases according to the arrangement.

Hence, to avoid empty verbalization and to establish mathematical ideas on a firm foundation, it is vitally necessary that the work in mathematics in the early grades be based on sound mathematical and psychological principles. Here our main concern will naturally be with the mathematics component but the psychological component will not be entirely neglected.

1.2 SETS AND SET LANGUAGE

Practically all current primary school mathematics texts begin with some discussion of the fundamental concept of a set. No attempt is made in the primary school, nor will any be made here, to provide a formal definition of a set. Examples serve to make the concept clear to children and, at a more advanced level, "set" is simply taken as a basic undefined term whose intuitive meaning is evident, as it is for the child.

2 Some Basic Concepts

There is no doubt that some primary-school texts and teachers have overemphasized the *formal* aspects of set language. Such overemphasis can be easily avoided, however, by basing beginning work with sets almost entirely on sets of *objects*. Thus sets of sticks, of stones, of children, etc., should be used in the early grades, rather than pictures of or symbols for sets. Unfortunately, however, books cannot include in them actual objects such as sticks and stones! Thus we find, in first-grade books, pictures of sets such as those shown in Fig. 1.1. Note that in the first group of pictures the membership of a set is indicated by circling the members, whereas in the second group of pictures the more standard notation involving "braces" is used. In most primary-school mathematics series, however, the use of braces is deferred until the upper grades.

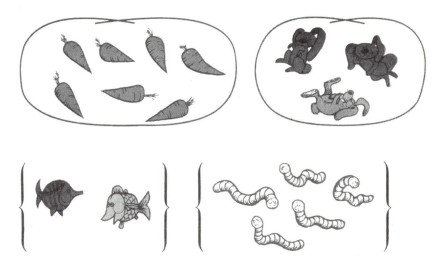

Figure 1.1

As the child develops maturity, he can be expected to make a greater use of abstraction and symbolism and thus to consider sets such as $\{1, 2, 3\}$ and $\{a, b, c\}$. I repeat, however, that early work with sets should emphasize *objects* rather than symbols.

For our purposes we will need, in this chapter, only the bare bones of set concepts and notation. Other aspects of sets will be considered when they are needed. These bare bones (which may be familiar to you from your previous work in mathematics) can be summarized as follows:

1. The **concept** of a set, as illustrated by such examples as a set of dishes, a herd of cows, a flock of sheep, etc., including the concept of the **empty** or **null** set with no members as, for example, the set of unicorns.

2. The use of **braces** (sometimes called "curly brackets") to indicate a set, as $\{a, b, c\}$, the set whose **members** or **elements** are a, b, c, and $\{\ \}$ for the empty set.

1.2   Sets and Set Language   3

3. We write, for example, $a \in \{a, b, c\}$ to indicate that a is a member of $\{a, b, c\}$, and $d \notin \{a, b, c\}$ to indicate that d is not a member of $\{a, b, c\}$. Note that since $\{\ \}$ has no members, it is always true that, for any a, $a \notin \{\ \}$. We read "$x \in A$" as "x is a member of A" or as "x is an element of A." (Normally we will indicate sets by capital letters and elements of sets by lower-case letters.) Similarly, "$x \notin A$" is read as "x is not a member of A" or as "x is not an element of A."

4. If we have two sets A and B such that for every $x \in A$ we also have $x \in B$ and for every $x \in B$ we also have $x \in A$, then we say that $A = B$. That is, A and B are **equal** if and only if every element of A is also an element of B and every element of B is also an element of A. For example, the set A of solutions to the equation $x^2 - 4 = 0$ is equal to the set $B = \{2, -2\}$.

Note that the order in which we write the elements of a set is not significant. Thus, for example,

$$\{a, b, c\} = \{c, b, a\} = \{b, c, a\}, \qquad \text{etc.}$$

What about a set such as $\{a, a, b\}$, where we have two "a's"? The general agreement is to consider the double listing as, so to speak, a sort of mistake, and thus to consider $\{a, a, b\} = \{a, b\}$. Similarly, $\{a, a, b, b, b, c\} = \{a, b, c\}$. (If we are really thinking of two *different* "a's"—say two persons both named Alfred—we should write, for example, $\{a_1, a_2, b\}$, which is not equal to either $\{a_1, b\}$ or $\{a_2, b\}$. Notice, by the way, that this somewhat abstract point simply never arises if you use objects to form sets: two popsicle sticks held in the hand are clearly two distinct sticks, no matter how alike they appear.)

5. If $x \in A$ implies that $x \in B$, we say that A is a **subset** of B. Thus, for example, $A = \{a, b\}$ is a subset of $B = \{a, b, c\}$ because the elements of A (a and b) are also elements of B. Note that (contrary to one's intuitive feeling for the prefix "sub"), every set is a subset of itself. Thus $\{a, b, c\}$ is a subset of $\{a, b, c\}$ simply because every element of $\{a, b, c\}$ is certainly an element of $\{a, b, c\}$.

6. We write $A \subseteq B$ to indicate that A is a subset of B and read "$A \subseteq B$" as "A is a subset of B." Thus $\{a, b\} \subseteq \{a, b, c\}$ and also $\{a, b, c\} \subseteq \{a, b, c\}$. (Note that we also have $\{a, b, c\} = \{a, b, c\}$.) As usual we use the "/" to indicate "not." Thus, for example, we write $\{a, d\} \nsubseteq \{a, b, c\}$ to indicate that $\{a, d\}$ is not a subset of $\{a, b, c\}$, and read "$\{a, d\} \nsubseteq \{a, b, c\}$" as "the set whose members are a and d is not a subset of the set whose members are a, b, and c." (Compare "$\subseteq$" with "$\leq$" as used for numbers where $a \leq b$ means that a is less than *or* equal to b.)

7. We also make the agreement that the null set will be considered as a subset of any set. That is, $\{\ \} \subseteq A$ for any set A. Another symbol for the null set is $\emptyset$ and so we can write $\emptyset \subseteq A$ for any set A. Note that, with this agreement, $A \subseteq A$ for any set A including the case when $A = \emptyset$.

4   Some Basic Concepts

We can picture sets and relations between sets by means of **Venn** diagrams, as shown in Fig. 1.2 (which includes an excellent picture of the empty set). Here the elements of the sets are considered as the points inside the curves.

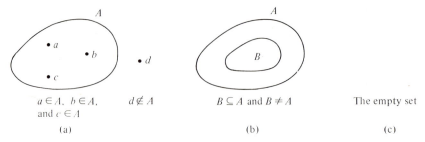

Figure 1.2

Programed Lesson 1.2

1. Use braces or ∅ (if appropriate) to indicate the following sets:
    a) The set of the first five letters of the English alphabet.
    b) The set of letters in the word "and."
    c) The set of letters in the word "Mississippi."
    d) The set of whole numbers between but not including 5 and 7.
    e) The set of whole numbers between but not including 5 and 6.
    f) The set of (real!) pink elephants.

    > a) {a, b, c, d, e}   b) {a, n, d}
    > c) {m, i, s, p}   (Note that {m, i, s, s, i, s, s, i, p, p, i} would be technically correct but, by agreement, this is equal to {m, i, s, p}.)
    > d) {6}   e) ∅   f) ∅

2. What difficulties would be encountered in writing down the set whose members are the great presidents of the United States?

    > The question would arise as to the meaning of "great." No two people are likely to agree on this. We require that a set be **well-defined**, i.e., described *unambiguously*, as would be the case if we said "presidents of the United States who have died in office."

3. Place a "∈" or "∉" as appropriate, to make true statements of the following:
    a) a___{b, c, d}
    b) a___{a}
    c) a___∅
    d) 0___∅
    e) □___{△,□,○}

    > a) ∉   b) ∈   c) ∉   d) ∉   (0 is not "nothing"!)   e) ∈

4. Place "⊆" or "⊈" as appropriate to make true statements of the following:
    a) {a, b}___{a, b, c, d}   b) ∅___{a}

1.3   **One-to-one Correspondence of Sets**   **5**

c) $\{a\}$___$\emptyset$ d) $\{\ \}$___$\{\ \}$

e) $\{a, b, c\}$___$\{b, a, c\}$ f) a___$\{a\}$

a) $\subseteq$ b) $\subseteq$ c) $\nsubseteq$ d) $\subseteq$ e) $\subseteq$ f) $\nsubseteq$

5. In which of the statements of Exercise 4 can "$\subseteq$" be replaced by "$=$" to make another true statement?

(d) and (e) only

6. If you have a quarter, a dime, and a nickel in your pocket, how many sets can you make with these coins?

8. They are $\{\ \}$, $\{q\}$, $\{d\}$, $\{n\}$, $\{q, d\}$, $\{q, n\}$, $\{d, n\}$, $\{q, d, n\}$.

7. How many different sets are there in the following list? Explain.

a) $\{1, 2, 3\}$ b) $\{3, 2, 1\}$

c) $\{1, 1 + 1, 1 + 1 + 1\}$ d) $\{1 \times 1, 1 \times 2, 1 \times 3\}$

e) $\{\frac{1}{1}, \frac{4}{2}, \frac{6}{2}\}$ f) $\{1 \div 1, 2 \div 1, 3 \div 1\}$

Only one; 1×1, $\frac{1}{1}$, $1 \div 1$ are all equal to 1; $1 + 1$, 1×2, $\frac{4}{2}$, and $2 \div 1$ are equal to 2; and $1 + 1 + 1$, 1×3, $\frac{6}{2}$, and $3 \div 1$ are all equal to 3. We have simply used different symbols for the numbers one, two, and three or, in the case of (a) and (b), listed the elements in a different order.

8. If A and B are sets and $A = B$, then whenever $x \in A$ it must also be true that x___ ___, and whenever $x \in B$ it must be true that x___ ___.

$\in$, B, $\in$, A

9. If A and B are sets and $A \neq B$, then it must be true that either there is some $x \in A$ such that x___ ___, or there is some $x \in B$ such that x___ ___.

$\notin$, B, $\notin$, A

10. If A and B are sets such that $A \subseteq B$, then if $x \in A$ it must be true that x___.

$\in$, B

1.3 ONE-TO-ONE CORRESPONDENCE OF SETS

Long before man learned how to count, and even today among some primitive peoples, large collections of things were kept track of by *matching*. Thus for example, by dropping a pebble into a pot for each animal leaving a corral in the morning, and then removing a pebble as each animal returned at night, a person could easily establish whether or not the same number, more, or fewer animals returned at night than left in the morning.

6 Some Basic Concepts

This idea of **matching,** or **one-to-one correspondence** (of animals and pebbles in our example) is absolutely basic to an understanding of arithmetic. It should be heavily stressed in early childhood mathematics education rather than emphasizing rote counting.

Thus, for example, a lesson early in the first grade might involve giving Mary and John each a set of pebbles and having them find out whether the two sets match. If they do, the children in the class can conclude that John has as many pebbles as Mary (and that Mary has as many pebbles as John). If, on the other hand, John runs out of pebbles in the matching process before Mary does, the children can conclude that Mary had more pebbles than John and that John had fewer pebbles than Mary. Note that in such activities even the use of the "number" is avoided and certainly there is no counting involved. What *is* happening is that the children are acquiring a solid basis for an understanding of the concept of a whole number.

Figure 1.3 shows an illustration from a first-grade text where the children are to draw lines between the objects pictured in the two sets (frogs and lily pads) to show that the sets can be matched.

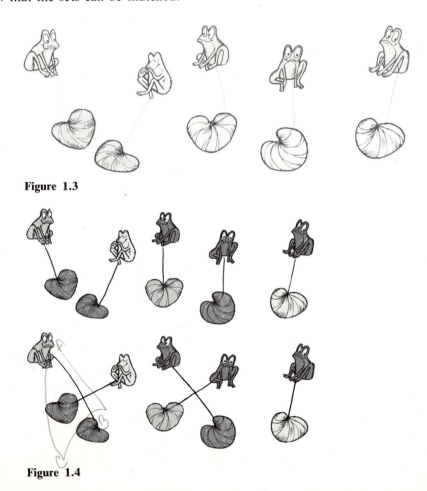

Figure 1.3

Figure 1.4

1.3 One-to-one Correspondence of Sets **7**

When we have two sets A and B such that we can pair each member of A with a single member of B, and each member of B with a single member of A, we say that A and B are **matching** sets (also called **equivalent** sets) and write $A \sim B$ (read "A matches B"). Any particular matching of the elements of the two sets is called a **one-to-one correspondence.**

Except in the case when A and B each have only one element (**singleton** sets), more than one way of matching is always possible. Figure 1.4 shows two ways in which a child might respond to the problem of matching frogs and lily pads. (The first way is probably the most natural one but the second way—as well as others—is entirely correct.)

The basic properties concerning the matching of sets are: For all sets A, B, and C,

1. $A \sim A$ (i.e., any set can be matched with itself). This is called the **reflexive** property of matching.

2. If $A \sim B$, then $B \sim A$ (i.e., if A can be matched with B, then B can be matched with A). This is called the **symmetric** property of matching.

3. If $A \sim B$ and $B \sim C$, then $A \sim C$ (i.e., if A can be matched with B and B can be matched with C, then A can be matched with C.) This is called the **transitive** property of matching.

The first property is certainly obvious if we make the additional agreement that $\emptyset \sim \emptyset$. Certainly, if $A \neq \emptyset$, every element of A can be matched with itself. And, for the second property, we simply note that if the elements of A can be matched with the elements of B, then certainly the elements of B can be matched with the elements of A.

We illustrate the reflexive and symmetric properties in Fig. 1.5.

$$A = \{a, b, c\} \qquad\qquad A = \{a, b, c\}$$
$$\updownarrow \updownarrow \updownarrow \qquad\qquad\quad \updownarrow \updownarrow \updownarrow$$
$$A = \{a, b, c\} \qquad\qquad B = \{x, y, z\} \qquad \textbf{Figure 1.5}$$

The transitive property is illustrated in Fig. 1.6, where the "curved" arrows indicate the desired one-to-one correspondence between A and C. That is, $a \leftrightarrow m$ and $m \leftrightarrow x$ so we let $a \leftrightarrow x$. Similarly, $b \leftrightarrow n$ and $n \leftrightarrow y$ so we let $b \leftrightarrow y$; $c \leftrightarrow o$ and $o \leftrightarrow z$ so we let $c \leftrightarrow z$.

$$A = \{a, \quad b, \quad c\}$$
$$B = \{m, \quad n, \quad o\}$$
$$C = \{x, \quad y, \quad z\} \qquad \textbf{Figure 1.6}$$

Practical applications of the transitive property are not difficult to find. Often there is a seating chart located at the entrance to the waiting area for a plane, with one tearout tag for each seat in the plane. Call the set of these tags B, let A be the set of prospective passengers, and C be the set of seats in the plane. Now if all the tags are picked up by the passengers we have A ~ B and, of course, B ~ C. Then we have A ~ C; that is, one passenger for each seat—again with no counting involved.

Note that if A = B, then certainly A ~ B. But, if A ~ B it may or may not be true that A = B. (See Fig. 1.5.)

Programed Lesson 1.3

1. In how many ways can a one-to-one correspondence be made between the sets S = {x, y} and T = {a, b}?

   Two.

   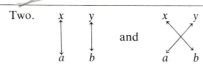

2. Show two one-to-one correspondences of S = {x, y, z} and T = {a, b, c} in which x ↔ a.

x ↔ a	x ↔ a
y ↔ b	y ↔ c
z ↔ c	z ↔ b

3. Which, if any, of the following diagrams show one-to-one correspondences between sets?

   (a) only

4. For all sets A and B, if A ~ B, then B ~ ____.

   A

5. For all sets A, A ~ ____.

   A

6. For all sets A, B, and C, if A ~ B and B ~ C, then A ~ ____.

   C

7. The properties described in the statements of Exercises 4, 5, 6 are called, respectively, the _____, _____, and _____ properties of matching.

   symmetric, reflexive, transitive

8. True or false?
   a) For all sets A and B, if $A = B$, then $A \sim B$.
   b) For all sets A and B, if $A \sim B$, then $A = B$.
   c) $\emptyset \sim \emptyset$

   $a)$ T $b)$ F $c)$ T

1.4 INFINITE SETS

The sets we have considered so far have been **finite** sets. But many of the sets considered in mathematics, such as various sets of numbers and the set of points in a plane, are examples of **infinite** sets. We write, for example, $\{0, 1, 2, 3, \ldots\}$ for the infinite set of **whole numbers,** where the three dots (ellipsis) inside the braces indicate that the listing is to be understood as continuing without end in the indicated pattern, in contrast to, for example, $\{0, 1, 2, 3, \ldots, 10\}$—the finite set of whole numbers from and including 0 to and including 10. Here the dots indicate only the omission of 4, 5, 6, 7, 8, and 9.

Now almost every adult has an intuitive idea of finite and infinite—(although the latter word is frequently misused as in the poet's "infinite stars in the heavens" to mean simply a very large number of stars!).

To make the notion of infinite sets more precise, we observe first that, if A is any *finite* set and B is any subset of A other than A itself, we certainly cannot have $B \sim A$. Whatever matching we attempt will leave at least one member of A unmatched with a member of B, as illustrated in Fig. 1.7. With infinite sets,

$A = \{a, b, c, d\}$
$\updownarrow \updownarrow \updownarrow \; \updownarrow$
$B = \{x, y, z\} \; ?$

Figure 1.7

however, the situation is quite different: it is perfectly possible to have a subset B of an infinite set A be in one-to-one correspondence with A and yet have $B \neq A$. We illustrate this in Fig. 1.8 for the set A of all whole numbers and the subset B ($\neq A$) of A consisting of all **even** numbers (multiples of 2). That is, to every $n \in A$ corresponds $2n \in B$ and to every $m \in B$ corresponds $\frac{m}{2} \in A$. Thus $B \sim A$ and $B \subseteq A$ but $B \neq A$.

$A = \{0, 1, 2, 3, 4, \ldots\}$
$\updownarrow \updownarrow \updownarrow \updownarrow \updownarrow$
$B = \{0, 2, 4, 6, 8, \ldots\}$

Figure 1.8

As another example of such a matching, consider the infinite set A of points from 0 to 3 (including 0 and 3) on a **number line,** as shown in Fig. 1.9. Each point of A can be matched with a point of the set B of points from 0 to 1 (including 0

10 Some Basic Concepts

and 1) on the number line. (In Fig. 1.9, B has been redrawn distinct from A to show the correspondence more clearly.) Here

$0 \in A \leftrightarrow 0 \in B, \quad \frac{3}{4} \in A \leftrightarrow \frac{1}{4} \in B, \quad 1 \leftrightarrow A \in \frac{1}{3} \in B,$
$2 \in A \leftrightarrow \frac{2}{3} \in B, \quad 3 \in A \leftrightarrow 1 \in B$

and, in general,

$a \in A \leftrightarrow \frac{a}{3} \in B$ and $b \in B \leftrightarrow 3b \in A$

Thus we again have $B \sim A$, $B \subseteq A$, and $B \neq A$, so that A is an infinite set.

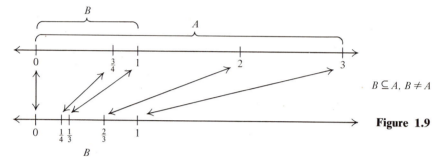

Figure 1.9

From these observations we are led to a formal definition of finite and infinite sets:

1. A set A is a **finite** set if no subset of A other than A is equivalent to A. (That is, if $B \subseteq A$ and $B \neq A$, then $B \nsim A$.)
2. A set A is an **infinite** set if there exists a subset B of A such that $B \subseteq A$, $B \neq A$, and $B \sim A$.

Note. The statement that $B \subseteq A$ and $B \neq A$ can also be expressed by saying that B is a **proper** subset of A; i.e., a proper subset of a set A (finite or infinite) is any subset of A other than A itself.

Programed Lesson 1.4

1. Show a one-to-one correspondence between the set $N = \{1, 2, 3, \ldots\}$ and the set $S = \{1, 4, 9, \ldots\}$ of the squares of the elements of N.

    $N = \{1, 2, 3, 4, \ldots\}$
    $\quad\quad\downarrow \downarrow \downarrow \downarrow \quad\quad n \in N \leftrightarrow n^2 \in S$
    $S = \{1, 4, 9, 16, \ldots\}$

2. (Refer to Exercise 1.) Since S is a _____ of N, $S \neq N$, and S__N, we conclude that N is an _____ set.

    subset, $\sim$, infinite

3. Let A be the set of points from 0 to 4 (including 0 and 4) on a number line and B be the set of points from 0 to 1 (including 0 and 1) on this number line. Then we can match $3 \in A$ with $\frac{3}{4} \in B$, $\frac{8}{5} \in A$ with _____ $\in B$, $\frac{2}{5} \in B$ with _____ $\in A$, and $\frac{3}{4} \in B$ with _____ $\in A$.

$\frac{2}{5}, \frac{8}{5}, 3$

4. In general, for A and B as in Exercise 3, we can match $a \in A$ with _____ $\in B$ and $b \in B$ with _____ $\in A$.

$\frac{a}{4}, 4b$

5. Since, for the sets A and B of Exercise 3, we have $B \sim A$, $B \subseteq A$, and B _____ A, we can conclude that A is an _____ set.

$\neq$, infinite

6. Which of the following sets are finite and which are infinite?
   a) The set of points on a line.
   b) The set of grains of sand on all the beaches of the world.
   c) The set of all pine needles in the forests of the world.
   d) The set of all whole numbers greater than 1,000,000,000,000.
   e) The set of subsets of an infinite set.

   (b) and (c) are finite sets; the others are infinite

1.5 CONCEPT OF A WHOLE NUMBER

Imagine a set of children, a set of cups, and a set of elephants, as portrayed in Fig. 1.10. What do these three sets have in common? Unless we already have the notion of number, it is difficult to imagine anything they have in common except that each one matches the others. With the notion of number at hand we are likely to say, of course, that the "commonness" lies in the fact that these sets each consist of two elements. But what do we mean by "two"? It is easy enough to show two *children* or two *cups* and we can, at least, show a picture of two *elephants*. But this is somewhat like showing a blue car, a blue flower, and a blue dress to communicate the idea of "blue". Such procedures *do* communicate—but only informally.

Figure 1.10

Now a more precise definition of blue is fairly easy to give (although probably less understandable than "definition by examples"!). Thus the second edition of Webster's Third International Dictionary begins its definition of "blue" with an example, "blue sky", and then says "the hue of a color normally evoked by radiant energy of wavelength 478.5 millimicrons."

For "two", the same dictionary says "one and one; twice one; the number next greater than one; the sum of one and one; two units of objects". All these attempts to define "two" simply avoid the issue—as in "two units of objects"—or shift the burden to a definition of "one"! (Indeed, a completely adequate formal definition of two was not made until 1923 by John von Neumann.)

Historical evidence that our ancestors struggled with the abstract idea of number is furnished by the Tsimchian Indians of Northwest Canada who once used seven different sets of number words. One was used only for counting flat objects or animals, another only for canoes, a third only for men, etc. Thus, for example, the two sets in Fig. 1.11 would be described numerically by the two different words shown with the sets.

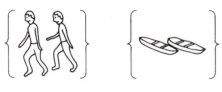

tepkadal   kalpeyeltik   **Figure 1.11**

Fortunately, whatever mathematical or psychological difficulties may underlie the concept of number, there is no great difficulty in persuading children that all the sets shown in Figs. 1.10 and 1.11 have associated with them the number *two*, and we say that each set has two members—and similarly for other matching sets. We need not attempt a formal definition of number but can rely on examples to develop the idea of the set of whole numbers, $\{0, 1, 2, 3, \ldots\}$, where 0 is the (**cardinal**) number of the empty set; 1 is the number of elements of any set that matches the set $\{a\}$; 2 is the number of elements of any set that matches the set $\{a, b\}$, etc. Note, in particular, that 0 does not mean "nothing". Thus, for example, 0 is the solution of the equation $x + 1 = 1$, whereas the equation $x + 1 = x$ has no solution; i.e., its **solution set** is $\emptyset$.

We will write, for any *finite* set A, $n(A)$ for the *number* of elements of A. Thus, for example,

$$n(\emptyset) = 0, \quad n(\{a\}) = 1, \quad \text{and} \quad n(\{a, b\}) = 2$$

In addition to the set $\{0, 1, 2, 3, \ldots\}$ of whole numbers, we will often have occasion to speak of the set $\{1, 2, 3, \ldots\}$ of **counting** or **natural** numbers.

Note that if A and B are any two matching sets, then $n(A) = n(B)$ and, conversely, if $n(A) = n(B)$, then A and B are matching sets. In symbols:

$$\text{If } A \sim B, \text{ then } n(A) = n(B), \quad \text{and} \quad \text{if } n(A) = n(B), \text{ then } A \sim B$$

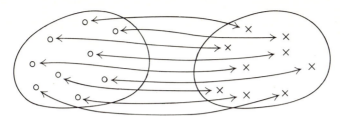

Figure 1.12

Thus, *without counting*, we know that the two sets shown in Fig. 1.12 have the same number of elements since, as the arrows show, they are matching sets.

Nonmatching sets can be used to introduce the concepts of less than and greater than for whole numbers, as illustrated by Fig. 1.13, taken from a first grade book.

Later on, in Chapter 3, we will consider, in some detail, the naming of, and symbols for, the whole numbers. You, of course, are thoroughly familiar with such names and symbols. Children, however, have a lot to learn at this point—including developing skill in writing the **numerals** (symbols for numbers) 0, 1, 2, 3, 4, 5, 6, 7, 8, and 9, and learning how to combine these numerals to make new numerals, such as 10, 11, 23, 435, etc.

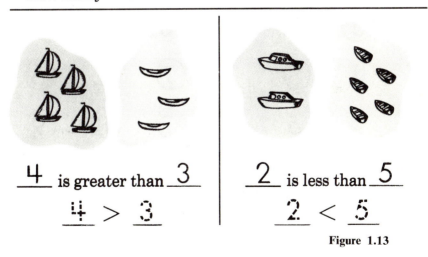

Figure 1.13

Programed Lesson 1.5

1. If $A = \{a, b, c, d\}$, then $n(A) = $ ___.

   4

2. $n(\emptyset) = $ ___.

   0

14   Some Basic Concepts

3. If A and B are any finite sets such that $A \sim B$, then $n(A)$ ____ $n(B)$.

   =

4. If A and B are any finite sets such that $n(A) = n(B)$, then A ____ B.

   $\sim$

5. The set $\{0, 1, 2, 3, \ldots\}$ is called the set of _____ numbers.

   whole

6. The set $\{1, 2, 3, \ldots\}$ is called the set of _____ numbers and also the set of _____ numbers.

   counting, natural (either order)

7. If A and B are any two finite sets such that $A \subseteq B$ and $A \neq B$, how do $n(A)$ and $n(B)$ compare?

   $n(A)$ is less than $n(B)$

1.6 GEOMETRY

At one time, the mathematics program of the early primary grades consisted almost entirely of arithmetic. Contemporary programs, however, include a good deal more, particularly in the area of geometry. Thus, sometimes as early as the first grade and certainly by the third grade, children are introduced to the concepts of line, line segment, congruent figures, simple closed curves, triangles, rectangles, and circles, without, of course, formal definitions being stressed.

In these introductions, of course, *activities* should be emphasized rather than memorization of definitions. Children should, for example, be given cutouts of geometric figures to match with congruent figures on the board, asked to draw their own examples of open and closed curves, etc.

Our purpose here is also not to emphasize formal definitions but simply to refresh your memory so that you will feel comfortable with some of the terminology and symbolism commonly used in both the early and later primary grades.

1. **Line segment.**   When we use a ruler and a pencil to join two points A and B we are drawing a line segment. Denoted by $\overline{AB}$.

2. **Ray.**   If we imagine we are looking along a line segment from A towards B and imagine a continuation of the line segment indefinitely far beyond B, we are visualizing the ray $\overrightarrow{AB}$, as suggested by Fig. 1.14.

Figure 1.14

1.6   **Geometry**   **15**

3. **Line.** If we imagine looking from A to B and beyond, and then from B to A and beyond, we are visualizing the line $\overleftrightarrow{AB}$, as suggested by Fig. 1.15.

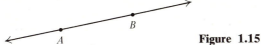

Figure 1.15

4. **Curve.** Figure 1.16 shows some curves which we can consider as any figure formed by marking a point A on a piece of paper, placing the tip of a pencil on it, and then tracing out any figure we like without lifting our pencil from the paper. Notice, by the way, that in the ordinary use of the word "curve" as distinguished from the mathematical use of the word here, we would probably not regard (b), (e), or (f) as curves.

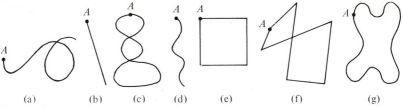

Figure 1.16

5. **Simple curve.** A curve which never crosses itself is called a simple curve. Thus (b), (d), (e), and (g) are simple curves.

6. **Closed** and **open curves.** A curve that goes back to its starting point is called a closed curve; one that does not go back to its starting point is called an open curve. Thus (c), (e), (f), and (g) are closed curves, and (a), (b), and (d) are open curves.

The other terms we have mentioned are ones which we will consider more formally in Chapter 8 but which are also covered briefly for review purposes in the following programed lesson. Notice, by the way, that all geometric figures can be considered as sets of points—emphasizing again the importance of the concept of a set.

Figure 1.17 shows some material from a third-grade book concerning the definitions given above.

Using the Ideas

1.

Figure 1.17

16 Some Basic Concepts 1.6

● *Can you name the simplest geometric figures?*

Discussing the Ideas

1. These figures suggest points
   Can you think of others?

2. These figures suggest line segments •————•
   Can you think of others?

3. A beam of light suggests a ray •————→. A ray has one endpoint and "goes on and on" in one direction. Can you think of other examples of rays?

4. The boy looking in opposite directions suggests a line ←————→. A line has no endpoints and "goes on and on" in both directions. What other situations suggest lines?

Figure 1.17 (continued)

Programed Lesson 1.6

1. Which of the following are:
   i) Simple curves?
   ii) Open curves?
   iii) Closed curves?
   iv) Simple open curves?
   v) Simple closed curves?

(a)   (b)   (c)   (d)

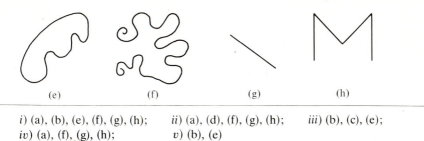

i) (a), (b), (e), (f), (g), (h);   ii) (a), (d), (f), (g), (h);   iii) (b), (c), (e);
iv) (a), (f), (g), (h);   v) (b), (e)

2. Congruent figures are figures that have the same size and shape. List the pairs of congruent figures shown below.

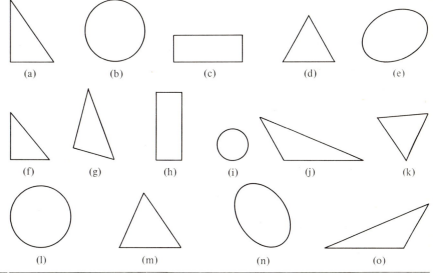

(a) and (g); (b) and (l); (c) and (h); (d) and (k); (e) and (n); (j) and (o)

3. Identify the following as triangle, rectangle, square, circle, or none of these.

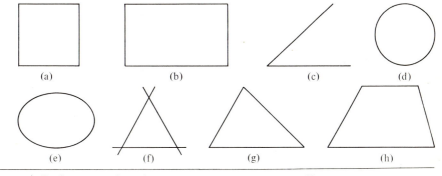

a) Both a rectangle and a square;   *b)* rectangle;
c) none of these;   *d)* circle;   *e)* none of these;
f) none of these;   *g)* triangle;   *h)* none of these.

18 Some Basic Concepts   **1.6**

4. True or false?
   a) $\overleftrightarrow{AB} = \overleftrightarrow{BA}$
   b) $\overrightarrow{AB} = \overleftarrow{BA}$
   c) $\overrightarrow{AB} = \overrightarrow{BA}$

---

a) T   *b*) T   *c*) F

Chapter Test

In preparation for this test you should review the meaning of the following terms and symbols, referring when necessary to the indicated pages:

Set (2)

Braces (2)

Equal sets (3)

Subset (3)

$\in$, $\notin$ (3)

$\subseteq$, $\not\subseteq$ (3)

{ }, Ø (3)

Venn diagram (4)

One-to-one correspondence (6)

Matching sets (6)

$\sim$ (7)

Reflexive, symmetric, and transitive properties of matching (7)

Finite set (9)

Infinite set (9)

Counting number (12)

Natural number (12)

Numeral (13)

Line segment (14)

Ray (14)

Line (15)

$\overline{AB}$ (14)

$\overrightarrow{AB}$ (14)

$\overleftrightarrow{AB}$ (15)

Curve (15)

Simple curve (15)

Closed curve (15)

Open curve (15)

1. True or false?
   a) For all sets A, $A \subseteq A$.
   b) For all sets A, $\emptyset \subseteq A$.
   c) For all sets A and B, if $A \sim B$, then $A = B$.
   d) For all finite sets A and B, if $n(A) = n(B)$, then $A = B$.
   e) For all sets A and B, if $A = B$, then $A \sim B$.
   f) For all sets A, $A \sim A$.
   g) A circle is a simple closed curve.

2. Use $\in$ or $\subseteq$ as appropriate to make a true statement of the following.
   a) x_____$\{x, y, z\}$
   b) $\{x\}$_____$\{x, y, z\}$
   c) $\{x, y, z\}$_____$\{z, y, x\}$
   d) x_____$\{x\}$

3. In how many ways can one establish a one-to-one correspondence between the sets $\{a, b, c, d\}$ and $\{w, x, y, z\}$ if $a \leftrightarrow z$ and $d \leftrightarrow w$?

References for Further Reading 19

4. If there exists a subset B of a set A such that $B \sim A$ and $B \neq A$, then A is a
_____ set.

5. If A and B are sets such that whenever $x \in A$, then $x \in B$, and whenever $x \in B$,
then $x \in A$, we say that A____B.

6. If there is a one-to-one correspondence between the elements of two sets A and B we
say that A and B are_____sets and write A____B.

7. The matching of a set of tickets with the set of seats in a theater results, for a soldout
performance, in the matching of the audience with the set of seats. This illustrates the
_____property of matching of sets.

8. If A and B are any finite matching sets, then $n(A)$____$n(B)$.

9. $\overrightarrow{AB}$, where A and B are points, is a symbol for a_____.

10. The set of natural numbers is the set $\{$_____$\}$.

TEST ANSWERS

1. *a)* T, *b)* T, *c)* F, *d)* F, *e)* T, *f)* T, *g)* T

2. *a)* $\in$, *b)* $\subseteq$, *c)* $\subseteq$, *d)* $\in$

3. Two $\{a, b, c, d\}$ $\{a, b, c, d\}$

        $\{w, x, y, z\}$ $\{w, x, y, z\}$

4. Infinite   *5.* $=$ *6.* Matching, $\sim$

7. Transitive   *8.* $=$ *9.* Ray

10. $\{1, 2, 3, \ldots\}$

REFERENCES FOR FURTHER READING*

Some articles on Piaget's theories and their relation to the teaching of arithmetic are:

1. Adler, I., "Mental growth and the art of teaching," *AT,* **13** (1966), pp. 576–584.
2. Coxford, A. F., "Piaget: Number and measurement," *AT,* **10** (1963), pp. 419–427.
3. Duckworth, E., "Piaget rediscovered," *AT,* **11** (1964), pp. 496–499.
4. Inskeep, J. E., Jr., "Building a case for the application of Piaget's theory and research in the classroom," *AT,* **19** (1972), pp. 255–260.
5. Sawada, D., "Piaget and pedagogy: Fundamental relationships," *AT,* **19** (1972), pp. 293–298.
6. Weaver, J. F., "Some concerns about the application of Piaget's theory and research to mathematical learning and instruction," *AT,* **19** (1972), pp. 263–269.

The use and misuse of sets in the primary school are discussed in the following article:

7. Vaughan, H. E., "What sets are not," *AT,* **17** (1970) pp. 55–60.

---

* Here, and elsewhere, *AT* means *The Arithmetic Teacher.*

20   Some Basic Concepts

Some problems regarding the teaching of the concept of set equality are discussed in the next article.

8. Dubisch, R., "Set equality," *AT*, **13** (1966), pp. 388–391.

The use of Venn diagrams in the primary school is discussed in the following article:

9. Smith, L. B., "Venn diagrams strengthen children's mathematical understanding," *AT*, **13** (1966), pp. 92–99.

Further details regarding the Tsimchian number language may be found on page 97 of the following book:

10. Allendoerfer, C. B., *Principles of Arithmetic and Geometry for Elementary School Teachers*, The Macmillan Co., N.Y. (1971).

Suggestions in regard to the teaching of geometry in the primary school are given in the following references:

11. *Geometry* (*K to 13*). Ontario Institute of Education, Toronto, Canada (1969).

12. Black, J. M., "Geometry alive in primary classrooms," *AT*, **14** (1967), pp. 90–102.

13. Brune, I., "Geometry in the grades," Chapter 9 of *Enrichment Mathematics for the Grades*, National Council of Teachers of Mathematics, Reston, Va. (1963).

14. Henderson, G. L., and C. P. Collier, "Geometry activities for later childhood education," *AT*, **20** (1973), pp. 444–453.

15. Immerzeel, G., "Geometry activities for early childhood education," *AT*, **20** (1973), pp. 438–443.

16. Kelley, S. J., *Learning Mathematics Through Activities*. James E. Freel and Associates, Inc., Cupertino, Calif. (1973).

17. Robinson, G. E., "The role of geometry in elementary school mathematics," *AT*, **13** (1966), pp. 3–10.

18. Rutland, L., and M. Hosier, "Some basic geometric ideas for the elementary teacher," *AT*, **8** (1961), pp. 357–362.

Chapter 2

Basic Ideas of Addition and Subtraction of Whole Numbers

2.1 THE UNION OF SETS

When discussing how to introduce addition in the first grade, the teacher's guide for the pupil's book may say something like this: "Ask a child to make a set of four bottle caps and another child to make a set of one stick. As the sets are made, draw them on the chalkboard with the numeral for each set written underneath. Now ask the two children to put the sets together to form a new set. Draw the new set with its corresponding numeral. Put in the plus and equal sign and have the class read the sentence 'Four plus one equals five'."

The accompanying pictures may look like Fig. 2.1

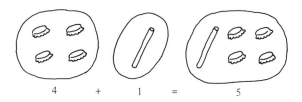

Figure 2.1

or like Fig. 2.2

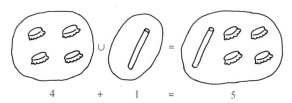

Figure 2.2

22   Basic Ideas of Addition and Subtraction of Whole Numbers   **2.1**

or like Fig. 2.3

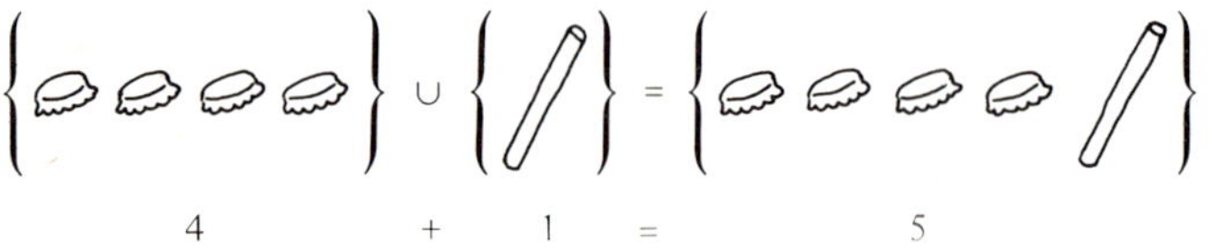

$$4 \qquad + \qquad 1 \qquad = \qquad 5$$

Figure 2.3

In the second two pictures, the symbol "∪" for **union** of sets is used. It is a useful symbol, but whether it should be introduced as early as the first grade is a debatable point. For our purposes it is worth using; and we write, for any sets A and B,

$$A \cup B$$

to mean the set whose elements are the elements of A together with the elements of B and we read "$A \cup B$" as "A union B." More formally:

$$x \in A \cup B \quad \text{if and only if} \quad x \in A \quad \text{or} \quad x \in B \quad \text{(or both)}$$

Thus, for example, if $A = \{a, b, c\}$ and $B = \{e, f\}$, then

$$A \cup B = \{a, b, c\} \cup \{e, f\} = \{a, b, c, e, f\} \qquad (a, b, c \in A; \quad e, f \in B)$$

and if $A = \{a, b, c\}$ and $B = \{a, b, e\}$, then

$$A \cup B = \{a, b, c\} \cup \{a, b, e\} = \{a, b, c, e\}$$
$$(c \in A \text{ only}; e \in B \text{ only}; a, b \in A \text{ and also } a, b \in B)$$

(Recall from Section 1.2 that $\{a, a, b, b, c, e\} = \{a, b, c, e\}$.)

A more compact way of defining $A \cup B$ is to use **set-builder** notation and write

$$A \cup B = \{x : x \in A \quad \text{or} \quad x \in B \quad \text{(or both)}\}$$

which we read as "the set of all x such that x is in A or x is in B (or both)".

This set-builder notation is commonly used in defining sets. Thus, for example,

$$\{x : x \text{ is a whole number}\} = \{0, 1, 2, 3, \ldots\}$$

and, using the symbols "$<$" for "less than" and "$>$" for "greater than",

$$\{x : x \text{ is a whole number}, \quad x < 6, \text{ and } x > 2\} = \{3, 4, 5\}$$

It is particularly useful in specifying infinite sets such as the set of whole numbers and finite sets such as $\{x : x \text{ is a whole number and } x < 1000\}$ that have a large number of members.

2.1 The Union of Sets 23

Figure 2.4 shows a Venn diagram picturing the union of the sets A and B as the shaded region. Note that, in this figure, $a \in A$ but $a \notin B$, $b \in B$ but $b \notin A$, whereas $c \in A$ and also $c \in B$.

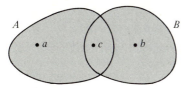

Figure 2.4

Programed Lesson 2.1

1. $\{\square, \bigcirc\} \cup \{\triangle\} = $ _____.

   $\{\square, \bigcirc, \triangle\}$

2. $\{x, y, z\} \cup \{x, y\} = $ _____.

   $\{x, y, z\}$

3. For all sets A, $A \cup \emptyset = $ ____.

   A

4. For all sets A, $A \cup A = $ ____.

   A

5. In the special case of two sets A and B such that $B \subseteq A$, what can you say about $A \cup B$?

   If $B \subseteq A$, then $A \cup B = A$.

6. Write a description of the set $\{x: x$ is a whole number and $x > 2\}$.

   The set of all x such that x is a whole number and x is greater than two.

7. Write in set-builder notation the set of all even numbers.

   $\{x: x$ is an even number$\}$

8. Write the set $\{1, 3, 5, 7, 9, \ldots\}$ in set-builder notation.

   $\{x: x$ is an odd number$\}$

9. What are the elements of the set $\{x: x$ is a president of the U.S.A. who was killed while in office$\}$?

   Lincoln, Garfield, McKinley, Kennedy

10. Write the set $\{$New Year's Day, Washington's Birthday, Memorial Day, Fourth of July, Labor Day, Veteran's Day, Christmas$\}$ in set-builder notation.

    $\{x: x$ is a legal holiday in the U.S.A.$\}$

24   **Basic Ideas of Addition and Subtraction of Whole Numbers**   **2.3**

2.2 SOME REMARKS ON THE LANGUAGE OF LOGIC

Before considering further the union of sets, let us look at the use of the word "or" and the phrase "if and only if" in the definition of union. The word and phrase are frequently used in mathematics and often misunderstood. First, the word "or". As used in ordinary speech it almost always is the *exclusive* "or", as in "I am going to a rock concert or to the opera tonight," meaning one or the other but not both. As used in mathematics, however, (unless stated otherwise), it always means the *inclusive* "or" (sometimes indicated in legal documents by "and/or"), meaning one or the other or *both*. So, when we agree to the inclusive meaning of "or", there is no need for the "(or both)" in the definition of the union of sets; we can and do write

$$x \in A \cup B \quad \text{if and only if} \quad x \in A \quad \text{or} \quad x \in B$$

and

$$A \cup B = \{x : x \in A \quad \text{or} \quad x \in B\}$$

What about "if and only if"? This phrase, as used in our definition of $x \in A \cup B$, is a shorthand way of combining the following two statements:

1. If $x \in A$ or $x \in B$, then $x \in A \cup B$.
2. If $x \in A \cup B$, then $x \in A$ or $x \in B$.

In more informal discourse, when "if and only if" refers to a definition, we can replace it by "means." Thus:

$$x \in A \cup B \quad \text{means} \quad x \in A \quad \text{or} \quad x \in B$$

2.3 DEFINITION OF THE ADDITION OF WHOLE NUMBERS

To define the addition of whole numbers in terms of sets, we must use **disjoint** sets—sets that have no elements in common. Thus

$\{a, b\}$ and $\{c, d\}$ are disjoint sets;
$\{a, b\}$ and $\{b, c\}$ are not disjoint sets (they have
    the element b in common);
$\emptyset$ and A are disjoint sets for any set A.

Note, by the way, that if sets of objects are used in the early primary-school classroom, rather than pictures or symbols of sets, disjointness is automatic. Thus a set of popsicle sticks held in one hand will always necessarily be disjoint with a set of popsicle sticks held in the other hand—no matter how much alike the sticks are. Once again, this illustrates the advantage of using objects rather than *pictures* of objects. Furthermore, in the work with the actual bottle caps and sticks, as discussed in Section 2.1, the children see only 5 objects—whether divided into sets of 4 and 1 elements, or combined to make a set of 5 elements. In the *picture* of the experiment, however, the operation of putting the objects together is not shown and thus the children see a total of *8* bottle caps and *2* sticks—*10* objects in all. This can be confusing to young children encountering the concept of addition for the first time.

2.3　　　　　　　　　　　**Definition of the Addition of Whole Numbers**　　**25**

We are now ready to formalize the relation between the addition of whole numbers and the union of sets, by making the following definition:

If A and B are finite disjoint sets,　　then $n(A) + n(B) = n(A \cup B)$

For example, to find $2 + 3$, we note that if $A = \{a, b\}$ and $B = \{c, d, e\}$, then A and B are disjoint sets with $n(A) = 2$ and $n(B) = 3$. Thus, by our definition,

$$n(A) + n(B) = n(A \cup B)$$

But $A \cup B = \{a, b, c, d, e\}$ so that $n(A \cup B) = 5$. Thus we have $2 + 3 = 5$.

Note the importance of using disjoint sets. If, for example, we were to take A as before but $B = \{b, d, e\}$, we would have $A \cup B = \{a, b, d, e\}$, a set of 4 elements, and conclude that $2 + 3 = 4$! Indeed, $n(A) + n(B) = n(A \cup B)$ only if A and B are disjoint sets.

From the definition of set union, we can conclude that, for all sets A and B,

$$A \cup B = B \cup A \qquad \textbf{(Commutative property of set union)}$$

and

$$A \cup (B \cup C) = (A \cup B) \cup C \qquad \textbf{(Associative property of set union)}$$

For example, if $A = \{a, b\}$, $B = \{c\}$, and $C = \{x, y, z\}$, we have $A \cup B = \{a, b, c\} = B \cup A$. (Remember that the order of the listing of the elements of the set is immaterial.) Also,

$$A \cup (B \cup C) = A \cup \{c, x, y, z\} = \{a, b\} \cup \{c, x, y, z\}$$
$$= \{a, b, c, x, y, z\}$$

and

$$(A \cup B) \cup C = \{a, b, c\} \cup C = \{a, b, c\} \cup \{x, y, z\}$$
$$= \{a, b, c, x, y, z\}$$

From these set properties and the definition of addition in terms of set union, we get the corresponding properties of addition:

For all whole numbers a, b, and c,

$$a + b = b + a \qquad \textbf{(Commutative property of addition)}$$
$$a + (b + c) = (a + b) + c \qquad \textbf{(Associative property of addition)}$$

Also, from $A \cup \emptyset = \emptyset \cup A = A$ for all sets A, $n(\emptyset) = 0$, and

$$n(A) + n(\emptyset) = n(A \cup \emptyset)$$

(because, for any set A, $\emptyset$ and A are disjoint sets), we get

$$a + 0 = 0 + a = a \qquad \textbf{(Additive property of 0)}$$

We call the 0 the **additive identity.**

26   Basic Ideas of Addition and Subtraction of Whole Numbers

It is important to note that "∪" is used with *sets* and "+" with *numbers*. Thus if A and B are sets, we write their union as $A \cup B$ and not as $A + B$. Then, since $n(A)$ and $n(B)$ are numbers (for finite sets A and B), we write $n(A) + n(B)$ and not $n(A) \cup n(B)$.

Figure 2.5 shows material from a first-grade book dealing with the commutative and associative properties of addition (sometimes called, for children, the **ordering** and **grouping** properties, respectively).

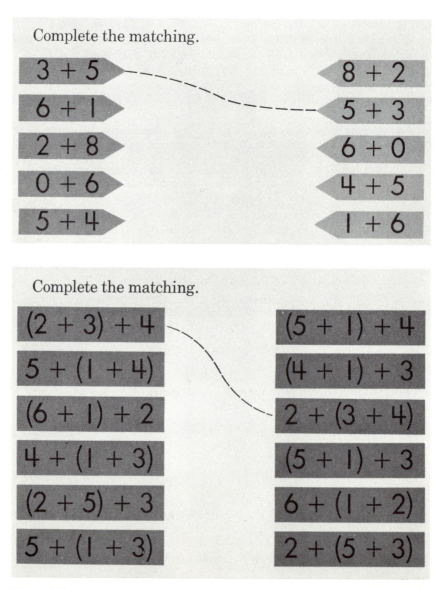

Figure 2.5

2.3   **Definition of the Addition of Whole Numbers   27**

Such illustrations of these properties, however, do not fully convey their power. Consider, for example, the following addition problem:

$$\begin{array}{r} 8 \\ 3 \\ 2 \\ +7 \\ \hline \end{array}$$

It is, of course, easy to do this addition in the "normal" way—"eight plus three equals eleven, plus two equals thirteen, plus seven is twenty". A quicker (or at least more elegant!) way, however, is to note that $8 + 2 = 10$, $3 + 7 = 10$, $10 + 10 = 20$. Such shortcuts become more useful in a problem such as:

$$\begin{array}{r} 175 \\ 220 \\ 125 \\ +180 \\ \hline \end{array}$$

where the observations that $175 + 125 = 300$, $220 + 180 = 400$, $300 + 400 = 700$, enable us to do the problem more easily than in the "normal" way.

The commutative and associative properties provide the justification for such procedures. Thus in the first problem, computing the answer in the "normal" way is equivalent to

$$[(8 + 3) + 2] + 7 = (11 + 2) + 7 = 13 + 7 = 20$$

whereas the shorter method computes

$$(8 + 2) + (3 + 7) = 10 + 10 = 20$$

The fact that

$$[(8 + 3) + 2] + 7 = (8 + 2) + (3 + 7)$$

follows from the commutative and associative properties. (In today's world of minicalculators, it would be silly to suggest that shortcuts in arithmetic computation are of any great importance per se! But some shortcuts, such as the one just described, do illustrate basic properties of our number system. Also, many children find them interesting—and hence are motivated in their work.)

Programed Lesson 2.3

1. Which of the following pairs of sets are disjoint?

a) $\{a, b\}, \{x, y\}$ b) $\{a, b\}, \{a, c\}$ c) $\{a\}, \emptyset$

(a) and (c)

28   Basic Ideas of Addition and Subtraction of Whole Numbers   **2.3**

2. Two sets A and B are said to be disjoint if, whenever $x \in A$, then x____B, and whenever $x \in B$, then x____A.

$\notin, \notin$

3. If $A = \{a, b, c\}$ and $B = \{d, e, f, g\}$, then $n(A) =$____and $n(B) =$____.

3, 4

4. With A and B as in Exercise 3, $A \cup B = \{$____$\}$ and A and B are____sets.

a, b, c, d, e, f, g; disjoint

5. From Exercises 3 and 4 and the fact that $n(A \cup B) =$ ____ we conclude that $3 +$____$=$____.

7, 4, 7

6. In general, if A and B are any finite ____ sets, then $n(A) + n(B) =$ ____.

disjoint, $n(A \cup B)$

7. The statement that for all sets A and B, $A \cup B = B \cup A$ is called the ____ property of set ____.

commutative, union

8. The associative property of set union states that for all sets A, B, and C, $A \cup (B \cup C) =$____.

$(A \cup B) \cup C$

9. Which properties of addition are illustrated by the following statements?
a) $2 + 3 = 3 + 2$
b) $2 + (3 + 4) = (3 + 4) + 2$
c) $2 + (3 + 4) = 2 + (4 + 3)$
d) $5 + 0 = 5$
e) $2 + (3 + 4) = (2 + 3) + 4$

(a), (b), and (c) commutative; (d) additive property of 0; (e) associative

10. Use the shortcuts suggested in the text to perform the following additions.

a) 28	**b)** 39	**c)** 125
47	62	368
53	41	275
+32	+28	+432

$a)$ $(28 + 32) + (47 + 53) = 60 + 100 = 160$;
$b)$ $(39 + 41) + (62 + 28) = 80 + 90 = 170$;
$c)$ $(125 + 275) + (368 + 432) = 400 + 800 = 1200$

11. $a + b$ refers to the____of whole numbers and $A \cup B$ refers to the ____ of sets.

addition, union

2.4 INTERSECTION OF SETS

The **intersection,** $A \cap B$, of two sets A and B is defined as the set whose elements are in *both* A and B. We read "$A \cap B$" as "A intersect B". In symbols: If A and B are any sets, then

$$A \cap B = \{x: x \in A \text{ and } x \in B\}$$

For example,

$$\{a, b, c, d\} \cap \{a, b, x, y\} = \{a, b\}$$

and

$$\{a, b, c, d\} \cap \{x, y, z\} = \emptyset$$

Venn diagrams can be used to picture intersection of sets, as shown in Fig. 2.6.

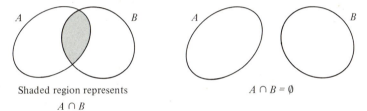

Shaded region represents $A \cap B$

$A \cap B = \emptyset$

Figure 2.6

Note that we can now describe disjoint sets more compactly: A and B are disjoint sets if and only if $A \cap B = \emptyset$. Hence we can write

For any finite sets A and B,

$$n(A) + n(B) = n(A \cup B) \text{ if and only if } A \cap B = \emptyset$$

Figure 2.7 (see top of page 30) is taken from a fifth-grade book and shows work on the union and intersection of sets.

Programed Lesson 2.4

1. {Alfred, Betty, Charles, Doris} ∩ {Alfred, Doris, Edith} = {_____}.

    Alfred, Doris

2. {○, □, △} ∩ {◇, ◯} = ___.

    ∅

3. If A and B are any sets, then $A \cap B = \{x: x \in A$ _____ $x \in B\}$.

    and

Basic Ideas of Addition and Subtraction of Whole Numbers 2.4

Union and intersection of sets

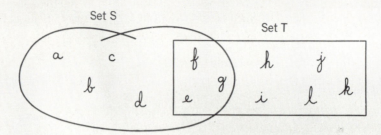

Set S contains all the letters in the red ring. Set T contains all the letters in the rectangle.

| The union of sets S and T | contains | the letters that are in one or the other or both sets. |

S ∪ T = {a, b, c, d, e, f, g, h, i, j, k, l}

| The intersection of sets S and T | contains | only the letters that are in both set S and set T. |

S ∩ T = {e, f, g}

The pictures below may help you to understand union and intersection of sets. In picture A the areas shaded pink show the **union** of the two sets. In picture B the area shaded a darker pink shows the **intersection** of the two sets.

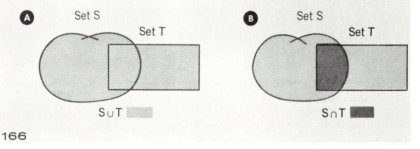

166

Figure 2.7

4. If A and B are any sets, then $A \cup B = \{x : x \in A$ _____ $x \in B\}$.

   or

5. If $A \cap B = \emptyset$, then $n(A \cup B) =$ ____ + ____ .

   $n(A), n(B)$

6. For all sets A, $A \cap A =$ ____.

    A

7. For all sets A, $A \cap \emptyset =$ ____.

    $\emptyset$

8. For all sets A and B, $A \cap B = B \cap$ ____.

    A

9. A natural name for the property that, for all sets A and B, $A \cap B = B \cap A$ is the
_____ property of set _____ .

    commutative, intersection

10. Verify that $A \cap (B \cap C) = (A \cap B) \cap C$ if $A = \{a, b, c\}$, $B = \{b, c, d\}$, and $C = \{c, d, e\}$.

    $A \cap (B \cap C) = A \cap \{c, d\} = \{a, b, c\} \cap \{c, d\} = \{c\};$
    $(A \cap B) \cap C = \{b, c\} \cap C = \{b, c\} \cap \{c, d, e\} = \{c\}$

11. It is true that, for all sets A, B, and C, $A \cap (B \cap C) = (A \cap B) \cap C$. A natural
name for this property is the _____ property of set _____ .

    associative, intersection

12. In the special case of two sets A and B such that $A \subseteq B$, what can you say about
$A \cap B$?

    If $A \subseteq B$, then $A \cap B = A$.

2.5 SUBTRACTION AND SET COMPLEMENTATION

Again, as for addition, the classroom situation is a simple one. We begin the study
of subtraction by taking, say, a set of 5 sticks and then removing from this set a
set of two sticks. What remains is a set of three sticks. The corresponding
arithmetic statement is, of course, $5 - 2 = 3$.

It is possible to formalize in set language this process of removing sticks. To
do so, we define the **relative complement** of a set B with respect to a set A, where
B is a subset of A, as the set of all elements of A which are *not* in B. Writing
$A \backslash B$ for this relative complement, and taking $A = \{a, b, c, d, e\}$ and $B = \{d, e\}$,
we have $A \backslash B = \{a, b, c\}$. We read "$A \backslash B$" as "$A$ less B." As another example,
let A be the set of whole numbers and $B = \{x: x \text{ is a whole number and } x < 10\}$.
Then $B \subseteq A$ and $A \backslash B = \{x: x \text{ is a whole number and } x = 10 \text{ or } x > 10\}$. In
set-builder notation we have, for all sets A and B with $B \subseteq A$,

$$A \backslash B = \{x: x \in A \text{ and } x \notin B\}$$

32 Basic Ideas of Addition and Subtraction of Whole Numbers 2.5

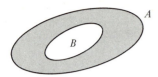

$A \backslash B$ is the shaded region

Figure 2.8

A Venn diagram for $A \backslash B$ is shown in Fig. 2.8.

Now we can define subtraction for $n(B)$ less than or equal to $n(A)$ (written $n(B) \leq n(A)$) by

$$n(A) - n(B) = n(A \backslash B) \quad \text{where } B \subseteq A$$

Thus, in our example above, where $A = \{a, b, c, d, e\}$ and $B = \{d, e\}$, we have $B \subseteq A$, $A \backslash B = \{a, b, c\}$, $n(A) = 5$, $n(B) = 2$, $n(A \backslash B) = 3$ so that

$$n(A) - n(B) = n(A \backslash B) \quad \text{yields} \quad 5 - 2 = 3$$

Figure 2.9 shows the use of sets in a first-grade book to illustrate subtraction.

Figure 2.9

Programed Lesson 2.5

1. For all sets A, $A \backslash A = $ ____.

   $\emptyset$

2. The fact that $A \backslash A = \emptyset$ for all sets A corresponds to the fact that, for all whole numbers a, $a - a = $ ____.

   0

3. For all sets A, $A \backslash \emptyset = $ ____.

   A

2.6   **Subtraction as Related to Addition**   **33**

4. The fact that $A \setminus \emptyset = A$ for all sets A corresponds to the fact that, for all whole numbers a, $a - 0 =$ _____.

   a

5. For all sets A and B such that B _____ A, $A \setminus B$ is the set of all elements of _____ that are _____ in _____.

   $\subseteq$, A, not, B

6. If $A = \{x, y, z, w\}$ and $B = \{x\}$, then $n(A) =$ _____ and $n(B) =$ _____.

   4, 1

7. With A and B as in Exercise 6, $A \setminus B = \{$_____$\}$ and so $n(A \setminus B) =$ _____.

   y, z, w;   3

8. Since, in Exercises 6 and 7, B _____ A, $n(A) = 4$, $n(B) = 1$, and $n(A \setminus B) = 3$, we can conclude that $4 -$ _____ $=$ _____.

   $\subseteq$, 1, 3

9. Let A be the set of students in a class and B be the set of women in this class; then $A \setminus B$ is the set of _____ in the class.

   men

10. Let A be the set of whole numbers and B be the set of even numbers; then $A \setminus B$ is the set of _____ numbers.

   odd

2.6 SUBTRACTION AS RELATED TO ADDITION

In teaching children the basic ideas of subtraction of whole numbers, it is very useful, as well as mathematically significant, to relate each subtraction problem to an addition problem. This is especially true if we look ahead to subtraction of negative integers or rational numbers. In such situations we still have the same relation of addition and subtraction, but the relative complement approach does not apply at all.

In the primary school classroom the situation in which we removed a set of 2 sticks from a set of 5 sticks to illustrate $5 - 2 = 3$ can then be used to illustrate $3 + 2 = 5$ simply by putting back the set of 2 sticks with the set of 3 sticks. On the other hand, if we begin with a set of 3 sticks and combine this set with a set of 2 sticks, we illustrate $3 + 2 = 5$. Now if we remove the second set of sticks we illustrate $5 - 2 = 3$. That is, to put it briefly, if a, b, and c are any whole numbers with $c \geq b$ (i.e., $c = b$ or $c > b$), then

$$c - b = a \quad \text{if and only if} \quad c = a + b$$

This fact is, of course, widely used as a check for subtraction: $5 - 2 = 4$ is incorrect because $5 \neq 4 + 2$.

34   Basic Ideas of Addition and Subtraction of Whole Numbers   **2.7**

It should be noted that, to every subtraction statement, there correspond two addition statements, and to every addition statement there correspond two subtraction statements. Thus to the subtraction statement $5 - 2 = 3$, there correspond the addition statement $2 + 3 = 5$ and also $3 + 2 = 5$. Similarly, to the addition statement $2 + 3 = 5$, there correspond the subtraction statement $5 - 3 = 2$ and also $5 - 2 = 3$.

Programed Lesson 2.6

1. The two addition statements corresponding to the subtraction statement $4 - 1 = 3$ are _____ and _____ .

   $1 + 3 = 4,$ $3 + 1 = 4$ (either order)

2. The two subtraction statements corresponding to the addition statement $5 + 4 = 9$ are _____ and _____ .

   $9 - 4 = 5,$ $9 - 5 = 4$ (either order)   .

3. 0 is the additive identity. Why cannot 0 be regarded as the subtractive identity?

   Although, for all whole numbers a, $a - 0 = a$, it is not true that $0 - a = a$ unless $a = 0$. (We could call 0 a *righthand* identity for subtraction.)

4. Is subtraction a commutative operation? Give a reason for your answer.

   No. For example, $3 - 2 \neq 2 - 3$.

5. Is subtraction an associative operation? Give a reason for your answer.

   No. For example, $6 - (4 - 2) = 6 - 2 = 4$ but $(6 - 4) - 2 = 2 - 2 = 0$.

6. If a and b are whole numbers and $a - b = b - a$, what can you say about a and b?

   $a = b$

7. If a, b, and c are whole numbers and $(a - b) - c = a - (b - c)$, what can you say about a, b, and c?

   $c = 0$; a and b can be any whole numbers such that $a \geq b$.

2.7 NUMBER LINES

An alternative (or supplementary) way of considering addition and subtraction is to utilize a **number line** which, unlike sets, can also be utilized to illustrate addition and subtraction for other kinds of numbers than whole numbers.

Thus, for example, Fig. 2.10 shows the addition $2 + 3 = 5$ on a number line. That is, beginning at 2, we make 3 "unit" jumps to the right. (For children, a cutout of a frog or a grasshopper adds interest—or the children themselves can jump on a number line drawn on the floor of the classroom!)

Figure 2.10

Similarly 5 − 3 = 2 is shown on a number line in Fig. 2.11 by beginning at 5 and making three "unit" jumps to the left.

Figure 2.11

Figure 2.12, from a first-grade book, shows, at top, a frog (purposefully!) jumping on a number line to illustrate 2 + 2 = 4 and, at bottom, to show 5 − 3 = 2.

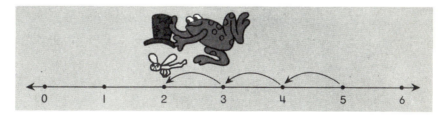

Figure 2.12

A number line can also be used to picture the relations of greater than and less than, as is illustrated by Fig. 2.13, taken from a second-grade book: ">" corresponds to "to the right of" and "<" corresponds to "to the left of".

36 Basic Ideas of Addition and Subtraction of Whole Numbers　　　　　　　　　2.8

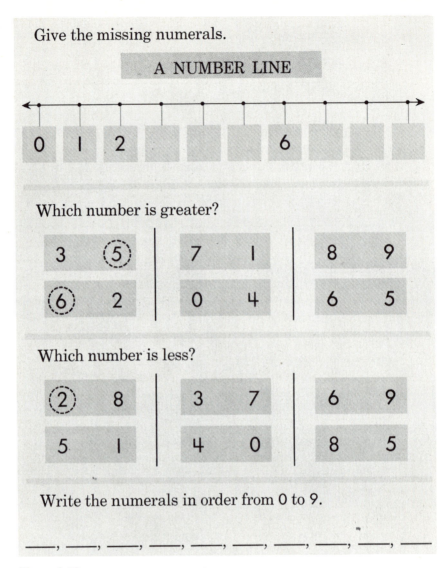

Figure 2.13

2.8 THE USE OF FRAMES

One might well ask at this point to what extent contemporary mathematics programs in the early grades differ from older programs. Certainly good teachers, from time immemorial, have used sticks and stones (therefore, sets!) to make concrete and vivid the idea of whole numbers, as well as addition and subtraction of whole numbers. Certainly the added features we have mentioned: some geometry, a little set language, use of a number line, an emphasis (perhaps overemphasis in some primary school programs) on properties such as the commutative property, do not add up to an enormous change in the early grades.

Here, and in the next section, we discuss other aspects of the newer mathematics programs which may be, in the long run, more significant than the ones mentioned above.

One of these innovations is the use of "frames" or "boxes" to phrase questions about arithmetic in a semialgebraic form. This procedure provides both for a painless transition to algebra later on and also for a variety of stimulating exercises.

Very early in a first-grade book we find material like that shown in Fig. 2.14, where the children are instructed by the teacher to write the correct numerals in the "boxes" so as to make a true statement. Clearly, one could just as well, from an adult point of view, write

$$1 + 2 = x \quad \text{and} \quad 2 + 2 = y,$$

and ask what x and y are. (Indeed, this transition to letters is sometimes made as early as the third grade. For the young child, however, the "box" is much more concrete and appealing.)

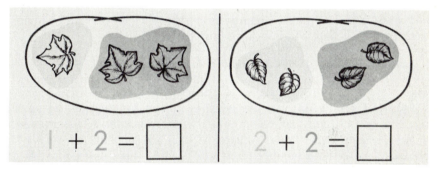

Figure 2.14

Now the problems given above may look like just a "fancy" way of asking for the sum of 1 and 2, and 2 and 2! But variations soon appear, as

$$2 + \square = 5$$

and, later,

$$2 + \square + \square = 8$$

(The agreement is that the same numerals must be put into each frame in the second example—compare with $2 + x + x = 8$.) Or we can use different-shaped frames where the agreement is that we may put different numerals in different-shaped frames, as in

$$6 = \square + \triangle$$

38   Basic Ideas of Addition and Subtraction of Whole Numbers   **2.8**

(But the same numeral *may* be put into each frame—compare $6 = x + y$, where one solution is $x = y = 3$.) Finding all possible answers for $6 = \square + \triangle$— including using 0—is an excellent way of reviewing the "six family" of basic addition facts. And a problem such as

$$\square + \triangle = \triangle + \square$$

is an excellent way to make children think about the commutative property of addition: For *all* whole numbers a and b, $a + b = b + a$. Even further (and there is no algebraic analogue for this), one can pose problems such as

$$6\,\square\,2 = 8 \qquad \text{and} \qquad 5\,\square\,3 = 2$$

where a "+" or a "−" is to be put into the frame to make a true statement. Or, for a more difficult problem,

$$(5\,\square\,3)\,\triangle\,2 = 6$$

(Later, multiplication and division symbols can be allowed as permissible substitutions and, as you will see in Chapter 5, quite difficult problems can be so constructed.)

Since in contemporary mathematics programs the symbols "<" and ">" are introduced early, we can also have problems such as

$$6 + 3\,\square\,9, \qquad 6 + 3\,\triangle\,10, \qquad 6 + 3\,\bigcirc\,8,$$

where "=", "<", or ">" are to be put into the frames.

Problems such as these are much more fun for the children than drill on "number facts." (Not that drill *may* not have a place in the classroom—although there are many who would replace all drill by such devices as we have just discussed and various number "activities" such as those that we will discuss in the next section.)

Programed Lesson 2.8

1. Fill in the frames with the proper numeral or numerals to make a true statement.
   a) $10 + \square = 15$
   b) $6 + \square + \square = 14$
   c) $2 \times \square + 3 = 11$

   ---
   a) 5 *b)* 4 *c)* 4

---

2. Filling in the frames for
   $$\square + (\triangle + \bigcirc) = (\square + \triangle) + \bigcirc$$
   illustrates the _____ property of addition.

   ---
   associative

3. Filling in the frames for

   $\square + 0 = \square$

   illustrates that 0 is an _____ _____ .

   additive identity

4. Fill in the frames with "+" or "−" to make a true statement.
   a) $(5 \square 3) \triangle 6 = 2$ b) $(5 \square 3) \triangle 6 = 8$
   c) $5 \square (2 \triangle 1) = 2$ d) $8 \square (3 \triangle 2) = 7$

   a) +, −   b) −, +   c) −, +   d) −, −

5. Fill in the frames with "=", ">" or "<" to make a true statement.
   a) $2 + 3 \square 4$ b) $3 + 4 \square 7$
   c) $5 - 2 \square 1$ d) $0 + 1 \square 2$

   a) >   b) =   c) >   d) <

2.9 NUMBER ACTIVITIES

This section could be a book in itself! However, any series of contemporary primary school mathematics texts will have a considerable number of such activities suggested in their teacher's guides, so there is no point in trying here to give any kind of exhaustive collection. Rather, we want you to get the flavor of such activities and for this reason, in both text and exercises, we will give some that involve arithmetic of the higher grades. In this way you, too (hopefully!) can experience some of the excitement that children feel when they participate in similar activities.

A popular activity for children that can be used as early as the first grade (and generalized for later work—see exercises) involves what is sometimes called an "arithmetic rabbit" and sometimes "Tom's Fantastic Mathematical Machine." (You can make up your own name or, better yet, have the children make up a name.) This device can be drawn like Fig. 2.15 to suggest the ears and tail of a rabbit, or like Fig. 2.16 to suggest a machine.

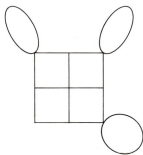

Figure 2.15

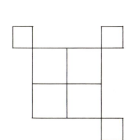

Figure 2.16

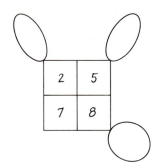
Figure 2.17

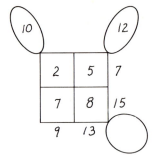
Figure 2.18

Now let's put some numerals for whole numbers into the inside squares, as shown in Fig. 2.17. We then begin to "complete" the rabbit in Fig. 2.18. This is done by calculating $2 + 7$, $5 + 8$ (vertically), $2 + 5$, $7 + 8$ (horizontally), and $2 + 8$, $7 + 5$ (diagonally). Now what about the "tail"? Observe that $9 + 13 = 22$, $7 + 15 = 22$, and (the "sum of the ears"), $10 + 12 = 22$. So, our completed rabbit is as shown in Fig. 2.19. It's fun for children—and there's lots of arithmetic practice involved!

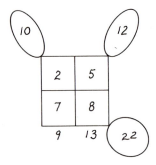
Figure 2.19

When subtraction has been introduced (or even before, as a prelude to subtraction), one can vary the rabbit activity with such problems as that shown in Fig. 2.20. This is by no means an easy problem for most children who have mastered the "standard" rabbit! But, from $9 - 5 = 4$, $16 - 9 = 7$, $8 - 5 = 3$, and $7 - 3 = 4$, the rabbit can be completed as shown in Fig. 2.21.

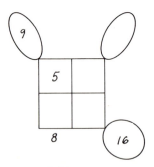
Figure 2.20

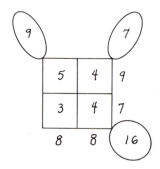
Figure 2.21

2.9 **Number Activities** 41

Here is another example of an activity—this time obviously at a higher level than first grade. In this activity the teacher asks the children each to pick four digits from the set $\{1, 2, 3, 4, 5, 6, 7, 8, 9\}$ with not all digits the same. Now the children are asked to form from these digits the largest and smallest four-digit numbers possible and to subtract the smaller from the larger. For example, if a child chose 4, 8, 9, and 2, his work at this point would look as follows:

$$
\begin{array}{r}
9842 \\
-2489 \\
\hline
7353
\end{array}
$$

Now the teacher tells the children to rearrange the digits of their answer to again make the largest possible number and the smallest possible number and then again subtract the smaller from the larger. Our "sample" child would then have:

$$
\begin{array}{r}
7533 \\
-3357 \\
\hline
4176
\end{array}
$$

Continuing in this way, the child would get:

$$
\begin{array}{r}
7641 \\
-1467 \\
\hline
6174
\end{array}
$$

and then see that the next step would be exactly the same, since he has, again, the digits 4, 1, 7, and 6. Is this an accident? A classroom of children will soon produce convincing evidence (but not a formal *proof*) that it is not an accident. Depending on the original choice of digits, a varying number of steps may be needed but (as has been proved) one always arrives at the number 6174—sometimes called **Kapreker's constant,** after the man who first discovered and proved the result. (*Note.* If a 0 appears at any time, it must be used in the next step. For example, if the digits chosen were 9, 9, 9, and 8, the first subtraction would give us $9998 - 8999 = 999$. But, to continue, we must think of 999 as 0999 so that the next step would be to calculate $9990 - 0999 = 8991$, then $9981 - 1899 = 8082$; then $8820 - 0288 = 8532$, etc.

Trying out various sets of digits to see whether 6174 is always obtained obviously provides lots of subtraction practice. It can also lead to further experimentation and possible discovery-type activities, as discussed in the next section.

Exercises 2.9

1. Complete the work begun above for the number 9998.

2. Carry out at least two other examples of a choice of four digits in the repeated subtraction activity.

3. Complete the following addition rabbit:

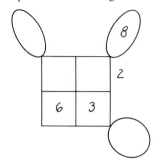

4. Explain why the number in the "tail" turns out to be the sum of the numbers in the "ears" as well as the horizontal and vertical "border" sums.

5. Construct a multiplication rabbit.

6. Can you construct a subtraction rabbit? A division rabbit?

7. Pick any whole number, square each of its digits, and add these squares to get a new number. Now take this number, square each of its digits, and add these squares to get another number. Continue in this way until you discover something interesting. (For example, if you began with 324, your first three steps would be):

   1. $324 \rightarrow 3^2 + 2^2 + 4^2 = 9 + 4 + 16 = 29$
   2. $29 \rightarrow 2^2 + 9^2 = 4 + 81 = 85$
   3. $85 \rightarrow 8^2 + 5^2 = 64 + 25 = 89$

2.10 DISCOVERY

The directions given in the last exercise of the preceding section may have seemed rather strange to you: "Continue in this way until you discover something interesting." We hope that you did discover something interesting and so see that open-ended questions of this kind can be meaningful.

Having children learn to *discover* things about mathematics, rather than regarding mathematics as just something *given* to them, is a basic goal of contemporary mathematics programs. And it is a goal which demands a great deal of the teacher. Even the best textbook, dedicated to the discovery approach, cannot fully exploit the ability of children to discover things about mathematics. The poorer texts, on the other hand, often make a mockery of the "discovery approach." Thus, for example, they may have a page looking something like this:

"You know that

$4 + 5 = 5 + 4$
$2 + 3 = 3 + 2$
$6 + 0 = 0 + 6$

2.10 **Discovery** **43**

and

$$7 + 3 = 3 + 7$$

Do you see that, for all whole numbers, the order of addition does not matter? We call this property the *commutative property of addition.*"

Now the "do you see" is supposed to convey the idea that the child has now discovered the commutative property of addition. Do *you* see, however, that the child is really being *told* the property? (A writing group I know of, dedicated to a *genuine* discovery approach, fined any author who used a "do you see"!)

How could we attempt to make the commutative property of addition more of a genuine discovery? Well, suppose that you ask the children to help you construct an addition table which, in final form, would look like Fig. 2.22 (without the diagonal line and the circles around the numerals).

+	0	1	2	3	4	5	6	7	8	9
0	0	1	2	③	4	5	6	7	8	9
1	1	2	3	4	5	6	7	8	9	10
2	2	3	4	5	6	⑦	8	9	10	11
3	③	4	5	6	7	8	9	10	11	⑫
4	4	5	6	7	8	⑨	10	11	12	13
5	5	6	⑦	8	⑨	10	11	12	13	14
6	6	7	8	9	10	11	12	13	14	15
7	7	8	9	10	11	12	13	14	15	16
8	8	9	10	11	12	13	14	15	16	17
9	9	10	11	⑫	13	14	15	16	17	18

Figure 2.22

You then ask "What interesting things can you observe about this table?" Now (and this is why no text can really be of much assistance in developing the discovery approach) you must expect the unexpected! In the context of what *you* are interested in, you are hoping that some child will call attention to the fact (in his or her own language, or by pointing) that the part of the table above the diagonal is the same as the part below, as indicated by the line and the circled items. (This fact can, of course, be easily rephrased to yield the commutative property.) But, alas, the first comment you may get is, "The numbers increase by one in each row as you go from left to right." Please do *not* say, "That's not what I wanted you to observe!" Instead, compliment the child on the correct observation and ask for additional observations. You may be surprised—as I have been—at the number of observations children are capable of making! Indeed, you may never get to the commutative property as an observation by a child. No

44   Basic Ideas of Addition and Subtraction of Whole Numbers   **2.10**

matter—the children have been busily engaged in developing mathematical ideas on their own. Nothing is lost and much is gained from such an activity even if, at the end, you have to call their attention to the commutative property yourself. (See the article at the end of this chapter for further remarks on this activity.)

In this example, we have provided the children with a starting point for a discovery activity. In an open classroom, however, where discovery activity is encouraged, you will find, at least occasionally, children excitingly announcing their own unaided discoveries. How about a child discovering that if you add two odd numbers, you always get an even number? Beautiful! And that child is not likely to write, for example, $7 + 9 = 15$. (He might, of course, write $7 + 9 = 18$—but at least 18 is an even number!)

How about a child who tells you that every perfect square other than 1 ($4 = 2^2$, $9 = 3^2$, etc.) has exactly three divisors? Yes, 4 has the set of divisors $\{1, 2, 4\}$ and 9 has the set of divisors $\{1, 3, 9\}$. But, how about 16 ($16 = 4^2$)? Whoops!—the set of divisors of 16 is $\{1, 2, 4, 8, 16\}$—considerably more than three. No matter; the child is trying to discover mathematics on his own and made a reasonable—even if incorrect—conjecture. And perhaps, with a little encouragement, he will go back to his problem and find that it is true that the squares of *prime* numbers (i.e., numbers with exactly two divisors: 2, 3, 5, 7, 11, . . .) all have exactly three divisors.

Exercises 2.10

1. Vary Exercise 7 of Section 2.9 by first adding the digits and then squaring. Thus

    $$113 \rightarrow (1 + 1 + 3)^2 = 5^2 = 25 \rightarrow (2 + 5)^2 = 7^2 = 49, \qquad \text{etc.}$$

    What can you discover that is interesting?

2. Take any number, multiply it by 2, add 16 to this new number, then divide the number you get by 2 and subtract the number you started with. What is interesting about this process? Make up other activities like this.

3. Section 2.9 dealt with the process of successive subtraction that produced Kapreker's constant. Apply the process to several three-digit numbers. What can you conclude?

Chapter Test

In preparation for this test, you should review the meaning of the following terms and symbols, referring when necessary to the indicated pages:

Union of sets (22)

∪ (22)

Commutative and associative properties
 of set intersection (31)

Commutative and associative properties
 of addition (25)

2.10   **Chapter Test**   **45**

Set-builder notation (22)	Additive property of 0 (25)
If and only if (24)	Additive identity (25)
Intersection of sets (29)	Relative complement (31)
∩ (29)	Number line (34)
Disjoint sets (24)	>, < (35)
Commutative and associative properties of set union (25)	

You should also review how to:

1. Show addition of whole numbers in terms of the union of sets (25);
2. Show subtraction of whole numbers in terms of set complementation (32);
3. Relate a subtraction problem to an addition problem (33);
4. Fill in frames with the correct numerals or signs of operation (37);
5. Complete addition "rabbits" (40).

1. $\{a, b, c, d\} \cup \{a, c\} = $ _____ and $\{a, b, c, d\} \cap \{a, c\} = $ _____ .

2. $\{a, b, c, d\} \cup \{e, f\} = $ _____ and $\{a, b, c, d\} \cap \{e, f\} = $ _____ .

3. For all sets A, $\emptyset \cup A = $ _____ and $\emptyset \cap A = $ _____ .

4. For all sets A, $A \cup A = $ _____ and $A \cap A = $ _____ .

5. For all sets A and B, if $A \subseteq B$, then $A \cup B = $ _____ and $A \cap B = $_____ .

6. Write, by listing the elements between braces, the set

$$\{x: x \text{ is a whole number} \quad \text{and} \quad x < 5\}.$$

7. Write in set-builder notation the set of all odd numbers.

8. Write the set $\{0, 2, 4, 6, 8, \ldots\}$ in set-builder notation.

9. Write a definition of $A \cup B$ in set-builder notation, using the symbol "$\in$".

10. The commutative property of set intersection states that, for all sets A and B, _____ = _____ .

11. The _____ property of set union states that for all sets A, B, and C, $A \cup (B \cup C) = (A \cup B) \cup C$.

12. The number 0 is called the _____ _____ .

13. If A and B are finite sets and $n(A) + n(B) = n(A \cup B)$, then $A \cap B = $ _____ .

14. Which properties of addition are illustrated by the following statements?
    a) $0 + 6 = 6$
    b) $2 + (5 + 7) = 2 + (7 + 5)$
    c) $2 + (5 + 7) = (5 + 7) + 2$
    d) $2 + (5 + 7) = (2 + 5) + 7$

46   Basic Ideas of Addition and Subtraction of Whole Numbers

15. $a + b$ refers to the addition of _____ and $A \cup B$ refers to the union of _____.

16. For all sets A, $A \backslash A =$ ____.

17. The subtraction fact $a - a = 0$, for all whole numbers a, corresponds to what fact about set complementation?

18. The subtraction fact $a - 0 = a$, for all whole numbers a, corresponds to what fact about set complementation?

19. For all sets C and D such that D ____ C, $C \backslash D$ is the set of all elements of ____ which are ____ in ____.

20. The two addition statements corresponding to $6 - 4 = 2$ are _____ and _____.

21. The two subtraction statements corresponding to the addition statement $6 + 3 = 9$ are _____ and _____.

22. $5 - 3 \neq 3 - 5$ shows that subtraction is not a _____ operation.

23. $10 - (5 - 3) \neq (10 - 5) - 3$ shows that subtraction is not an _____ operation.

24. Show the addition $3 + 5 = 8$ on a number line.

25. Show the subtraction $7 - 5 = 2$ on a number line.

26. Fill in the frames with the proper numerals to make a true statement.
   a) $6 + \square = 8$
   b) $5 + \square + \square = 7$
   c) $4 + \square = 4$

27. Filling in the frames for $\square + \triangle = \triangle + \square$ illustrates the _____ property of _____.

28. Fill in the frames with "+", or "−" to make a true statement.
   a) $(6 \square 2) \triangle 4 = 8$
   b) $(6 \square 2) \triangle 4 = 12$
   c) $5 \square (2 \triangle 1) = 6$

29. Fill in the frames with "=", ">", or "<" to make a true statement.
   a) $5 + 7 \square 11$
   b) $3 + 2 \square 5$
   c) $5 + 8 \square 15$

2.10 Chapter Test 47

30. Complete the following addition "rabbit."

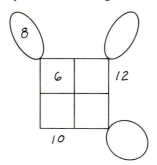

TEST ANSWERS

1. $\{a, b, c, d\}, \{a, c\}$
2. $\{a, b, c, d, e, f\}, \emptyset$
3. $A, \emptyset$
4. A, A
5. B, A
6. $\{0, 1, 2, 3, 4\}$
7. $\{x: x \text{ is an odd number}\}$
8. $\{x: x \text{ is an even number}\}$
9. $\{x: x \in A \text{ or } x \in B\}$
10. $A \cap B, B \cap A$
11. Associative
12. Additive identity
13. $\emptyset$
14. a) Additive property of zero,
    c) Commutative property,
    b) Commutative property,
    d) Associative property.
15. Numbers, sets
16. $\emptyset$
17. For all sets A, $A \setminus A = \emptyset$
18. For all sets A, $A \setminus \emptyset = A$
19. $\subseteq$, C, not, D
20. $6 = 4 + 2, 6 = 2 + 4$
21. $6 = 9 - 3, 3 = 9 - 6$
22. Commutative
23. Associative
24.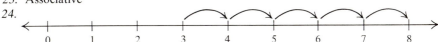
25.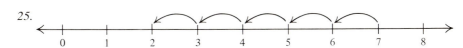
26. a) 2   b) 1   c) 0
27. Commutative, addition
28. a) $-, +$ b) $+, +$ c) $+, -$
29. a) $>$ b) $=$ c) $<$
30.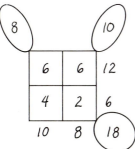

The addition table:
Experiences in practice–discovery

by Margaret A. Hervey and Bonnie H. Litwiller

Reprinted from *The Arithmetic Teacher*, Vol. 19 (1972), pp. 179–181.

Mathematics is often referred to as a study of relations and patterns. Many elementary teachers would agree with this idea, but they may have difficulty in finding materials that will provide opportunities for their students to discover patterns. The addition table may be used as an effective vehicle for the discovery of some number patterns. While searching for patterns, the students are engaging in purposeful practice, which results in a "practice–discovery" activity.

Given an addition table like the one shown in Fig. 1, students in our elementary methods classes observed such patterns as the following:

1. Each row and column is a sequence of alternating even–odd or odd–even numbers.

2. The entries in any diagonal are either all odd or all even.

3. The entries in any diagonal from upper right to lower left are constant, whereas the entries in the diagonals from upper left to lower right are either consecutive even numbers or consecutive odd numbers.

4. The consecutive sums of the entries in the rows or columns have a constant difference of 10.

5. The sums of the diagonals parallel to the main diagonal (the diagonal containing the entries 0, 2, 4, 6, . . . , 18) are consecutive multiples of 9.

6. In any square drawn on the table, the sums of the two diagonals are equal.

+	0	1	2	3	4	5	6	7	8	9
0	0	1	2	3	4	5	6	7	8	9
1	1	2	3	4	5	6	7	8	9	10
2	2	3	4	5	6	7	8	9	10	11
3	3	4	5	6	7	8	9	10	11	12
4	4	5	6	7	8	9	10	11	12	13
5	5	6	7	8	9	10	11	12	13	14
6	6	7	8	9	10	11	12	13	14	15
7	7	8	9	10	11	12	13	14	15	16
8	8	9	10	11	12	13	14	15	16	17
9	9	10	11	12	13	14	15	16	17	18

Fig. 1. Addition table.

In addition to these patterns, others may be found by drawing polygons on the addition table. In the table shown in Fig. 2, isosceles right triangles have been drawn so that the legs of the triangles lie in the rows and columns.

The Addition Table: Experiences in Practice–Discovery 49

Our students worked through the following set of instructions for each triangle A, B, and C in Fig. 2. Each student had copies of the addition table so that triangles could be sketched on the table.

1. Find the sum of the numbers that represent the vertices and call this sum SV.

   In $\triangle A$, the sum is $1 + 4 + 7 = 12$.
   In $\triangle B$, the sum is $5 + 9 + 13 = 27$.
   In $\triangle C$, the sum is $8 + 13 + 18 = 39$.

2. Find the sum of the numbers that are in the interior of the triangle and call this sum SI.

   In $\triangle A$, the sum is 4.
   In $\triangle B$, the sum is $8 + 9 + 10 = 27$.
   In $\triangle C$, the sum is $11 + 12 + 13 + 13 + 14 + 15 = 78$.

3. Find the ratios of the sums found in steps 1 and 2, that is, SV/SI.

   From $\triangle A$, the ratio is 12/4, or 3/1.
   From $\triangle B$, the ratio is 27/27, or 3/3.
   From $\triangle C$, the ratio is 39/78, or 3/6.

+	0	1	2	3	4	5	6	7	8	9
0	0	1	2	3	4	5	6	7	8	9
1	1	2	3	4	5	6	7	8	9	10
2	2	3	4	5	6	7	8	9	10	11
3	3	4	5	6	7	8	9	10	11	12
4	4	5	6	7	8	9	10	11	12	13
5	5	6	7	8	9	10	11	12	13	14
6	6	7	8	9	10	11	12	13	14	15
7	7	8	9	10	11	12	13	14	15	16
8	8	9	10	11	12	13	14	15	16	17
9	9	10	11	12	13	14	15	16	17	18

Fig. 2. Addition table with triangles superimposed.

The students observed that they had all used the same set of isosceles right triangles. They then drew triangles at random on the addition table so that the lengths of the legs measured 3 units, 4 units, and 5 units, as was the case in $\triangle A$, $\triangle B$, and $\triangle C$. It was found that the ratios 3/1, 3/3, and 3/6 held regardless of the placement of the triangles on the addition table. Note, however, that the legs of the triangles lie in the rows and columns of the table.

Students then discovered that if right triangles are drawn so that the successive lengths of the legs are increased by 1, the sequence of ratios SV/SI that results is 3/1, 3/3, 3/6, 3/10, 3/15, 3/21, 3/28. What could this sequence represent? The constant 3 may represent the three vertices of a triangle. The sequence 1, 3, 6, 10, 15, 21, 28 may represent the number of interior numbers in the triangles and may be recognized as the first eight triangular numbers. The arrangement of the set of interior numbers of the right triangles is similar to a geometric representation of the triangular numbers (see Fig. 3).

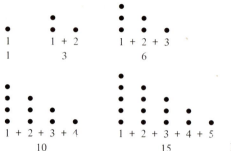

Figure 3

Students were then given the following instructions to follow for each triangle in Fig. 3:

1. Find the sum of the numbers that lie on the sides of the triangle and call this sum *SP*.

    In △*A*, the sum is 1 + 2 + 3 + 4 + 5 + 6 + 7 + 5 + 3 = 36.
    In △*B*, the sum is 108.
    In △*C*, the sum is 195.

2. Find the ratios of *SP/SI*, where *SI* is the sum of the interior points.

    The resulting sequence is 9/1, 12/3, 15/6, 18/10, 21/15, 24/21, 27/28. How could these ratios be interpreted? The sequence 9, 12, 15, 18, 21, 24, 27 may represent the number of numbers that lie on the triangles. It may also be noted that there is a constant increase of 3 in the sequence. Each time a new triangle is drawn with an additional number in each leg, the hypotenuse also contains an additional number, giving a total of three additional numbers. Again the triangular numbers may be seen.

3. Find the ratios *SV/SP* from *SV/SI* ÷ *SP/SI* = *SV/SP*.

    The resulting sequence is 1/3, 1/4, 1/5, 1/6, 1/7, 1/8, 1/9, where 3, 4, 5, 6, 7, 8, and 9 represent the number of units in the successive legs of the triangles.

    In △*A* the length of each leg is 3 units, in △*B* the length is 4 units, and in △*C* the length is 5 units.

Next the students were asked to draw squares on the addition table so that the sides of the squares lay on the rows and columns of the table as shown in Fig. 4.

They were given the following instructions to follow for each square:

1. Find the sum of the numbers that represent the vertices and call this sum *SV*.

    In square *D*, the sum is 12.
    In square *E*, the sum is 28.
    In square *F*, the sum is 36.

2. Find the sum of the numbers that are in the interior of the square and call this sum *SI*.

    In square *D*, the sum is 3.
    In square *E*, the sum is 28.
    In square *F*, the sum is 81.

The Addition Table: Experiences in Practice–Discovery   51

3. Find the ratios of the sums found in steps 1 and 2, that is, SV/SI.

From square D, the ratio is 12/3, or 4/1.
From square E, the ratio is 28/28, or 4/4.
From square F, the ratio is 36/81, or 4/9.

+	0	1	2	3	4	5	6	7	8	9
0	0	1	2	3	4	5	6	7	8	9
1	1	2	3	4	5	6	7	8	9	10
2	2	3	4	5	6	7	8	9	10	11
3	3	4	5	6	7	8	9	10	11	12
4	4	5	6	7	8	9	10	11	12	13
5	5	6	7	8	9	10	11	12	13	14
6	6	7	8	9	10	11	12	13	14	15
7	7	8	9	10	11	12	13	14	15	16
8	8	9	10	11	12	13	14	15	16	17
9	9	10	11	12	13	14	15	16	17	18

Fig. 4. Addition table with squares superimposed.

If squares are drawn on the addition table so that the successive lengths are increased by 1, the sequence of ratios SV/SI = 4/1, 4/4, 4/9, 4/16, 4/25, 4/36, 4/49, 4/64, 4/81. The constant 4 may represent the four vertices of a square, and the sequence 1, 4, 9, 16, 25, 36, ..., 81 is easily recognized as the first nine square numbers. The arrangement of the set of interior numbers of the square is similar to the geometric representation of the square numbers (see Fig. 5).

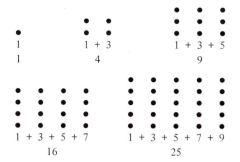

Figure 5

The ratio SP/SI, where SP is the sum of the numbers that lie on the sides of the square and SI is the sum of the numbers that are in the interior of the square, results in the sequence 8/1, 12/4, 16/9, 20/16, 24/25, 28/36, 32/49, 36/64, 40/81 where 8, 12, 16, 20, ..., 40 may represent the number of numbers that lie on the sides of the square. A constant increase of 4 may be noted. Also, $SV/SI \div SP/SI = SV/SP$, or 1/2, 1/3, 1/4, 1/5, 1/6, 1/7, 1/8, 1/9, 1/10, where 2, 3, 4, 5, ..., 10 represent the number of units in each side of the squares.

52　Basic Ideas of Addition and Subtraction of Whole Numbers

An additional practice–discovery activity is to find the sequences when hexagons and octagons are drawn on the addition table. The ratios exhibit patterns that are similar to the relations previously shown for the triangles and squares.

It may be noted that one characteristic of the addition table is that the consecutive entries in the rows have a constant increase and the consecutive entries in the columns also have a constant increase. Since it is this idea of constant increases on which the table is built, the patterns for the polygons should hold on any table with a similar design, such as the 100 table, a page of a calendar, a table of consecutive odd numbers or even numbers, a table of multiples of n where n is a whole number, and a table formed by "skip counting" (beginning with 1).

REFERENCES FOR FURTHER READING

1. Steinberg, Z., "Will the set of children . . .?" *AT,* **18** (1971), pp. 105–108. Describes work in a first-grade class on the concepts of set, set union, and set intersection.

2. "Arithmetic with frames," Chapter 3 of *Enrichment Mathematics for the Grades,* National Council of Teachers of Mathematics, Reston, Va. (1963), or *AT,* **4** (1957), pp. 119–124. Both of these articles are written by some of the original users of frames—the staff of the University of Illinois Committee on School Mathematics project.

3. Folsom, M., "Frames, frames, and more frames," *AT,* **10** (1963), pp. 484–485.

The following three references discuss the general ideas of teaching by the discovery method.

4. Ausubel, D. P., "Some psychological and educational limitations of learning by discovery," *AT,* **11** (1964), pp. 290–301.

5. Jones, P. S., "Discovery teaching—from Socrates to modernity," *AT,* **17** (1970) (unnumbered pages).

6. Kersh, B. Y., "Learning by discovery", *AT,* **11** (1964), pp. 226–231. Contains many other additional references.

7. Dubisch, R., "Generalizing a property of prime numbers," *AT,* **21** (1974), pp. 93–94. A discussion of the remarks in the text on numbers that have exactly three divisors, and extended to 4, 5, . . . divisors.

The next two articles deal with various number patterns that can lead to discovery.

8. Carman, R. A., and M. J. Carman, "Number patterns," *AT,* **17** (1970), pp. 637–639.

9. Hashisaki, J., and J. A. Peterson, "Patterns in arithmetic," *AT,* **13** (1966), pp. 209–212.

10. Goldenberg, E. P., "Scrutinizing number charts," *AT,* **17** (1970), pp. 645–653. Describes a far-reaching discovery-type activity originally developed by David Page for the University of Illinois Arithmetic Project under the name "Maneuvers on Lattices."

"Magic squares" are described in the following two articles.

11. Adkins, B. E., "Adapting magic squares to classroom use," *AT,* **10** (1963), pp. 498–500.

12. Cappon, J. Sr., "Easy construction of magic squares for classroom use," *AT,* **12** (1965), pp. 100–105.

Many games such as dominoes and those involving dice provide excellent practice in arithmetic as well as some opportunities for discovery of strategies. One very good game, not well known in the U.S.A. but played throughout Africa under various names, is described in the following article.

13. Haggerty, J. B., "KALAH—An ancient game of mathematical skill," *AT*, **11** (1964), pp. 326–330.

There are many books available on activities and games for children. Two of them are:

14. Kelley, S. J., *Learning Mathematics through Activities*, James E. Freel and Associates, Cupertino, Calif. (1973).

15. Spitzer, H., *Enrichment of Arithmetic*, McGraw-Hill Book Co., New York (1964).

Chapter 3

Naming Whole Numbers

3.1 EARLY BEGINNINGS

Consider the problem of counting the number of members of the set $\{\bigcirc, \triangle, \square\}$. We count one, two, three. But Germans count eins, zwei, drei; the French un, deux, trois; and the Japanese ichi, ni, san. The words are arbitrary and vary from language to language. However, we, the Germans, the French, and the Japanese all write 1, 2, 3. But the Romans wrote I, II, III (and we still sometimes use these symbols on clocks and inscriptions on buildings) and the Ethiopians (for ceremonial purposes) still often use $\breve{\delta}$, $\breve{\epsilon}$, $\ddot{\imath}$.

What we (and the Germans, the French, the Japanese, and the Ethiopians) do is to use the same set of words and the same set of symbols to describe the number of members of any kind of finite set. Whether we have a set of houses, a set of people, or a set of marbles, we use, in English, the words one, two, three, and so on, to state the number of elements of the set. (But, as was mentioned in Section 1.5, the example of the Tsimschian Indians shows that this wasn't always done. Furthermore, there is some possible evidence of the same situation once having applied in English in that there are several ways of indicating *twoness*: a *pair* of shoes, a *brace* of ducks, a *duet*, etc. As Bertrand Russell once wrote: "It must have required many ages to discover that a brace of pheasants and a couple of days were both instances of the number two.")

3.2 CONCEPT OF A BASE

Ignoring such exceptions as that of the Tsimschian language, let us consider the problems involved in developing a number vocabulary. The first problem, from the point of view of the learner, is simply one of memorization. Suppose (and there is reason for this choice of language!) you were forced to memorize the Japanese names for the first one hundred numbers. Now the first ten are, in order,

56 Naming Whole Numbers 3.3

ichi, ni, san, shi, go, roku, shichi, hachi, ku, and ju. You would then, I am sure, be immensely relieved to know that with no further words to learn you could easily recite the next nine number words in a simple and obvious fashion. They are ju-ichi, ju-ni, ju-san, ju-shi, ju-go, ju-roku, ju-shichi, ju-hachi, and ju-ku.

Now if English enjoyed the same systematic structure for naming numbers as Japanese does, we would count, as we do now, one, two, . . . , ten but then ten and one, ten and two, . . . , ten and nine. (Actually, linguistic evidence exists to show that our present words eleven, twelve, and so on, are corruptions of words that indicated their meaning more clearly. Thus, for example, "eleven" is the modern form of the Anglo-Saxon word *einlufon* meaning "one left over," i.e., ten plus one.)

Continuing, twenty is related to twain ten (where "twain" can still be found in the dictionary as a poetic form of "two," as in Kipling's "never the twain shall meet"). Thus number language in English (and French, German, etc.) becomes more obviously understandable as we name larger numbers. For example, "two thousand, five hundred, and sixty-two" is quite a reasonable "description" of the number 2562. (Japanese, by the way, continues its systematic development with ni-ju-ni for twenty two, san-ju for thirty and so on. All of this is quite an asset in beginning mathematics instruction.)

What we have been saying is that all languages we have mentioned, including Japanese and Amharic (the official language of Ethiopia), base their number-naming systems firmly upon ten (whether or not the language clearly reveals this). We have ten and one, two tens, ten tens (= one hundred), and so on. Why? Although no proof can be provided that it all happened because of the anatomy of humans with their ten fingers, the conjecture seems eminently reasonable. (A discovery of an eight-fingered race on Mars who based their counting system on eight would certainly lend support to this conjecture!)

We say that our system of naming numbers is a **base ten** system. Other bases have been used and, indeed, are still in occasional use in some parts of Africa today. In particular, base five and (among the Mayans) base twenty have been used. (Such choices of base can still, however, be construed as related to anatomy—with base twenty likely to be restricted to warm climates!) We will consider base five in some detail in the next section.

3.3 BASE FIVE NUMBER LANGUAGE

The purpose of introducing a different base here (and elsewhere in this book) is simply to help you to think again about base ten. You are so thoroughly familiar with base ten that it is often not easy for you to see the possibility of any difficulty that a child might have with it—especially in such a "simple" matter as counting. Learning to think in a base other than ten will soon convince you that things aren't as simple as they may seem! (Some series of primary school texts also give, in the *upper* grades, work with other bases to get students to think again about material that they have studied earlier in base ten.)

Base Five Number Language 57

Now the child, of course, first has to learn the words one, two, three, etc., which are, at the beginning, new to him. So, to put yourself in his place to some extent, you must begin your base five experience by learning some new words. (But only five of them!) To make life easier for you, however, we will use words rather closely related to our number words in base ten:

{□}	{□△}	{□△○}	{□△○◇}	{□△○◇✳}
ome	**tow**	**tree**	**fur**	**fit**

What's next? Why fit and ome, of course. (Which, if our new language evolved, might become "fome" to confuse later generations of children—as our eleven does now!) Then, fit and tow, fit and tree, fit and fur. Then what: fit and fit? No! In our base ten language we don't say "ten and ten" (although we could). It's two tens. So, here we say "tow fits." Then tow fits and ome (you continue) to tow fits and fur. Now you should certainly know that next comes tree fits, and then be able to continue easily until you come to fur fits and fur. What now? Fit fits? Think a moment. When we get, in base ten language, to nine tens and nine, do we say next "ten tens"? We could, of course, but the generally accepted language is "hundred". That is, a new word is needed. So, as our base five language developed, we would need a new word for "fit fits"—say "tef". Then tef and ome, and we are off again.

An actual base five language that is used by the Kuanyan tribe of southeastern Africa is mwe, vali, tatu, ne, tano, tano na mwe, tano na vali, tano na tatu, tano na ne, for the numbers one, two, . . . , nine, respectively.

Exercises 3.3

1. Beginning with fur tefs (4×25 in base ten), continue counting until you get to fur tefs, fur fits, and fur, that is, to

   $$(4 \times 25) + (4 \times 5) + 4, \quad \text{in base ten}$$

   (You will undoubtedly find that the words do not exactly flow trippingly off your tongue—and so you will better appreciate children's difficulties in learning to count!) What does naming the next number involve? (*Hint*. Think of nine hundred, ninety (i.e., nine tens), and nine in base ten.)

2. Make up a base four language and write numbers in it up to twenty.

3. The Bine tribe of the western district of Papua count iepa, neneni, nesae, nesae iepa, nesae neneni, neneni sesae, neneni nesae iepa, etc. What base are they using?

4. In the Yoruba language of West Africa, the word for "four" is "errin", the word for twenty is "logun", and the words for "sixteen" are "errin din logun". Explain!

5. In the Ateso language of a Kenya tribe, the words for the first fifteen numbers are idope, iyarei, iuni, iwongon, ikany, ikany kape, ikany karei, ikany kauni, ikany kaangon, itomon, itomon adiope, itomon aarei, itomon auni, itomon aangon, itomon akany. What conclusions can you draw about their choice of base?

3.4 NUMERALS IN BASE TEN

Numerals are simply names of numbers. In that sense "two" or "zwei", for example, are just as much numerals as the symbols "2" or "II". Customarily, however, the word "numeral" is reserved for the latter kind. Once again there is a long history of their development, which varies from culture to culture. The end result, however, is now in common use almost all over the world. We have as symbols for the first nine counting numbers,

1, 2, 3, 4, 5, 6, 7, 8, 9

Then, as you know, we introduce the marvelous symbol "0" for the number zero. With this added to the numerals for the first nine counting numbers we can proceed, with the utmost ease, to write a numeral for any whole number. This is not so for Roman or Amharic numerals, however. Even though a separate symbol is not needed for every number in the Roman system (XI, for example, simply combines the symbols X for ten and I for one, to get a symbol for eleven), a never-ending succession of symbols is needed to prevent long repetitions of symbols. Thus D = XXXXX, C = DD, etc.

Similarly, the Ethiopians introduce a new symbol, $\bar{\mathit{I}}$, for ten and then write $\bar{\mathit{I\!E}}$, $\bar{\mathit{I\!\bar{e}}}$, and $\bar{\mathit{I\!\!\bar{I}}}$ for eleven, twelve, and thirteen, respectively. Separate symbols $\bar{\mathit{\bar{\lambda}}}$, $\bar{\mathit{\bar{\vartheta}}}$, and $\bar{\mathit{\bar{\vartheta}}\!}$, etc., are then introduced for twenty, thirty, forty, etc., so that considerably more memorization of symbols is needed for this system than for the Roman system.

All of the simplicity of what are called **Hindu-Arabic numerals** (thus named because they reached Europe from the Hindus via the Arabs) is a result of the brilliant invention (inventor or inventors unknown) of the concept of **place value.** Thus, for example, 111 means 1 hundred, 1 ten, and 1 one, as shown in Fig. 3.1, in contrast to the Roman III, which simply means three ones. The numerals 0, 1, 2, ... , 9, that are used to write a numeral for any number, are called the **digits** of the numeral. Thus 2143 has 2, 1, 4, and 3 for its digits.

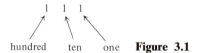

Figure 3.1

3.5 PICTURING BASE TEN NUMERALS FOR CHILDREN

In teaching the concept of place value to children, considerable use should be made of "stick bundles" and the abacus (plus other devices such as number trays, bean sticks, etc., which we'll not discuss here). Collections of individual sticks can and should, of course, be used to illustrate the numbers 1, 2, ... , 9. But now we bundle sticks into sets of ten to show the numbers from 10 to 99, as illustrated in Fig. 3.2 taken from a first-grade text. (It is, of course, possible to tie together ten bundles of ten each to show one hundred, but this gets too cumbersome.)

3.5 Picturing Base Ten Numerals for Children 59

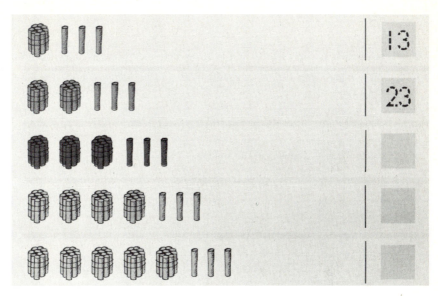

Figure 3.2

Figure 3.3, taken from a fifth-grade text, shows the representation of 562,342 on an abacus. Clearly, additional rods can be used on the left for still larger numbers, and, to the right, to show tenths, hundredths, etc. (Note that a rod with no beads on it represents the number 0.)

2. It is difficult to show thousands and millions with sets of objects, but we can show larger numbers easily on the abacus. Write the number shown on this abacus.

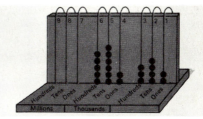

Figure 3.3

Programed Lesson 3.5

1. How can 34 be represented by sticks and stick bundles?

   3 bundles of ten sticks each and 4 single sticks

2. How can 60 be represented by sticks and stick bundles?

   6 bundles of ten sticks each and no single sticks

3. What number is represented by the sticks and stick bundles shown below?

60 Naming Whole Numbers 3.5

36

4. What number is represented by the sticks and stick bundles shown below?

21

5. Show 2370 on a drawing of an abacus.

6. Show 1035 on a drawing of an abacus.

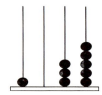

7. What number is represented on the abacus shown below?

2302

8. What number is represented on the abacus shown below?

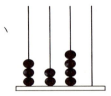

3240

3.6 NUMERALS IN BASE FIVE

Again, to fully appreciate the problems experienced by children in learning to write numerals in base ten, you should really go through the process of learning new symbols for numbers in base five—just as you were asked to learn (somewhat) new words. I will spare you this task, however, by "borrowing" a sufficient number of symbols from the customary base ten numerals, i.e., 0, 1, 2, 3, 4 for zero, ome, tow, tree, and fur, respectively. Then what is the numeral for fit? Just as 10 is the numeral for ten in base ten, so 10 is the numeral for fit in base five (fit). That is, in base ten, 10 stands for one ten and zero ones, whereas in base five 10 stands for one five and zero ones. To distinguish between numerals in different bases, we can use subscripts. Thus 10_{ten} means the numeral for ten in base ten, and 10_{five} means the numeral for five in base five. Similarly 125_{ten} is a base ten numeral and 243_{five} is a base five numeral. When no subscript is used, we shall agree that base ten is meant. Thus, for example, 125 will mean 125_{ten}. (Generally, however, as shown in the next section, it helps to avoid confusion to use the subscript "ten" when working with more than one base.)

The first ten numerals in base five are thus 0, 1, 2, 3, 4, 10, 11, 12, 13, and 14. Now what? There is a temptation to write 15. But "5" is not a symbol in base five, just as there is no separate symbol for ten in base ten. Instead we continue with

  20, 21, 22, 23, 24,

where 20 means two fives and no ones, 21 means two fives and one one, etc. Then

  30, 31, 32, 33, 34, 40, 41, 42, 43, 44

What is next? (Think of the next base ten numeral after 99.)

Be careful not to read, for example, 321_{five} as "three hundred twenty-one." "Three hundred twenty one" is base ten language. We can read 321_{five} as "three twenty-fives, two fives, and one" or, in the language introduced in Section 3.3, as "tree tefs, tow fits, and ome." In practice, however, we would normally simply say "three, two, one, base five."

Note that we can also use stick bundles and abaci to represent numbers in bases other than ten. Our "basic" bundles for base five would, of course, consist of five sticks, and the next "higher" bundle would consist of five of these basic bundles. Thus the sticks and stick bundles shown in Fig. 3.4 would represent 23_{five}. Similarly, the base five abacus shown in Fig. 3.5 represents 203_{five}.

Figure 3.4

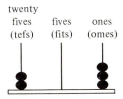

Figure 3.5

62 Naming Whole Numbers

Programed Lesson 3.6

1. How would 321_{five} be represented by stick bundles?

   By three (tree) bundles of twenty-five (tef) sticks, two bundles of five (fit) sticks, and one (ome) single stick.

2. How would 321_{five} be represented on a base five abacus?

   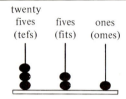

   twenty
   fives   fives   ones
   (tefs)   (fits)   (omes)

3. How would 321_{four} be represented by stick bundles?

   By three bundles of sixteen sticks, two bundles of four sticks, and one single stick.

4. How would 321_{four} be represented on a base four abacus?

   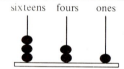

   sixteens   fours   ones

5. Write the base five numerals from 44_{five} to 200_{five}.

   100, 101, 102, 103, 104, 110, 111, 112, 113,
   114, 120, 121, 122, 123, 124, 130, 131, 132,
   133, 134, 140, 141, 142, 143, 144, 200

6. Write the four base five numerals following 443_{five}.

   444, 1000, 1001, 1002

7. Write the first twenty numerals in base four beginning with 1.

   1, 2, 3, 10, 11, 12, 13, 20, 21, 22, 23, 30, 31,
   32, 33, 100, 101, 102, 103, 110

3.7 CHANGING FROM ONE BASE TO ANOTHER

Suppose that we had 24_{ten} sticks in the form of two bundles of ten each and four single sticks. Imagine (or better, actually perform the experiment!) untying the bundles and having all the 24_{ten} sticks lying on a table. Now begin gathering these sticks up in bundles of five. If this is done, we will wind up with four bundles of five sticks each and four single sticks.

Conclusion: $24_{ten} = 44_{five}$.

Similarly, if we had 32_{five} sticks in the form of three bundles of five each and two single sticks, we could untie the bundles and then begin gathering these sticks into bundles of *ten*. The result will be one bundle of ten and seven single sticks.

Conclusion: $32_{five} = 17_{ten}$.

Now such a procedure for changing from one base to another is very meaningful—especially if done with actual sticks—but, on the other hand, it is not a very efficient procedure in converting, for example, 3249_{ten} to another base. Essential for a more efficient procedure is the use of **expanded notation,** where we use the fact that $10^2 = 10 \times 10 = 100$, $10^3 = 10 \times 10 \times 10 = 1000$, etc., to write, for example,

$$2708_{ten} = [(2 \times 10^3) + (7 \times 10^2) + (0 \times 10) + 8]_{ten}$$

and

$$324_{five} = [(3 \times 10^2) + (2 \times 10) + 4]_{five}$$

(Recall that 10_{five} is our symbol for five in base five and is read as "five" or "one, zero, base five" and not as "ten".) Thus, to convert 324_{five} to a numeral in base ten, we use the fact that $10_{five} = 5_{ten}$, so that

$$324_{five} = [(3 \times 10^2) + (2 \times 10) + 4]_{five}$$
$$= [(3 \times 5^2) + (2 \times 5) + 4]_{ten}$$
$$= (75 + 10 + 4)_{ten} = 89_{ten}$$

Here is another example:

$$1122_{three} = [(1 \times 10^{10}) + (1 \times 10^2) + (2 \times 10) + 2]_{three}$$
$$= [(1 \times 3^3) + (1 \times 3^2) + (2 \times 3) + 2]_{ten}$$
$$= (27 + 9 + 6 + 2)_{ten} = 44_{ten}$$

(Note the exponent "10" in 10^{10}. It would be technically incorrect to write 10^3 here since "3" is not a symbol in base three—although, in practice, there is no harm in using base ten symbols for *exponents*.)

Going from base ten to a base other than ten is not as simple as going from a base other than ten to base ten as we have been discussing. Suppose, for example, that we want to go from base ten to base five. Then what we need to do first is to write our base ten numeral in terms of powers of 5:

$$5^1 = 5, \quad 5^2 = 25, \quad 5^3 = 125, \quad 5^4 = 625, \quad \text{etc.}$$

64   Naming Whole Numbers   3.7

For example, given 729_{ten}, we first divide 729 by 625, the largest power of 5 less than 729:

$$
\begin{array}{r}
1 \\
625 \overline{\smash{)}729} \\
625 \\
\hline
104
\end{array}
$$

(All such divisions are in base ten, of course.) This shows us that

$$729_{\text{ten}} = [(1 \times 625) + 104]_{\text{ten}} = [(1 \times 5^4) + 104]_{\text{ten}}.$$

Now since $104 < 125$, we divide 104 by 25:

$$
\begin{array}{r}
4 \\
25 \overline{\smash{)}104} \\
100 \\
\hline
4
\end{array}
$$

Thus,

$$
\begin{aligned}
729_{\text{ten}} &= [(1 \times 5^4) + (0 \times 5^3) + (4 \times 5^2) + 4]_{\text{ten}} \\
&= [(1 \times 5^4) + (0 \times 5^3) + (4 \times 5^2) + (0 \times 5) + 4]_{\text{ten}} \\
&= [(1 \times 10^4) + (0 \times 10^3) + (4 \times 10^2) + (0 \times 10) + 4]_{\text{five}} \\
&= 10,404_{\text{five}}
\end{aligned}
$$

Note how we must be careful to insert the "0's":

$$
\begin{aligned}
144_{\text{five}} &= [(1 \times 10^2) + (4 \times 10) + 4]_{\text{five}} \\
&= [(1 \times 5^2) + (4 \times 5) + 4]_{\text{ten}} \\
&= (25 + 20 + 4)_{\text{ten}} \\
&= 49_{\text{ten}} \neq 729_{\text{ten}}.
\end{aligned}
$$

Here is another example: To convert 729_{ten} to base four, we first calculate the powers of 4 less than 729. We have $4^1 = 4$, $4^2 = 16$, $4^3 = 64$, $4^4 = 256$ (but $4^5 = 1024 > 729$). Now we divide 729 by 256 to get

$$
\begin{array}{r}
2 \\
256 \overline{\smash{)}729} \\
512 \\
\hline
217
\end{array}
$$

so that $729_{\text{ten}} = [(2 \times 256) + 217]_{\text{ten}}$. Next we divide 217 by 64 to get

$$
\begin{array}{r}
3 \\
64 \overline{\smash{)}217} \\
192 \\
\hline
25
\end{array}
$$

3.7   **Changing from One Base to Another**   **65**

so that $729_{ten} = [(2 \times 256) + (3 \times 64) + 25]_{ten}$. We now proceed to divide 25 by 16 to get

$$
\begin{array}{r}
1 \\
16\overline{)25} \\
\underline{16} \\
9
\end{array}
$$

so that $729_{ten} = [(2 \times 256) + (3 \times 64) + (1 \times 16) + 9]_{ten}$. Finally, we divide 9 by 4 to get

$$
\begin{array}{r}
2 \\
4\overline{)9} \\
\underline{8} \\
1
\end{array}
$$

so that

$$729_{ten} = [(2 \times 256) + (3 \times 64) + (1 \times 16) + (2 \times 4) + 1]_{ten}$$

Thus

$$
\begin{aligned}
729_{ten} &= [(2 \times 256) + (3 \times 64) + (1 \times 16) + (2 \times 4) + 1]_{ten} \\
&= [(2 \times 4^4) + (3 \times 4^3) + (1 \times 4^2) + (2 \times 4) + 1]_{ten} \\
&= [(2 \times 10^{10}) + (3 \times 10^3) + (1 \times 10^2) + (2 \times 10) + 1]_{four} \\
&= 23{,}121_{four}
\end{aligned}
$$

(Note that 4^4_{ten} is written as 10^{10}_{four} rather than 10^4_{four}—see earlier comment on this point.)

In practice, of course, the work just shown can be greatly condensed. In particular, we can write, for example,

$$(2 \times 10^2) + (3 \times 10) + 4$$

as

$$2 \times 10^2 + 3 \times 10 + 4,$$

omitting the parentheses. It is better, however, to write out enough detail so that the meaning of the process is clear, rather than to condense the work to such an extent that the meaning is obscure.

Programed Lesson 3.7

1. Write the following numerals in expanded form in the given base.
   a) 2314_{ten}

    $(2 \times 10^3 + 3 \times 10^2 + 1 \times 10 + 4)_{ten}$

66   Naming Whole Numbers   **3.7**

b) $20{,}532_{\text{ten}}$

$(2 \times 10^4 + 0 \times 10^3 + 5 \times 10^2 + 3 \times 10 + 2)_{\text{ten}}$

c) 232_{five}

$(2 \times 10^2 + 3 \times 10 + 2)_{\text{five}}$

d) 1033_{four}

$(1 \times 10^3 + 0 \times 10^2 + 3 \times 10 + 3)_{\text{four}}$

e) 256_{seven}

$(2 \times 10^2 + 5 \times 10 + 6)_{\text{seven}}$

f) 202_{three}

$(2 \times 10^2 + 0 \times 10 + 2)_{\text{three}}$

2. 10^4_{three} is better written as $10^{\underline{}}_{\text{three}}$.

11

3. 10^6_{four} is better written as $10^{\underline{}}_{\text{four}}$.

12

4. To convert 234_{five} to base ten we first write $234_{\text{five}} = (2 \times 10^2 + 3 \times \underline{} + \underline{})_{\text{five}}$.

10, 4

5. Now we write $(2 \times 10^2 + 3 \times 10 + 4)_{\text{five}} = (2 \times \underline{} + 3 \times 5 + \underline{})_{\text{ten}}$

5^2, 4

6. Since $(2 \times 5^2 + 3 \times 5 + 4)_{\text{ten}} = (50 + \underline{} + \underline{})_{\text{ten}}$, we conclude that $234_{\text{five}} = \underline{}_{\text{ten}}$.

15, 4, 69

7. To convert 763_{ten} to base five, we first note that the largest power of five that is less than 763 is $\underline{}$.

$5^4 \, (= 625)$

8. Now we divide 763 by 625 to get $763_{\text{ten}} = (\underline{} \times 625 + \underline{})_{\text{ten}}$.

1, 138

9. Since the largest power of five that is less than 138_{ten} is $\underline{}$, we divide 138 by $\underline{}$ to get $138_{\text{ten}} = (\underline{} \times 125 + \underline{})_{\text{ten}})$.

3.7   **Changing from One Base to Another**   **67**

$5^3 \ (= 125), \ 125, \ 1, \ 13$

10. Since the largest power of five that is less than 13_{ten} is ____, we divide 13 by ____ to get $13_{ten} = ($____ $\times \ 5 + $____$)_{ten}$.

$5^1 \ (= 5), \ 5, \ 2, \ 3$

11. Thus $763_{ten} = ($____ $\times \ 5^4 + $____ $\times \ 5^3 + $____ $\times \ 5^2 + $____ $\times \ 5 + $____$)_{ten}$.

$1, \ 1, \ 0, \ 2, \ 3$

12. But

$$(1 \times 5^4 + 1 \times 5^3 + 0 \times 5^2 + 2 \times 5 + 3)_{ten}$$
$$= (1 \times 10^4 + 1 \times \text{\_\_\_\_} + 0 \times \text{\_\_\_\_} + 2 \times \text{\_\_\_\_} + \text{\_\_\_\_})_{five}.$$

$10^3, \ 10^2, \ 10, \ 3$

13. Finally, since $(1 \times 10^4 + 1 \times 10^3 + 0 \times 10^2 + 2 \times 10 + 3)_{five} = $____$_{five}$, we conclude that $763_{ten} = $____$_{five}$.

$11{,}023, \ 11{,}023$

Exercises 3.7

1. Convert 3214_{five} to a base ten numeral.

2. Convert 645_{seven} to a base ten numeral.

3. Convert 1232_{four} to a base ten numeral.

4. Convert 1296_{ten} to a base five numeral.

5. Convert 1296_{ten} to a base four numeral.

6. Convert 1296_{ten} to a base seven numeral.

7. Convert 1232_{four} to a base three numeral.

8. Convert 1202_{three} to a base four numeral.

Chapter Test

In preparation for this test, you should review the concepts of the base of a numeration system (56), *of place value* (58), *and of expanded notation* (63). *You should also review how to:*

1. Count in bases other than ten (57);

68 Naming Whole Numbers

2. Use stick bundles to represent a whole number in various bases (59 and 61);
3. Use an abacus to represent a whole number in various bases (59 and 61);
4. Write numerals for whole numbers in various bases (61);
5. Change a numeral in a base other than ten to a numeral in base ten (63); and
6. Change a numeral in base ten to a numeral in a base other than ten (64).

1. Make up a base three language and write words for numbers in this language up to, and including, thirteen.
2. How would 102_{three} be shown by stick bundles?
3. Show 102_{three} on a drawing of a base three abacus.
4. Write the first thirty base three numerals.
5. Write 1022_{three} as a base ten numeral.
6. Write 64_{ten} as a base three numeral.

TEST ANSWERS

1. The choice of words is arbitrary, of course. Borrowing our base five language to some extent, we would have ome, tow, tree, tree and ome, tree and tow, tow trees, tow trees and ome, tow trees and tow. (1, 2, 3, 3 + 1, 3 + 2, 2 × 3, 2 × 3 + 1, 2 × 3 + 2). Now we must have a word for tree trees—say "forest". Then we have forest, forest and ome, forest and tow, forest and tree (3^2, $3^2 + 1$, $3^2 + 2$, $3^2 + 3$). Next would come forest, tree, and ome ($3^2 + 3 + 1$), and this would complete our list of thirteen.
2. One bundle of nine sticks (i.e., three bundles of bundles of three), no bundles of three, and two single sticks.
3. nines threes ones

4. 1, 2, 10, 11, 12, 20, 21, 22, 100, 101, 102, 110, 111, 112, 120, 121, 122, 200, 201, 202, 210, 211, 212, 220, 221, 222, 1000, 1001, 1002, 1010.
5. $(1022)_{three} = (1 \times 10^{10} + 0 \times 10^2 + 2 \times 10 + 2)_{three}$
   $= (1 \times 3^3 + 0 \times 3^2 + 2 \times 3 + 2)_{ten}$
   $= (27 + 0 + 6 + 2)_{ten} = 35_{ten}$.
6. The largest power of 3 less than 64 is $3^3 = 27$. We have $64 = 2 \times 27 + 10$. Then the largest power of 3 less than 10 is $3^2 = 9$ and we have $10 = 1 \times 9 + 1$. So

$64_{ten} = (2 \times 27 + 1 \times 9 + 0 \times 3 + 1)_{ten}$
$= (2 \times 3^3 + 1 \times 3^2 + 0 \times 3 + 1)_{ten}$
$= (2 \times 10^{10} + 1 \times 10^2 + 0 \times 3 + 1)_{three} = 2101_{three}$

Early Mayan mathematics

by Donald R. Byrkit

Reprinted from *The Arithmetic Teacher*, Vol. 17 (1970), pp. 387–390.

During the period 500 B.C. to 600 A.D., at a time when the rest of the world was, arithmetically speaking, in the dark ages, the Mayas of Central America built many buildings, temples, and courts that required considerable knowledge of mathematics. They developed a vigesimal system (base 20) of numbers with positional notation and a special symbol for zero. They wrote and used numbers as great as 12,489,781 and developed an intricate calendar that ran, never missing a day, for 2,148 years.

As early as 752 B.C., the Mayas were trying to develop a mathematical system to apply to calendrics in order to predict eclipses of the sun. A special cycle of 260 days running concurrently with the official year of 360 days and the solar year of 365 days enabled them to predict eclipses with great accuracy. The purpose of such prediction was to enable them to attempt to drive away the eclipse by magical means! So great was the accuracy of their predictions, however, that they were only occasionally able to pride themselves on the success of their magic. They would probably have been less happy if they had known that it was a mistake in their calculations instead.

The vigesimal system was employed by these barefoot Indians as the basis for their system. Two basic types of numeral systems (each with variations) were used. One, corresponding roughly to our arabic numerals, employed head-variant symbols for the numbers zero through thirteen inclusive. The symbol for ten was the death's head, and the rest of the symbols probably represented the Oxlahuntiku, the thirteen gods of the upper world. The numbers from fourteen through nineteen, and occasionally thirteen, were formed by affixing the fleshless lower jaw of the death's-head symbol for ten to the appropriate symbol for a number from three through nine. For example, if six were represented by the head of a pupil with a crew cut and ten were represented by a pupil wearing glasses, then a pupil with a crew cut and wearing glasses would represent sixteen. For numbers above nineteen, another system was introduced. Since the Mayas generally used numbers to indicate how many of a particular object were being discussed, the referent could be used in the number system. For twenty of anything, the object under discussion was drawn with a flag flying from it. Four hundred items were represented with a tree or hair affixed, while a copal pouch attached meant that there were eight thousand of the item. Thus 9,182 bracelets would be represented by a drawing of five bracelets: one with a copal pouch attached, one with two trees or hairs attached, one flying nineteen flags, and two standing alone.

The other and more commonly employed system consisted, like that of the Romans, of a symbol for one and a symbol for five, additively employed (only), to represent numbers from one through nineteen. The symbol for unity was a dot (pebble) and for five a horizontal bar (stick). Additions were made vertically so that thirteen would be represented as shown in Fig. 1. At twenty, however, positional notation was introduced, again

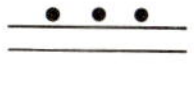

Figure 1

70   Naming Whole Numbers

employed vertically, with a special symbol for zero. This symbol is often called a "shell design," but may in fact have been a frontal view of a conventionalized closed fist.

Thus, 151 would be represented as in Fig. 2 and evaluated as in Fig. 3. Then 20, 251, 107, 125 would be as in Fig. 4.

$$
\begin{array}{ll}
7 \times 20 = 140 \\
11 \times 1 = 11 \\
\hline
151
\end{array}
$$

Figure 2   **Figure 3**

$$
\begin{array}{l}
1 \times 20 = 20 \\[4pt]
0 \times 1 = 0 \\
\hline
20 \\[4pt]
12 \times 20 = 240 \\
11 \times 1 = 11 \\
\hline
251 \\[4pt]
5 \times 20 = 100 \\
7 \times 1 = 7 \\
\hline
107 \\[4pt]
6 \times 20 = 120 \\
5 \times 1 = 5 \\
\hline
125
\end{array}
$$

Figure 4

In everyday business or civil affairs, the vigesimal system was used as a complete positional system with each higher (literally) position representing a higher power of twenty.

Thus this symbol would represent 8,428 and could be evaluated as in Fig. 5.

$$
\begin{array}{l}
1 \times 20^3 = 1 \times 8{,}000 = 8{,}000 \\
1 \times 20^2 = 1 \times 400 = 400 \\
1 \times 20 = 20 \\
8 \times 1 = 8 \\
\hline
8{,}428
\end{array}
$$

Figure 5

Since the solar year contains approximately 365 days, the straight vigesimal system was modified slightly when used in computing time, in order to simplify calculations in calendrics. Whereas the normal system represented ascending powers of 20, the modified system substituted $18 \cdot 20$ for 20^2, and $18 \cdot 20^{n-1}$ for 20^n in succeeding places. A number, then, could be represented in two different ways, and a numeral (or number representation) could represent two different numbers, depending on whether it dealt with time or not. Thus the ascending positions determined the number as in Table 1, and the numeral representing 400, 8,000, 160,000 in business or civil use would represent 360, 7,200, 144,000 in calendric use. The number represented in Fig. 6 would, on the one hand, in normal use, be read as *uuc pic, oxlahun hunbak, uac kal, uaxaclahun*—or 7 eight-thousands, 13 four-hundreds, 6 twenties, and 18—and would total out to 61,338; on the other hand, in calendric use, it would be read as *uuckatun, oxlahun tun, uac uinal, uaxaclahun kin*—or seven 7200-day periods, thirteen 360-day periods (or official years), six 20-day periods (or months), and 18 days—and would total out to 54,138 days. Proponents of the theory that the *tun*, or official year, was the basic unit of the system hold that the vigesimal system was used throughout and that the above would be simply 7 *katun*, 13 *tun* (or 153 *tun*) plus some fractional parts.

Table 1

Position	Normal	Modified
$n+1$	20^n	$18 \cdot 20^{n-1}$
.	.	.
.	.	.
.	.	.
4	20^3	$18 \cdot 20^2$
3	20^2	$18 \cdot 20$
2	20	20
1	1	1

Figure 6

The mathematical ingenuity of the Maya is made more striking by the way in which the 260-day solar eclipse or religious year and the 365-day civil year were incorporated into each other. With 18 months of 20 days each, plus the "five unlucky days," the day names would pop up at the same position every four years. Thus *Chuen 10 Tzec* could be Friday, April 1, this year, and again four years hence. Each day, however, had two designations. The number in front of the day gave its position in the 13-day religious cycle. Thus *6 Ik 1 Pop* could be Wednesday, January 2 (January 1 would be *0 Pop*—the Mayans didn't count a day until it was over). But if *8 Chuen 10 Tzec* were Friday, April 1, this year, four years hence would be *12 Chuen 10 Tzec*, and *8 Chuen 10 Tzec* would not occur again until 52 civil years (or 73 religious years) had elapsed. This period was called the "calendar round" and represented the smallest number of days (18,980) that would give a whole number of years in both systems. 18,980 is the smallest number that is a multiple of both 260 and 365. An even more encompassing system, called the "long count," placed the starting date around 3372 B.C. and counted time from that date.

Although the Mayas made no great accomplishments in algebra, their achievements in arithmetic and calendrics will long stand as a monument to their intellect.

72   Naming Whole Numbers

The Mayan number system could be a great help to elementary and junior high school teachers who are teaching their students to understand the positional number system. Most bases other than ten are artificially constructed by using the arabic numerals; the Mayan system offers a nontrivial example of a positional number system in a base other than ten. Furthermore, the system does not use the arabic numerals. Thus a teacher could use the Mayan system (the normal one) to provide such an example to intermediate grades and above. Children, particularly in junior high school, might enjoy working out addition and multiplication tables in the Mayan system and attempting to formulate rules for the operations with large numbers.

An interesting way to illustrate the Mayan calendar in the classroom would be to make four clock faces, with one hand each. The first could have the numerals 0–12; the second could have the twenty day names; the third, the numerals 0–19; the fourth, the 18 Mayan months plus "unlucky days." It would be instructive to have the first and third reversible, each with the two types of Mayan numerals, one type on each side. Thus the "wheels-within-wheels" involvement of the Mayan civil year with the Tzolkin (religious year) would be illustrated. Perhaps the intricacy of such a system would appeal to some youngsters.

The Mayan numeration system also is a good subject for class discussion. Why was twenty used as a base? What illustrations can you give where we use an elapsed-time system (such as 20 minutes past 2 o'clock)? Why did the Mayas use 18 months of 20 days instead of 20 months of 20 days to conform to the vigesimal system? For more advanced classes, calculate today's date in Mayan, assuming January 1, 3372 B.C., as the starting date. The possibilities are great, and the [literature can] provide excellent information for the resourceful. `

REFERENCES FOR FURTHER READING

The first five references discuss various numeration systems—both words and symbols.

1. Cowle, I. M., "Ancient systems of numeration—stimulating, illuminating," *AT*, **17** (1970), pp. 413–416. Egyptian, Roman, Babylonian.

2. Willerding, M. F., "Other number systems—Aids to understanding mathematics," *AT*, **8** (1961), pp. 350–356. Egyptian, Roman, Mayan.

3. Wolfers, E. P., "The original counting systems of Papua and New Guinea," *AT*, **18** (1971), pp. 77–83.

4. Zaslavsky, C., *Africa Counts*, Prindle, Weber, and Schmidt, Boston (1973).

5. Peterson, J. A., and J. Hashisaki, *Theory of Arithmetic* (Third Edition), John Wiley and Sons, Inc., New York (1971). Egyptian, Roman, Ionic (Early Greek), and Japanese.

The following references have to do with numeration in bases other than ten.

6. Hess, A. L., "A critical review of the Hindu-Arabic numeration system," *AT*, **17** (1970), pp. 493–497. Includes a discussion of base twelve.

7. Rudd, L. E., "Nondecimal numeration systems," Chapter 2 of *Enrichment Mathematics for the Grades*, National Council of Teachers of Mathematics, Reston, Va. (1963).

8. Ranucci, E. R., "Tantalizing ternary," *AT*, **15** (1968), pp. 718–722. Shows how numerals in base three can be used to solve various common puzzles.

9. Smith, K. J., "Inventing a numeration system," *AT*, **20** (1973), pp. 550–553. Shows how children might develop a numeration system in a base other than ten.

Chapter 4

Addition and Subtraction Algorithms for Whole Numbers

4.1 INTRODUCTION

The various procedures (rules or **algorithms**) for adding and subtracting whole numbers are, of course, developed gradually in the primary school. Here, however, we will consider the complete story. Once again, there is no problem in regard to your knowing the algorithms. Rather, the problem is to make sure that you understand why the algorithms work, so that you can teach them to children intelligently rather than simply giving them rules to follow blindly.

To dramatize the all too frequent plight of children who are asked to do this and to do that without any explanation of the reasons why, let us imagine that you know how to add whole numbers but have no idea of subtraction except for knowing how to subtract from 9. Suppose, then, the lesson went something like this:

"Today you are going to learn how to subtract any whole number from any whole number larger than it. For example, let us subtract 2148 from 6192. First put down the problem like this:

$$6192$$
$$-2148$$

Now copy the top number over again on the right like this:

$$6192 \qquad 6192$$
$$-2148$$

Next, subtract each digit of the bottom number of our problem from 9 (going from left to right) and write your answers, in order from left to right, underneath the number 6192. Since $9 - 2 = 7$, $9 - 1 = 8$, $9 - 4 = 5$, and $9 - 8 = 1$, our work will now look like this:

$$6192 \qquad 6192$$
$$\underline{-2148} \qquad \underline{7851}$$

73

74   Addition and Subtraction Algorithms for Whole Numbers   4.1

Now *add* the two numbers on the right to get

$$
\begin{array}{r}
6192 \\
-2148 \\
\hline
\end{array}
\qquad
\begin{array}{r}
6192 \\
+7851 \\
\hline
14043
\end{array}
$$

Finally, cross out the first digit on the left of the sum and add 1 to the remaining number to obtain the answer to our subtraction problem. So our work looks like this:

$$
\begin{array}{r}
6192 \\
-2148 \\
\hline
4044
\end{array}
\qquad
\begin{array}{r}
6192 \\
+7851 \\
\hline
\cancel{1}4043 \\
+1 \\
\hline
4044
\end{array}
$$

That is how we do subtraction problems. Now open your book to page X and do the problems there in exactly this way."

Now if such a lesson had been presented to you, I hope that you would have been able to follow the instructions. But I also hope you would have been upset and angry about the lesson so that you would have asked "What does subtraction mean?" and then "Why does this method work?" (It *does* always work—except for trivial problems such as 6192 − 6192—providing that, when the number of digits in the number we are subtracting is less than the number of digits in the number we are subtracting from, we introduce 0's for the "missing" digits. For example, we rewrite

$$
\begin{array}{r}
324 \\
-48 \\
\hline
\end{array}
\qquad \text{as} \qquad
\begin{array}{r}
324 \\
-048 \\
\hline
\end{array}
$$

This then gives us

$$
\begin{array}{r}
324 \\
+951 \\
\hline
\cancel{1}275 \\
+1 \\
\hline
276
\end{array}
$$
\qquad for a correct answer of 276.

Now the point of this imaginary lesson is *not* to advocate teaching this procedure for subtraction to children, in place of the usual one, but simply to point out that the standard algorithms for the operations of arithmetic are all too frequently presented in a similar "do this because it works" fashion. Thinking about how ridiculous the above presentation sounded will, I hope, help to ensure that you will not present arithmetic in such a way to children.

To see why the subtraction method described above works (called the **method of complements**) do the following programed lesson.

4.1   **Introduction**   **75**

Programed Lesson 4.1

1. The method of complements applied to the subtraction

$$\begin{array}{r} 8 \\ -2 \\ \hline \end{array} \quad \text{gives us} \quad \begin{array}{r} 8 \\ +7 \\ \hline 15 \\ +1 \\ \hline 6 \end{array} \quad \text{where} \quad 7 = \underline{} - 2$$

so we conclude that $8 - 2 = 6$.

9

2. The step

$$\begin{array}{r} 8 \\ +7 \\ \hline 15 \end{array} \quad \text{is equivalent to subtracting} \underline{} \text{ from } 8 + (9 - 2).$$

10

3. The step

$$\begin{array}{r} 8 \\ +7 \\ \hline 15 \\ +1 \end{array} \quad \text{is equivalent to adding} \underline{} \text{ to } 8 + (9 - 2).$$

1

4. So we are saying that $8 - 2 = [8 + (9 - 2)] - \underline{} + \underline{}$.

10, 1

5. To show that the statement in Exercise 4 is correct, we can write

$$[8 + (9 - 2)] - 10 + 1 = [8 + (9 - 2)] - 9$$
$$= (8 - 2) + (9 - \underline{}) = (8 - 2) + 0 = 8 - 2$$

9

6. Similarly

$$\begin{array}{r} 6192 \\ -2148 \\ \hline \end{array} \quad \text{done by first writing} \quad \begin{array}{r} 6192 \\ +7851 \\ \hline 14043 \end{array} \quad \text{says that}$$

we first add ($\underline{} - 2148$) to 6192.

9999

7. Then striking out "1" on the left and adding 1 to 4043 is the same as subtracting ____ from 14,043 and then adding ____.

   10,000, 1

8. So we are saying that

   6192 − 2148 = [6192 + (9999 − 2148)] − ____ + ____

   10,000, 1

9. To show that the statement in Exercise 8 is correct, we can write

   [6192 + (9999 − 2148)] − 10,000 + 1
   = [6192 + (9999 − 2148)] − 9999
   = (6192 − 2148) + (9999 − ____)
   = (6192 − 2148) + 0 = 6192 − 2148

   9999

Exercises 4.1

In Exercises 1 through 5, do the subtractions by the method of complements.

| 1. | 46
−13 | 2. | 428
−192 | 3. | 428
−39 | 4. | 2342
−1158 | 5. | 4385
−324 |

6. Explain why the method of complements works in Exercises 1 and 2.

4.2 ADDITION WITH STICK BUNDLES AND ABACUS

The first step toward an understanding of the (standard) addition algorithm by children might well involve sticks and stick bundles. (See Section 3.5.) For example, 43 + 32 is easily shown with sticks and stick bundles, as illustrated in Fig. 4.1, taken from a second-grade book.

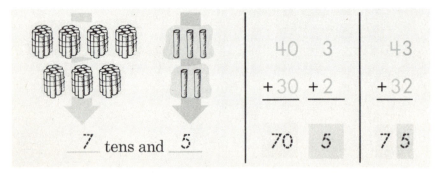

Figure 4.1

What about a problem such as 47 + 34 where "carrying" is involved? Figure 4.2, also from this second-grade book, shows how this problem can be done with sticks and stick bundles. (The next step, in which ten of the eleven single sticks are used to make another bundle of ten so that there are 8 bundles of ten and one single stick, is not shown in this figure. The teacher's guide, however, points out that this should be done.)

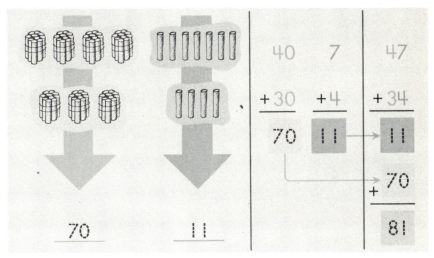

Figure 4.2

Now let us consider these two problems on an abacus. The steps involved in finding 43 + 32 are shown in Fig. 4.3.

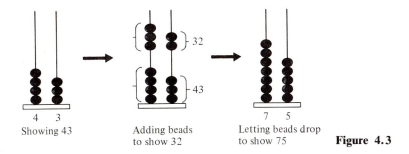

Figure 4.3

Figure 4.4 shows the first three steps in finding 47 + 34 with an abacus.

Now we use the basic idea of the abacus—namely that ten beads on any rod are "worth" one bead on the rod next to it on the left. So we remove ten beads from the ones rod and add 1 bead to the tens rod, to obtain the configuration shown in Fig. 4.5. This, in brief outline, shows how stick bundles and the abacus can be used to introduce the addition algorithm. Further details can be found in the teacher's guides for various primary-school mathematics series.

78 Addition and Subtraction Algorithms for Whole Numbers 4.3

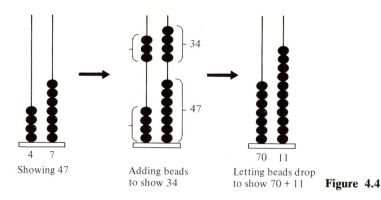

4 7			70 11	
Showing 47		Adding beads to show 34	Letting beads drop to show 70 + 11	**Figure 4.4**

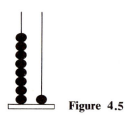

Figure 4.5

Note, by the way, the importance of having an "open ended" abacus on which we can place as many as 18 beads on a rod. Unfortunately, most of the abaci sold commercially have the rods closed at both ends and only nine beads on a rod.

Exercises 4.2

1. Describe how you would show the additions (a) 43 + 36 and (b) 26 + 37, by means of sticks and stick bundles.

2. Show how you would use an abacus to show the additions in Exercise 1.

3. Show how you would use an abacus to show the additions (a) 123 + 245 and (b) 456 + 874.

4.3 THE ADDITION ALGORITHM IN SYMBOLS

Figures 4.1 and 4.2 indicate, respectively, how we can write

$$43 + 32 = (40 + 3) + (30 + 2) = (40 + 30) + (3 + 2)$$
$$= 70 + 5 = 75$$

and

$$47 + 34 = (40 + 7) + (30 + 4) = (40 + 30) + (7 + 4)$$
$$= 70 + 11 = 70 + (10 + 1)$$
$$= (70 + 10) + 1 = 80 + 1 = 81$$

(Note the extensive use of the commutative and associative properties of addition.)

Figure 4.2 also indicates the use of the method of **partial sums,** where we write

```
   47
  +34
   11   (7 + 4 = 11)
  +70   (40 + 30 = 70)
   81
```

(Notice how this process eliminates "carrying.") Here are two other examples of the use of this method:

```
   229                              5859
   262                              6948
  +434                             +7404
    15   (9 + 2 + 4 = 15)            21   (9 + 8 + 4 = 21)
   110   (20 + 60 + 30 = 110)        90   (50 + 40 + 0 = 90)
  +800   (200 + 200 + 400 = 800)   2100   (800 + 900 + 400 = 2100)
   925                           +18000   (5000 + 6000 + 7000 = 18,000)
                                      1   (1 + 0 + 0 + 0 = 1)
                                    110   (20 + 90 + 0 + 0 = 110)
                                    100   (100 + 0 = 100)
                                  10000   (2000 + 8000 = 10,000)
                                 +10000   (10,000 = 10,000)
                                  20211
```

Note that, in our last example, two stages of partial addition are needed to avoid "carrying." Also note that the partial addition method works just as well from left to right as from right to left. For example,

```
   229
   262
  +434
   800
   110
  + 15
   925
```

The "standard" way of proceeding, finally, would look like this for the three problems we have just considered.

```
    1          11          212
   47         229         5859
  +34         262         6948
   81        +434        +7404
              925        20211
```

80   Addition and Subtraction Algorithms for Whole Numbers   **4.4**

Notice how the small numerals written at the top of the problems simply serve to abbreviate the process used in the method of partial sums.

A gradual development of the addition algorithms from stick bundles, to abacus, to partial sums (going back over the steps when necessary) should make the standard procedure meaningful to children. Again, details of such a development can be found in the teacher's guide of the children's texts and, of course, a much slower development than that given here is needed with children.

Exercises 4.3

In Exercises 1 through 5, perform the additions by the method of partial sums.

1.	57	**2.**	174	**3.**	28	**4.**	124	**5.**	3138
	+39		+539		34		289		1249
					+59		+514		+8229

6. A child makes the same kind of mistake in each of the following problems:

54	38	272	427
+9	+8	+88	+228
81	91	711	691

If he makes the same kind of mistake in the problem below, what will his answer be?

269
+56

7. In the following problems, each letter stands for one of the digits $0, 1, 2, \ldots, 9$. Determine which digit each letter stands for.

a)   48
     $+A$
     $\overline{BB}$

b) CC
     $+D$
     $\overline{50}$

c) $E7$
     $+EG$
     $\overline{95}$

4.4 SUBTRACTION WITH STICK BUNDLES AND ABACUS

Figure 4.6, from a second-grade book, shows a subtraction with sticks and stick bundles not involving "borrowing" and Fig. 4.7, also from a second-grade book, shows a subtraction involving "borrowing." [I have used the word "borrowing" here since it, like "carrying", is probably familiar to you. Current textbooks, however, avoid such terms in favor of "rewriting" or "regrouping." Thus $47 + 34$ is computed by rewriting $47 + 34$ as $(40 + 30) + (7 + 4)$, and $42 - 15$ by rewriting, as will be discussed in the next section, $42 - 15$ as $(30 + 12) - (10 + 5)$].

On an abacus, our first problem, the calculation of $48 - 23$, is done as shown in Fig. 4.8.

4.4 Subtraction with Stick Bundles and Abacus

Find the differences.

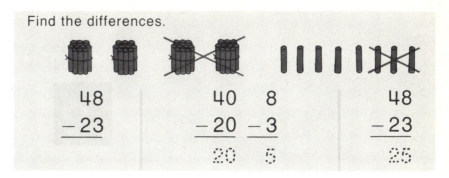

Figure 4.6

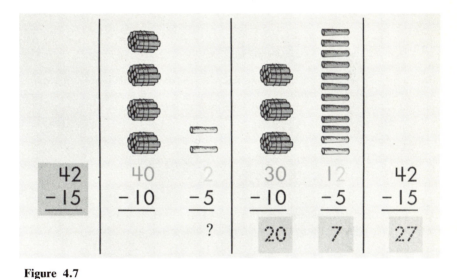

Figure 4.7

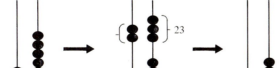

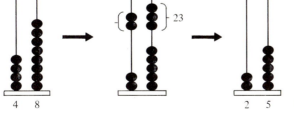

Figure 4.8

In our second problem, 42 − 15, our difficulty, of course, lies in the fact that in our abacus representation of 42, as shown in Fig. 4.9, we have fewer than 5 beads on the ones rod. What we do, then, is to "trade" one bead on the tens rod for 10 beads on the ones rod. Now our abacus appears as in Fig. 4.10, and we proceed as before.

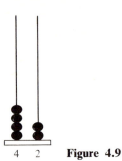
4   2   **Figure 4.9**

3   12   **Figure 4.10**

Exercises 4.4

1. Describe how you would show the subtractions (a) 64 − 23 and (b) 64 − 27, by means of sticks and stick bundles.

2. Show how you would use an abacus to show the subtractions in Exercise 1.

3. Show how you would use an abacus to show the subtractions:
   a) 347 − 125   **b)** 347 − 129   **c)** 347 − 159.

4.5 SUBTRACTION ALGORITHMS IN SYMBOLS

There doesn't seem to be any sort of method of "partial subtraction" to lead nicely into the customary procedure for subtraction! What we can do, of course, is to write, for the problems considered in the previous section,

$$\begin{array}{c} 48 \\ -23 \\ \hline \end{array} \quad \rightarrow \quad \begin{array}{r} 40 + 8 \\ (-)20 + 3 \\ \hline 20 + 5 = 25 \end{array}$$

and

$$\begin{array}{c} 42 \\ -15 \\ \hline \end{array} \quad \rightarrow \quad \begin{array}{r} 40 + 2 \\ (-)10 + 5 \\ \hline \end{array} \quad \rightarrow \quad \begin{array}{r} 30 + 12 \\ (-)10 + 5 \\ \hline 20 + 7 = 27 \end{array}$$

In the second example, considerable condensation normally takes place, so that the work looks like this:

$$\begin{array}{r} \overset{3}{\cancel{4}}{}^{1}2 \\ -1\ 5 \\ \hline 2\ 7 \end{array}$$

Obviously, if such condensation is pushed upon children too soon, the procedure will become a rote one performed without understanding.

4.5   **Subtraction Algorithms in Symbols   83**

Here is another example of this procedure and the explanation for it:

$$\begin{array}{r} 2\,{}^{1}1 \\ \not{3}\,\not{2}\,6 \\ -1\,3\,8 \\ \hline 1\,8\,8 \end{array}$$

$$\begin{array}{r} 326 \\ -138 \end{array} \rightarrow \begin{array}{r} 300 + 20 + 6 \\ (-)100 + 30 + 8 \end{array} \rightarrow \begin{array}{r} 300 + 10 + 16 \\ (-)100 + 30 + 8 \end{array} \rightarrow \begin{array}{r} 200 + 110 + 16 \\ (-)100 + 30 + 8 \\ \hline 100 + 80 + 8 = 188 \end{array}$$

We can dramatize these problems for children by having them think in terms of coinage. For example, in

$$\begin{array}{r} 42 \\ -15 \\ \hline \end{array}$$

we can think of the "top person" as having 4 dimes and 2 pennies (40 + 2). "Top" then exchanges one of the dimes for ten pennies, so "top" now has 3 dimes and 12 pennies (30 + 12). (This shows clearly that no "borrowing" is really involved!) Similarly, in the second example we can consider 326 in terms of 3 dollar bills, 2 dimes, and 6 pennies. Then 1 dime is exchanged for 10 pennies, giving 300 + 10 + 16 and then, 1 dollar bill is exchanged for 10 dimes giving 200 + 110 + 16 (2 dollar bills, 11 dimes, and 16 pennies).

Although the procedure just described is the one commonly used, the **equal additions** method (also called the **Austrian** method) is sometimes seen. In any event, this alternative method is worth knowing. It can be considered as based on the algebraic equality: For all numbers a, b, and c,

$$a - b = (a + c) - (b + c)$$

For example, $42 - 15$ is found by this method by first writing:

$$42 - 15 = (42 + 10) - (15 + 10)$$
$$(a = 42, b = 15, \text{ and } c = 10 \text{ in the equality})$$

Now we write $42 + 10$ as

$$(40 + 2) + 10 = 40 + (2 + 10) = 40 + 12$$

and $15 + 10$ as:

$$(10 + 5) + 10 = (10 + 10) + 5 = 20 + 5.$$

So we have

$$\begin{array}{r} 42 \\ -15 \\ \hline \end{array} \quad \rightarrow \quad \begin{array}{r} 40 + 12 \\ (-)20 + 5 \\ \hline 20 + 7 = 27 \end{array} \quad (\text{since } 12 - 5 = 7 \text{ and } 40 - 20 = 20)$$

84   Addition and Subtraction Algorithms for Whole Numbers   **4.5**

In abbreviated form, we have

$$
\begin{array}{r}
4^1 2 \\
2 \\
-\not{1}\,5 \\
\hline
2\,7
\end{array}
$$

Here, in the coinage interpretation, we can think of the "top person" as having 4 dimes and 2 pennies, and the "bottom person" as having 1 dime and 5 pennies. Each now receives 10¢, but "top" receives the 10¢ in the form of 12 pennies, so that "top" now has 4 dimes and 12 pennies, whereas the "bottom person" receives the 10¢ in the form of 1 dime, so that "bottom" now has 2 dimes and 5 pennies. The difference in their wealth in both cases is 27¢!

Here is another subtraction done by this method:

$$
\begin{array}{r}
3^1 2^1 4 \\
2\,9 \\
-\not{1}\,\not{8}\,8 \\
\hline
1\,3\,6
\end{array}
$$

Notice that we have done two "equal additions" here—first adding 10 to write $324 + 10 = 320 + 14$ and $188 + 10 = 198$, and then adding 100 to write $320 + 14 + 100 = 300 + 120 + 14$ and $190 + 8 + 100 = 298$. In the coinage interpretation, the "top person" has 3 dollar bills, 2 dimes, and 4 pennies; and the "bottom person" has 1 dollar bill, 8 dimes, and 8 pennies. Each is given $1.10 with "top" receiving this amount as 10 dimes and 10 pennies, and "bottom" receiving it as 1 dollar bill and 1 dime. Then "top" has 3 dollar bills, 12 dimes, and 14 pennies, and "bottom" has 2 dollar bills, 9 dimes, and 8 pennies. Thus:

$$
\begin{array}{ccc}
 & \text{3 dollar} & \\
 & \text{bills} & \text{12 dimes} \quad \text{14 pennies} \\
324 & 300 \;+\; 120 \;+\; 14 \\
-188 \quad\rightarrow & (-)200 \;+\;\; 90 \;+\;\;\; 8 \\
\hline
 & \uparrow \qquad\quad \uparrow \qquad\quad \uparrow \\
 & \text{2 dollar} \quad \text{9 dimes} \quad \text{8 pennies} \\
 & \text{bills}
\end{array}
$$

If you have some difficulty with this second method, stop and think that children initially have as much difficulty with the usual method, which, to you, is familiar and easy! (Still another algorithm for subtraction is described in the article at the end of this chapter.)

Programed Lesson 4.5

1. Complete the following expanded versions of the "borrowing" procedure.

$$
\textbf{a)} \quad
\begin{array}{r}
92 \\
-78
\end{array}
\;\rightarrow\;
\begin{array}{r}
90 + \underline{\quad} \\
(-)70 + \underline{\quad}
\end{array}
\;\rightarrow\;
\begin{array}{r}
\underline{\quad} + 12 \\
\underline{\quad} + \;\;8 \\
\hline
\underline{\quad} + \underline{\quad} = \underline{\quad}
\end{array}
$$

4.5   **Subtraction Algorithms in Symbols**   **85**

$$
\begin{array}{c}
90 + 2 \\
(-)70 + 8 \\
\hline
\end{array}
\quad \rightarrow \quad
\begin{array}{c}
80 + 12 \\
(-)70 + 8 \\
\hline
10 + 4 = 14
\end{array}
$$

b)

$$
\begin{array}{c}
426 \\
-198 \\
\hline
\end{array}
\quad \rightarrow \quad
\begin{array}{c}
400 + \underline{} + 6 \\
(-) \underline{} + 90 + \underline{} \\
\hline
\end{array}
\quad \rightarrow \quad
\begin{array}{c}
400 + 10 + \underline{} \\
\underline{} + 90 + \underline{} \\
\hline
300 + \underline{} + 16
\end{array}
$$

$$
\rightarrow \quad
\begin{array}{c}
\underline{} + 90 + \underline{} \\
\hline
\underline{} + \underline{} + \underline{} = \underline{}
\end{array}
$$

$$
\begin{array}{c}
400 + 20 + 6 \\
(-)100 + 90 + 8 \\
\hline
\end{array}
\quad \rightarrow \quad
\begin{array}{c}
400 + 10 + 16 \\
(-)100 + 90 + 8 \\
\hline
\end{array}
\quad \rightarrow \quad
\begin{array}{c}
300 + 110 + 16 \\
(-)100 + 90 + 8 \\
\hline
200 + 20 + 8 = 228
\end{array}
$$

2. Complete the following expanded versions of the equal additions method.

a)

$$
\begin{array}{c}
92 \\
-78 \\
\hline
\end{array}
\quad \rightarrow \quad
\begin{array}{c}
90 + \underline{} \\
(-)\underline{} + 8 \\
\hline
\end{array}
\quad \rightarrow \quad
\begin{array}{c}
90 + 12 \\
(-)\underline{} + 8 \\
\hline
\underline{} + \underline{} = \underline{}
\end{array}
$$

$$
\begin{array}{c}
90 + 2 \\
(-)70 + 8 \\
\hline
\end{array}
\quad \rightarrow \quad
\begin{array}{c}
90 + 12 \\
(-)80 + 8 \\
\hline
10 + 4 = 14
\end{array}
$$

b)

$$
\begin{array}{c}
426 \\
-198 \\
\hline
\end{array}
\quad \rightarrow \quad
\begin{array}{c}
400 + \underline{} + 6 \\
(-)100 + \underline{} + 8 \\
\hline
\end{array}
\quad \rightarrow \quad
\begin{array}{c}
400 + 20 + 16 \\
(-)100 + \underline{} + 8 \\
\hline
\end{array}
$$

$$
\rightarrow \quad
\begin{array}{c}
400 + 120 + 16 \\
(-)200 + \underline{} + 8 \\
\hline
\underline{} + \underline{} + \underline{} = \underline{}
\end{array}
$$

$$
\begin{array}{c}
400 + 20 + 6 \\
(-)100 + 90 + 8 \\
\hline
\end{array}
\quad \rightarrow \quad
\begin{array}{c}
400 + 20 + 16 \\
(-)100 + 100 + 8 \\
\hline
\end{array}
\quad \rightarrow \quad
\begin{array}{c}
400 + 120 + 16 \\
(-)200 + 100 + 8 \\
\hline
200 + 20 + 8 \\
= 228
\end{array}
$$

Exercises 4.5

1. Do the following subtractions by the "borrowing" method, and then give an explanation, as in the text, to show why this procedure works.

$$
Example: \quad
\begin{array}{c}
\overset{4}{\cancel{5}}\overset{1}{2} \\
-3\,7 \\
\hline
1\,5
\end{array}
\qquad
\begin{array}{c}
52 \\
-37 \\
\hline
\end{array}
\quad \rightarrow \quad
\begin{array}{c}
50 + 2 \\
(-)30 + 7 \\
\hline
\end{array}
\quad \rightarrow \quad
\begin{array}{c}
40 + 12 \\
(-)30 + 7 \\
\hline
10 + 5 = 15
\end{array}
$$

86   Addition and Subtraction Algorithms for Whole Numbers   **4.6**

a)   92   b)   232   c)   526   d)   3472
    -78 -118 -188 -1885

2. Do the problems of Exercise 1 using the equal additions method with explanations as to why this procedure works.

Example: $5\overset{1}{2}$ 52   50 + 2   50 + 12
        $\overset{4}{\underset{}{-3}}\,7$ $\underline{-37}$ $\rightarrow$ $\underline{(-)30 + 7}$ $\rightarrow$ $\underline{(-)40 + 7}$
                                                    10 + 5 = 15

3. Give a "coinage" explanation, as in the text, for Exercises 1(a) and (c).

4. Give a "coinage" explanation, as in the text, for Exercises 2(a) and (c).

5. In the following problems, each letter stands for one of the digits 0, 1, 2, ..., 9. Determine which digit each letter stands for.

a)   42   b)   7C   c)   EE.
    $-A$ $-CD$ $-1D$
    $\overline{B5}$ $\overline{26}$ $\overline{7D}$

4.6 ADDITION AND SUBTRACTION IN OTHER BASES

I commented before that learning to write numerals in bases other than ten can help you to better understand numeration in base ten and also to appreciate children's difficulties with base ten. Likewise, doing computations in bases other than ten will help you to better understand computations in base ten and to appreciate children's difficulties with such computations in base ten. It is *not* intended that you become very proficient in computation in bases other than ten, nor is it implied that you will be teaching computation in bases other than ten in the lower grades. (A small amount of work in other bases is sometimes included in grades 6 through 8 for the same purpose as here—namely, to review the ideas of numeration and computation in a different framework.)

If we were to demand of you real proficiency in addition and subtraction in base five, you would need to memorize the base five addition table shown in Fig. 4.11 and to know the corresponding basic "subtraction facts" in base five: $13 - 4 = 4$, $10 - 3 = 2$, $11 - 2 = 4$, etc. For the work here, however, you can utilize your knowledge of computation in base ten. Thus, for example, you can think of $13_{\text{five}} - 4_{\text{five}}$ as $8_{\text{ten}} - 4_{\text{ten}} = 4_{\text{ten}} = 4_{\text{five}}$. (Notice, by the way, that the table of basic "addition facts" for base five is considerably smaller than the corresponding one in base ten that children need to memorize!)

+	0	1	2	3	4
0	0	1	2	3	4
1	1	2	3	4	10
2	2	3	4	10	11
3	3	4	10	11	12
4	4	10	11	12	13

Figure 4.11

4.6   **Addition and Subtraction in Other Bases**   **87**

Now examine carefully how the following addition and subtraction problems in base five are done, and make sure that you know exactly what is going on. In particular, notice that the *basic* principles of the addition and subtraction algorithms remain unchanged as we go from base ten to base five. Thus, however difficult you may find problems in bases other than ten, note that you—unlike children working for the first time in base ten—can draw upon what you already know about similar computations in base ten.

Keep in mind that all numerals in the four examples below are in base five.

1.   121 (no "carrying" involved)   2. $\overset{1\,1}{134}$ ("carrying" involved)

    $\underline{+212}$ $\underline{+243}$

     333   432

In considering Example 2, you may appreciate the value of the method of partial sums better than you did when we were considering addition in base ten. Done by the method of partial sums, the second example becomes:

$$
\begin{array}{r}
134 \\
\underline{+243} \\
12 \quad (3 + 4 = 12) \\
120 \quad (30 + 40 = 120) \\
\underline{+300} \quad (100 + 200 = 300) \\
432
\end{array}
$$

3.   243 (No "borrowing" involved)   4. $\overset{2\,{}^1 0\,{}_1}{3\,\cancel{1}\,2}$ ("Borrowing" involved)

   $\underline{-112}$ $\underline{-1\,4\,3}$

    131   1 1 4

I am sure that you see, in this last example, that "borrowing" doesn't seem as simple and routine as it appears to you (but not to children!) in base ten. The same problem done in "slow motion" would appear as follows:

$$
\begin{array}{c}
312 \\
\underline{-143}
\end{array}
\quad \rightarrow \quad
\begin{array}{c}
300 + 10 + 2 \\
\underline{(-)100 + 40 + 3}
\end{array}
\quad \rightarrow \quad
\begin{array}{c}
300 + 0 + 12 \\
\underline{(-)100 + 40 + 3}
\end{array}
$$

$$
\rightarrow \quad
\begin{array}{c}
200 + 100 + 12 \\
\underline{(-)100 + 40 + 3} \\
100 + 10 + 4 = 114
\end{array}
$$

Also, by the method of equal additions we have:

$$
\begin{array}{r}
3\,{}^1 1\,{}^1 2 \\
\overset{2\ 10}{\underline{-\cancel{1}\,\cancel{4}\,3}} \\
1\ 1\ 4
\end{array}
\qquad \text{(Note that } 4 + 1 = 10 \text{ in base five.)}
$$

88   Addition and Subtraction Algorithms for Whole Numbers   **4.6**

This, in "slow motion," becomes

$$312 + 10 = (310 + 2) + 10 = 310 + (2 + 10)$$
$$= 310 + 12 \quad \text{(adding 10 to 312)}$$

and

$$(310 + 12) + 100 = (310 + 100) + 12 = 300 + (10 + 100) + 12$$
$$= 300 + 110 + 12 \quad \text{(adding 100 more to 312)}$$

Now, adding 10 to 143:

$$143 + 10 = (100 + 40 + 3) + 10 = 100 + (40 + 10) + 3$$
$$= 100 + 100 + 3 \quad \text{(since } 40 + 10 = 100 \text{ in base five)}$$

and now adding 100 more to 143:

$$(100 + 100 + 3) + 100 = (100 + 100) + 100 + 3$$
$$= 200 + 100 + 3$$

Or, condensing the work above,

$$
\begin{array}{ccccc}
312 & & 300 + 10 + 2 & & 300 + 10 + 12 \\
-143 & \rightarrow & (-)100 + 40 + 3 & \rightarrow & (-)100 + 100 + 3 \\
\hline
\end{array}
$$

$$
\begin{array}{cc}
 & 300 + 110 + 12 \\
\rightarrow & (-)200 + 100 + 3 \\
\hline
 & 100 + 10 + 4 = 114
\end{array}
$$

It may be helpful to you to think here in terms of quarters, nickels, and pennies (100_{five}, 10_{five}, and 1_{five}, respectively) used as dollars, dimes, and pennies (100_{ten}, 10_{ten}, and 1_{ten}, respectively) were used earlier in explaining subtraction procedures in base ten. For example, writing $300 + 10 + 2$ as $200 + 100 + 12$ in base five is equivalent to exchanging 1 nickel for 10_{five} ($= 5_{\text{ten}}$) pennies and also 1 quarter for 10_{five} ($= 5_{\text{ten}}$) nickels. Similarly,

$$
\begin{array}{ccc}
312 & & 300 + 110 + 12 \\
-143 & \rightarrow & (-)200 + 100 + 3 \\
\hline
\end{array}
$$

is equivalent to giving "top" 10_{five} ($= 5_{\text{ten}}$) pennies and 10_{five} ($= 5_{\text{ten}}$) nickels for a total of $5 \times 1¢ + 5 \times 5¢ = 30¢$ (in base ten), and "bottom" 1 nickel and 1 quarter (again, 30¢ in base ten). (Unfortunately, our coinage can be so used only for base ten and base five!)

The purpose of having you do some computations in bases other than ten is not, I repeat, to have you obtain proficiency in computations in these bases, but simply to make you realize the potential difficulties of children with base *ten* computation. By now you should realize that, for real comprehension of base five arithmetic, even you would profit from some work with sticks and bundles (of five,

4.6   Addition and Subtraction in Other Bases   89

of course!) and a base five abacus. The programed lesson below, preceding the exercises, suggests the value of work with these important teaching tools before taking on problems involving only the abstract symbols.

Programed Lesson 4.6

1. Describe how you would show 234_{five} by means of sticks and stick bundles.

    Two bundles of twenty-five sticks (i.e., five bundles of five sticks each), three bundles of five sticks, and four single sticks.

2. Show, by picturing stick bundles, the addition, $23_{\text{five}} + 14_{\text{five}}$.

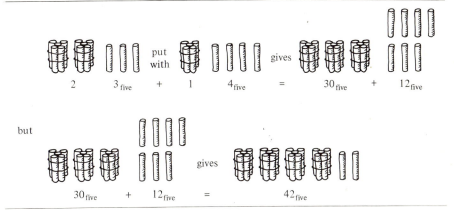

3. Show, by picturing stick bundles, the subtraction, $23_{\text{five}} - 14_{\text{five}}$.

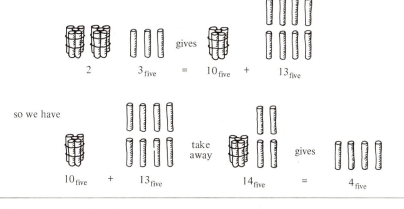

4. Show 423_{five} on a base five abacus.

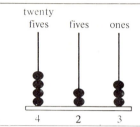

5. Show the addition $423_{five} + 134_{five}$ on a base five abacus.

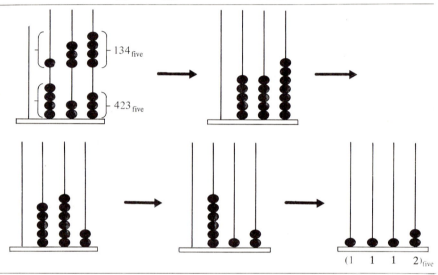

6. Show the subtraction $423_{five} - 134_{five}$ on a base five abacus.

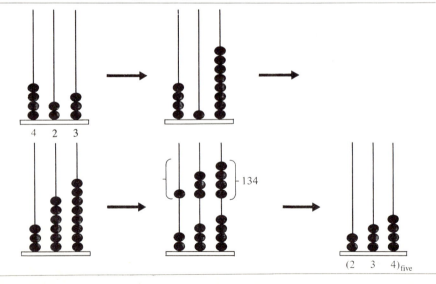

4.6   **Addition and Subtraction in Other Bases**   **91**

Exercises 4.6

In the following exercises try to avoid a purely mechanical procedure. Think through the problems and feel free to rephrase the problems, in your own mind at least, in terms of sticks and stick bundles, an abacus, or coinage.

1. Perform the following additions in base five, first by the method of partial sums, and then by the abbreviated method.

   a) 412
       +314

   b) 24
       13
       22
      +32

   c) 1322
       2401
     +1324

2. Perform the following subtractions in base five by the "borrowing" method, and then give an explanation, as in the text, as to why the method works.

   Example.

   $$\begin{array}{r} {}^{3}\!\!{}^{1}\!\!\!\not{4}\,1 \\ -2\,2 \\ \hline 1\,4 \end{array} \qquad \begin{array}{r} 41 \\ -22 \\ \hline \end{array} \rightarrow \begin{array}{r} 40 + 1 \\ (-)20 + 2 \\ \hline \end{array} \rightarrow \begin{array}{r} 30 + 11 \\ (-)20 + 2 \\ \hline 10 + 4 = 14 \end{array}$$

   a) 32
     −14

   b) 422
     −114

   c) 422
     −134

   d) 4241
     −1323

3. Do the problems in Exercise 2 using the equal additions method with explanations.

   Example.

   $$\begin{array}{r} {}^{4}\!{}^{1}\!1 \\ 3 \\ -\not{2}\,2 \\ \hline 1\,4 \end{array} \qquad \begin{array}{r} 41 \\ -22 \\ \hline \end{array} \rightarrow \begin{array}{r} 40 + 1 \\ (-)20 + 2 \\ \hline \end{array} \rightarrow \begin{array}{r} 40 + 11 \\ (-)30 + 2 \\ \hline 10 + 4 = 14 \end{array}$$

4. Do as in Exercise 1 for the following additions in base four.

   a) 312
       +313

   b) 23
       13
       20
      +32

   c) 323
       102
     +310

5. Do the following subtraction problems in base four by the "borrowing" method, with explanations as in Exercise 2.

   a) 322
     −133

   b) 1321
     −1232

   c) 3012
     −1233

6. Do the problems of Exercise 5 by the equal additions method, with explanations as in Exercise 3.

7. The following problems are done correctly if the numerals represent numbers in certain bases. What is the base in each case?

   a) 5
     +3
     ─
     11

   b) 32
     +43
     ───
     105

   c) 13
     −4
     ──
     4

   d) 123
     −34
     ───
     45

92   Addition and Subtraction Algorithms for Whole Numbers

8. In base ten, subtraction by the method of complements involves, as a first step, doing some subtractions from 9. Noting that, in base ten, $10 - 1 = 9$, what can you conjecture as to the first step in doing subtraction by the method of complements in other bases?

9. Apply your conjecture of Exercise 8 to do the subtraction problems of Exercises 2 and 5 by the method of complements.

Chapter Test

In preparation for this test, you should review how to perform computations with whole numbers as follows:

1. Add, using stick bundles (76) and (77);

2. Add, using an abacus (78);

3. Add, by the method of partial sums (79);

4. Subtract, using stick bundles (80);

5. Subtract, using an abacus (81);

6. Subtract, by the method of complements (73 and 74);

7. Subtract by the "borrowing" method, and provide a detailed explanation of the method (82 and 83);

8. Subtract by the equal additions method, and provide a detailed explanation of the method (83 and 84);

9. Add and subtract (by any method) in bases other than ten (86–89).

1. Describe how you would show, in base ten, the addition $48 + 37$ by means of sticks and stick bundles.

2. Describe how you would show, in base ten, the addition $324 + 968$ by means of an abacus.

3. Describe how you would show the addition $34_{six} + 12_{six}$ by means of sticks and stick bundles.

4. Describe how you would show the addition $134_{six} + 342_{six}$ by means of an abacus.

5. Describe how you would show, in base ten, the subtraction $47 - 29$ by means of sticks and stick bundles.

6. Describe how you would show, in base ten, the subtraction $503 - 147$ by means of an abacus.

7. Describe how you would show the subtraction $42_{six} - 14_{six}$ by means of sticks and stick bundles.

8. Describe how you would show the subtraction $301_{six} - 143_{six}$ by means of an abacus.

9. Do the following addition in base ten by means of the method of partial sums.

   2432
   7894
   +3217

10. Do the following addition in base six by means of the method of partial sums.

    132
    243
    +124

11. Do the following base ten subtraction (a) by the "borrowing" method, (b) by the equal additions method, and (c) by the method of complements. Give an explanation of your work for parts (a) and (b).

    742
    −347

12. Do the following base six subtraction by each of the three methods asked for in Problem 11. Give an explanation of your work for the "borrowing" method and the equal additions method.

    322
    −114

13. Fill in the frames in the following addition problems (base ten).

    a) 4 ☐   b) 3 ☐
       +○ 8   +○ 7
       ─────   ─────
        7 4   12 0

14. Fill in the frames in the following subtraction problems (base ten).

    a) 4 ☐   b) 5 ☐
       −○ 7   −○ 4
       ─────   ─────
        2 7   2 8

TEST ANSWERS

1.

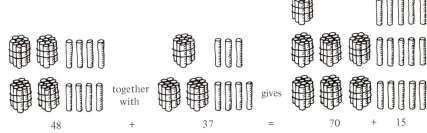

48   +   37   =   70   +   15

together with   gives

94 Addition and Subtraction Algorithms for Whole Numbers

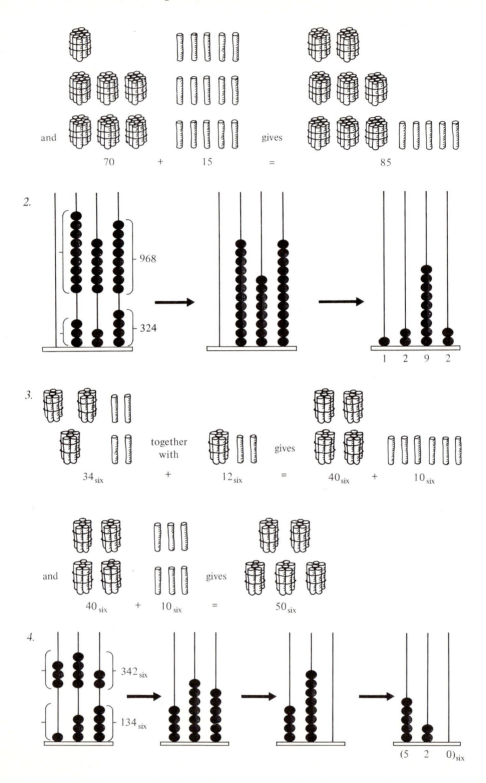

Chapter Test

5.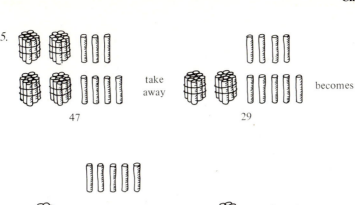

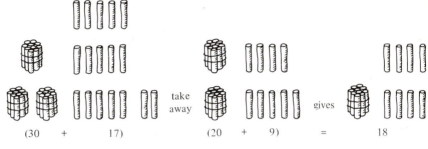

6.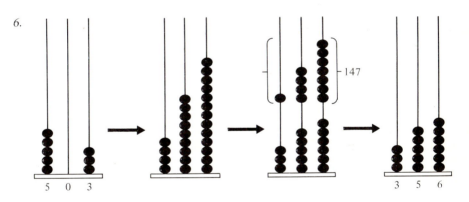

7.

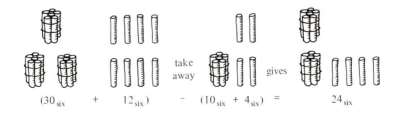

8.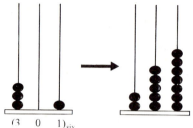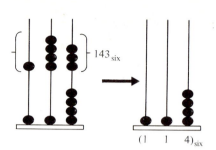

9. 2432
    7894
    +3217
    ─────
       13
      130
     1400
    +12000
    ──────
    13543

10. 132
    243
    +124
    ────
     13
    130
    +400
    ────
    543

11. a) "Borrowing"

$$\begin{array}{r} 6^13 \\ 7\,\not{4}^12 \\ -3\,4\,7 \\ \hline 3\,9\,5 \end{array}$$

$\begin{array}{r}742\\-347\\\hline\end{array}$ → $\begin{array}{r}700+40+2\\(-)300+40+7\\\hline\end{array}$ → $\begin{array}{r}700+30+12\\(-)300+40+\ 7\\\hline\end{array}$

→ $\begin{array}{r}600+130+12\\(-)300+\ 40+\ 7\\\hline 300+\ 90+\ 5 = 395\end{array}$

b) Equal additions

$$\begin{array}{r} 7\,4^1\!{}^12 \\ 4\,5 \\ -\not{3}\,4\,7 \\ \hline 3\,9\,5 \end{array}$$

$\begin{array}{r}742\\-347\\\hline\end{array}$ → $\begin{array}{r}700+40+2\\(-)300+40+7\\\hline\end{array}$ → $\begin{array}{r}700+40+12\\(-)300+50+\ 7\\\hline\end{array}$

→ $\begin{array}{r}700+140+12\\(-)400+\ 50+\ 7\\\hline 300+\ 90+\ 5 = 395\end{array}$

c) Complements

$\begin{array}{r}742\\-347\\\hline\end{array}$ $\begin{array}{r}742\\+652\\\hline \not{1}394\\+1\\\hline 395\end{array}$

Chapter Test 97

12. "Borrowing"

$$
\begin{array}{r}
3\,\overset{1}{2}\overset{1}{2} \\
-1\,1\,4 \\
\hline
2\,0\,4
\end{array}
\qquad
\begin{array}{r}
322 \\
-114 \\
\hline
\end{array}
\;\rightarrow\;
\begin{array}{r}
300 + 20 + 2 \\
(-)100 + 10 + 4 \\
\hline
\end{array}
$$

$$
\rightarrow\;
\begin{array}{r}
300 + 10 + 12 \\
(-)100 + 10 + 4 \\
\hline
200 + 0 + 4 = 204
\end{array}
$$

Equal additions

$$
\begin{array}{r}
3\,2\overset{1}{2} \\
-1\,\overset{2}{1}\,4 \\
\hline
2\,0\,4
\end{array}
\qquad
\begin{array}{r}
322 \\
-114 \\
\hline
\end{array}
\;\rightarrow\;
\begin{array}{r}
300 + 20 + 2 \\
(-)100 + 10 + 4 \\
\hline
\end{array}
$$

$$
\rightarrow\;
\begin{array}{r}
300 + 20 + 12 \\
(-)100 + 20 + 4 \\
\hline
200 + 0 + 4 = 204
\end{array}
$$

Complements

$$
\begin{array}{r}
322 \\
-114 \\
\hline
\end{array}
\qquad
\begin{array}{r}
322 \\
+441 \\
\hline
\cancel{1}203 \\
+1 \\
\hline
204
\end{array}
$$

13. *a)*
$$
\begin{array}{r}
46 \\
+28 \\
\hline
74
\end{array}
$$
b)
$$
\begin{array}{r}
33 \\
+87 \\
\hline
120
\end{array}
$$

14. *a)*
$$
\begin{array}{r}
44 \\
-17 \\
\hline
27
\end{array}
$$
b)
$$
\begin{array}{r}
52 \\
-24 \\
\hline
28
\end{array}
$$

A new algorithm for subtraction?

by Hitoshi Ikeda and Masue Ando

Reprinted from *The Arithmetic Teacher*, Vol. 21 (1974), pp. 716–719.

To suggest a new algorithm in place of one that has had popular use over many decades can be very upsetting to many teachers. Such was the case when the subtractive algorithm for division appeared. The new algorithm, however, gained popularity over the conventional division algorithm as it contributed to easier rationalization, especially in the early stages of learning long division. Such a painless transition makes one bold enough to suggest another change. This article proposes that a new algorithm replace the more commonly used forms in the subtraction of whole numbers.

The use of the decomposition algorithm in the subtraction of whole numbers has persisted in our country despite earlier studies which showed that on the criterion of efficiency, the equal additions algorithm was more successful than the decomposition method. Let it be assumed that teachers in our country have preferred the decomposition method because they have found it easier to rationalize and also because it is the only method most of them have been exposed to.

In recent years, there has been increasing emphasis on having students experience multiple embodiments of concepts through the use of appropriate manipulative materials.

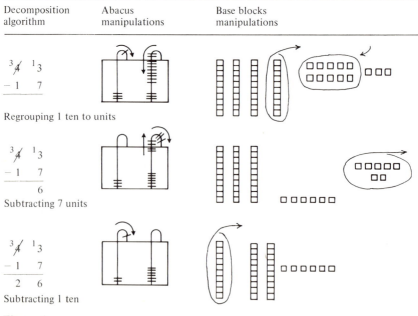

Figure 1

A New Algorithm for Subtraction

Manipulative aids to represent arithmetic operations should clarify the operational concept as well as the algorithm related to the operation. If the manipulations are to correspond to sequential steps in the algorithm, different algorithms are possible depending on the way the manipulations are processed. Consider the problem 43 − 17. Figure 1 shows the sequence of manipulations on the abacus and with the base blocks in relation to steps in the decomposition algorithm. (The manipulations pictured use the "take away" action of subtraction rather than the set separation idea.)

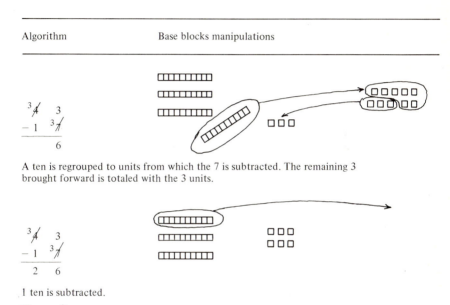

Figure 2

Since most abacuses of the type shown in Fig. 1 do not provide space for the 10 units regrouped from 1 ten to be brought forward, consider the manipulations with the base blocks. First, the ten rod is taken to the supply box; there it is exchanged for 10 units which are brought forward; finally, the number of units to be subtracted are counted and placed back in the supply box. To reduce the amount of back-and-forth manipulations, consider the sequence pictured in Fig. 2, using the same problem. If regrouping is not necessary subtraction proceeds as usual. If regrouping is necessary, this approach suggests that the amount to be subtracted be taken from the regrouped 10, regardless of which place the regrouping occurs. Figure 3 shows how the algorithm can be used with three other examples.

An advantage of this algorithm is that the computation involves no sums over ten and no differences from minuends over ten. Readiness activities must emphasize the components of ten. It differs from the so-called complementary method which adds the complement of the subtrahend to the minuend and then subtracts that power of ten from which the complement was determined. (The complement is the difference between that number and the next higher power of ten.) In a simplified approach, the complement of the subtrahend is usually determined by subtracting the unit's digit from 10 and all other digits from 9. (In the example in Fig. 4, 8 from 10 is 2 and 3 from 9 is 6.) After the addition of the complement, 1 is subtracted from the left of the leftmost digit of the subtrahend. For some

Example A:

$$\begin{array}{rrr} 3 & {}^{0}\not{1} & 2 \\ -1 & 6 & {}^{2}\not{8} \\ \hline & & 4 \end{array}$$

Subtracting 8 from the regrouped ten leaves 2 units for a total of 4 units.

$$\begin{array}{rrr} {}^{2}\not{3} & {}^{0}\not{1} & 2 \\ -1 & {}^{4}\not{6} & {}^{2}\not{8} \\ \hline 1 & 4 & 4 \end{array}$$

Subtracting 6 tens from the regrouped hundred leaves 4 tens for a total of 4 tens. Finally, subtracting 1 hundred leaves 144.

Example B: (middle zero, regrouping a hundred)

$$\begin{array}{rrr} {}^{4}\not{5} & 0 & 7 \\ -2 & {}^{2}\not{8} & 2 \\ \hline 2 & 2 & 5 \end{array}$$

Subtracting 2 units; subtracting 8 tens from the regrouped hundred leaves 2 tens. Finally, subtracting 2 hundreds leaves 225.

Example C: (middle zero, regrouping a ten and a hundred)

$$\begin{array}{rrr} {}^{3}\not{4} & 0 & 3 \\ -1 & {}^{7}\not{3} & 6 \\ \hline \end{array}$$

Since there are no tens available for regrouping, a hundred is regrouped first. 3 tens from the regrouped amount leaves 7 tens.

$$\begin{array}{rrr} {}^{3}\not{4} & 0 & 3 \\ -1 & {}^{6}\not{7}\not{3} & {}^{4}\not{6} \\ \hline 2 & 6 & 7 \end{array}$$

Now 6 units are subtracted from a regrouped ten leaving a total of 7 units and 6 tens. Finally, subtracting 1 hundred leaves 267.

Figure 3

Example:

$$\begin{array}{rrr} & 6 & 4 \\ - & 3 & 8 \\ \hline \end{array}$$

Since the power of ten higher than 38 is 100, the complement of 38 is 62. The 62 is written above the 38 in the problem.

$$\begin{array}{rrr} & 6 & 4 \\ + & {}^{6}\not{3} & {}^{2}\not{8} \\ \hline \not{1} & 2 & 6 \end{array}$$

The 1 crossed out indicates that 100 is subtracted from the sum.

Rationale: $64 - 38 = 64 + (100 - 38) - 100$
$= (64 + 62) - 100$

Figure 4

problems, this simply requires crossing out the 1 at the left as in the example in Fig. 4. For others, a bit more care is required, as in

$$\begin{array}{l} 9\,2\,3 \\ \;\;5\,3 \\ \underline{-\not{4}\not{7}} \\ 8 \\ \not{9}\,7\,6. \end{array}$$

The algorithm proposed in this article utilizes complements of ten only when regrouping is necessary and thus removes the necessity to subtract some power of ten from the result. Furthermore, while the complementary method changes a subtraction problem into an addition problem, the proposed method retains the subtractive notion in association with manipulative actions. Both the complementary method and the equal additions method are difficult to represent with manipulative materials.

In summary, the proposed algorithm:

1. Can be rationalized more easily than the equal additions or the complementary methods;

2. Can be represented by simpler manipulative actions than those representing the decomposition algorithm; and

3. Requires the knowledge of fewer number facts than when the decomposition or equal additions methods are used since the largest minuend used in each place is ten.

In view of these points the adoption of the proposed algorithm for subtraction of whole numbers should be given strong consideration.

REFERENCES FOR FURTHER READING

1. Cleminson, R. A., "Developing the subtraction algorithm," *AT*, **20** (1973), pp. 634–638. Gives details of the use of stick bundles and the abacus in teaching subtraction.

2. Cunningham, G. C., "Making a counting abacus," *AT*, **14** (1967), pp. 132–134.

3. Johnson, P. B., "Modern math in a toga," *AT*, **12** (1965), pp. 343–347. Shows how to do computations using Roman numerals.

Although the emphasis in teaching computations in the primary school should definitely be on the standard algorithms, some children will enjoy learning various shortcuts. The following three references deal with this topic.

4. Brumfiel, C., and I. Vance, "On whole number computation," *AT*, **16** (1969), pp. 253–257.

5. Freitag, H. T., and A. H. Freitag, "Shortcuts for the human computer," *AT*, **13** (1966), pp. 671–676.

6. Yates, W. E., "The Trachtenberg system as a motivational device," *AT*, **13** (1966), pp. 67–78.

Chapter 5

Basic Concepts of Multiplication and Division of Whole Numbers

5.1 PRIMARY-SCHOOL BEGINNINGS

The usual initial approach to multiplication in the primary school is, like addition, through the use of sets as is illustrated by Fig. 5.1, taken from a third-grade text.

Figure 5.1

Following this, as shown in Fig. 5.2, taken from the same third-grade text, the connection between multiplication and addition is made. (Some texts may regard, for example, 4 × 3 as representing 4 + 4 + 4 rather than 3 + 3 + 3 + 3 but the convention used here is the common one.)

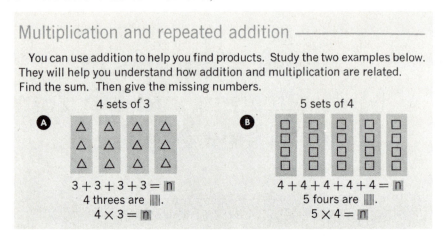

Figure 5.2

The only difficulties in the presentation of the concept of multiplication to children arise when the special products involving 0 are considered, e.g., 2 × 0 and 0 × 2. Actually, although we can't draw a picture of 2 × 0, we certainly can consider it as 0 + 0 and hence get 2 × 0 = 0 + 0 = 0. On the other hand, attempts to show that 0 × 2 = 0 as "if we have 0 sets of 2 members each, we have 0 members," etc., only serve to confuse children. Thus it is best, after getting 2 × 0 = 0, to say something such as "It is convenient and useful for us to agree that 0 × 2 = 0." (Among other reasons, taking 0 × 2 = 0 guarantees that the commutative property of multiplication, $a \times b = b \times a$, will hold for all whole numbers a and b.)

5.2 MULTIPLICATION IN TERMS OF CARTESIAN PRODUCTS OF SETS

Figure 5.3 shows an exercise from our third-grade text. You will certainly be easily able to figure out that the answer to (c) is, as suggested by (d), 4 × 3. This, however, will not be evident to children, as is indicated by the inclusion of several exercises in this third-grade text that involve the idea of pairing the elements of one set with the elements of another.

What is involved here is what mathematicians call the **Cartesian product** of two sets. The two sets pictured in Fig. 5.3 can be symbolized as

$A = \{a, b, p, o\}$ and $B = \{c, k, \ell\}$,

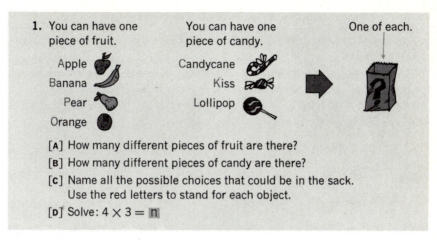

Figure 5.3

and we are concerned with the number of possible pairs of fruit and candy (in that order). These pairs form the set

$$C = \{(a, c), (a, k), (a, \ell), (b, c), (b, k), (b, \ell), (p, c), (p, k),$$
$$(p, \ell), (o, c), (o, k), (o, \ell)\}$$

C is the Cartesian product of A and B, and we write $C = A \times B$ (commonly read as "C equals A **cross** B").

In general, if A and B are any sets, we define $A \times B$ by

$$A \times B = \{(a, b): \quad a \in A \quad \text{and} \quad b \in B\}$$

We call (a, b) an **ordered pair** and consider that $(a, b) = (c, d)$ if and only if $a = c$ and $b = d$. Thus, for example, (apple, candy cane) ≠ (candy cane, apple) although {apple, candy cane} = {candy cane, apple}.

What is $A \times \emptyset$ for any set A? Since there are no elements in $\emptyset$, no ordered pairs (a, b) with $a \in A$ and $b \in \emptyset$ are possible and so $A \times \emptyset = \emptyset$ for all sets A. Similarly, $\emptyset \times A = \emptyset$ for all sets A.

We can use the concept of Cartesian product to make a formal definition of the product of two whole numbers a and b, as follows:

Let A and B be any sets such that $n(A) = a$ and $n(B) = b$. Then, using "·" for multiplication of whole numbers so as not to confuse multiplication of whole numbers with the Cartesian product of sets, we have

$$a \cdot b = n(A) \cdot n(B) = n(A \times B)$$

Thus, in our fruit and candy example, we have $n(A) = 4$, $n(B) = 3$, and $n(A \times B) = 12$, so that

$$4 \cdot 3 = n(A) \cdot n(B) = n(A \times B) = 12$$

(But we will continue to use "×" rather than "·" to indicate multiplication of whole numbers unless we are considering at the same time the Cartesian product of sets.)

A very nice way of picturing Cartesian products geometrically in relation to the product of whole numbers is by means of intersecting lines. Thus $2 \times 3 = 6$ can be represented as shown in Fig. 5.4, where the answer, 6, appears as the number of intersections of the lines.

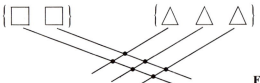

Figure 5.4

If a, b, and c are any whole numbers such that $c = a \times b$, we say that c is a **multiple** of a and also a multiple of b. Thus since $6 = 2 \times 3$, 6 is a multiple of 2 and also a multiple of 3. Note that, since $a = a \times 1$ for all whole numbers a, any whole number is a multiple of itself. Furthermore, since $0 = 0 \times a$ for all whole numbers a, 0 is a multiple of every whole number.

Programed Lesson 5.2

Take $A = \{a, b, c\}$ and $B = \{x, y\}$ throughout this lesson.

1. $A \times B = \{_____\}$.

   $\{(a, x), (a, y), (b, x), (b, y), (c, x), (c, y)\}$

2. $n(A) = \_\_$, $n(B) = \_\_$, and $n(A \times B) = \_\_$.

   3, 2, 6

3. Since $n(A) = 3$, $n(B) = 2$, $n(A \times B) = 6$, and $n(A) \cdot n(B) = n(A \times B)$, we can conclude that $\_\_\_\_ = \_\_$.

   3×2, 6

4. $B \times A = \{_____\}$.

   $\{(x, a), (x, b), (x, c), (y, a), (y, b), (y, c)\}$

5. Is it true that $A \times B = B \times A$? Why?

   No. For example, $(a, x) \in A \times B$ but $(a, x) \notin B \times A$.

6. Is it true that $A \times B \sim B \times A$? Why?

5.2 Multiplication in Terms of Cartesian Products of Sets

Yes. We can match $(a, x) \in A \times B$ with $(x, a) \in B \times A$, $(c, y) \in A \times B$ with $(y, c) \in B \times A$, etc.

7. Since $A \times B \sim B \times A$, it follows that $n(A \times B) \_\_ n(B \times A)$.

   $=$

8. Since $n(A \times B) = n(B \times A)$, $n(A) \cdot n(B) = n(A \times B)$, and $n(B) \cdot n(A) = n(B \times A)$, it follows that $n(A) \cdot n(B) \_\_ n(B) \cdot n(A)$; i.e., that $3 \times 2 = \_\_$.

   $=$, 2×3

9. $A \times \emptyset = \_\_$.

   $\emptyset$

10. $n(A \times \emptyset) = n(A) \cdot \_\_$.

    $n(\emptyset)$

11. $n(A) = \_\_$ and $n(\emptyset) = \_\_$.

    3, 0

12. Since $n(A) \cdot n(\emptyset) = n(A \times \emptyset)$, $A \times \emptyset = \emptyset$, $n(A) = 3$, and $n(\emptyset) = 0$, we can conclude that $\_\_ \cdot 0 = \_\_$.

    3, 0

13. Show the product 4×2 by use of sets and intersecting lines.

    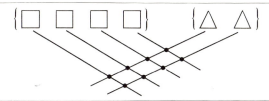

14. Anne, Mary, and Felicia play the piano; Susan and Jill play the violin. Show, by a Cartesian product, the number of duets that be played so that each pianist plays once with each violinist.

    $\{(A, S), (A, J), (M, S), (M, J), (F, S), (F, J)\}$

15. At a car sale, the models available are the Hurricane and the Typhoon; each is available in chartreuse, pink, violet, and orange. Show, by a Cartesian product, the number of different model–color choices available.

    $\{(H, c), (H, p), (H, v), (H, o), (T, c), (T, p), (T, v), (T, o)\}$

5.3 DIVISION AS PARTITION AND AS SUCCESSIVE SUBTRACTION

Figure 5.5 illustrates how we can think of division in terms of a **partition** of a set into equivalent subsets, and Fig. 5.6 illustrates how we can think of division in terms of successive subtractions (both illustrations from a third-grade book).

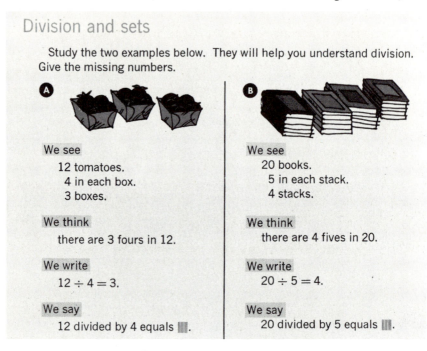

Figure 5.5

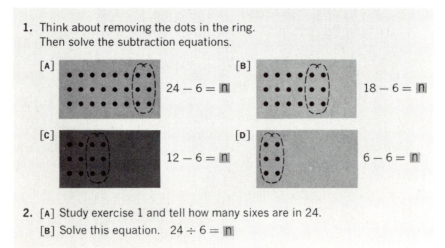

Figure 5.6

5.3 Division as Partition and as Successive Subtraction

Both points of view of division are useful. The partition approach gives a natural physical interpretation of division and, as we shall see later, the usual procedure for "long division" (e.g., dividing 2342 by 49) simply applies the idea of division as successive subtraction in a condensed form.

Of course, in either way of looking at a division we may obtain a **remainder**. Thus $7 - 2 - 2 - 2 = 1$, which gives us "7 divided by 2 is 3 with remainder 1" and the same conclusion is reached by the partition shown in Fig. 5.7.

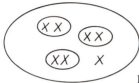

Figure 5.7

In general, if a and b are any whole numbers with $b \neq 0$, we have

$$a \div b = q \quad \text{with remainder } r,$$

where q and r are whole numbers and $0 \leq r < b$ (read "r is greater than or equal to zero and less than b"—a combination of the statements $r > 0$ or $r = 0$, and $r < b$). If $a = 6$ and $b = 2$, then $q = 3$ and $r = 0$; if $a = 7$ and $b = 2$, then $q = 3$ and $r = 1$; if $a = 1$ and $b = 3$, then $q = 0$ and $r = 1$.

Programed Lesson 5.3

1. Show the division $8 \div 2$ as the partition of a set of 8 objects into sets of 2 objects.

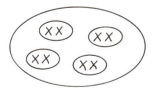

2. Show the division $8 \div 3$ (with remainder) as the partition of a set of 8 objects into sets of 3 objects.

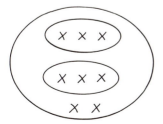

110   Basic Concepts of Multiplication and Division of Whole Numbers   5.4

3. Show the division $8 \div 2$ by successive subtractions.

$$8 - 2 - 2 - 2 - 2 = 0$$

4. Show the division $8 \div 3$ (with remainder) by successive subtractions.

$$8 - 3 - 3 = 2$$

5. The members of the Little Men's Marching and Chowder Society are to march five abreast in a parade. If 32 members are present, what is the least number of additional members needed to form lines of 5 apiece?

3

5.4 DIVISION AS THE INVERSE OF MULTIPLICATION

In Section 2.6 we related subtraction of whole numbers to addition of whole numbers by stating that

$$c - b = a \quad \text{if and only if } c = a + b$$

for any whole numbers a, b, and c. Likewise, we can relate division of whole numbers to multiplication of whole numbers by stating that:

1. $a \div b = q \quad$ if and only if $a = b \times q$

for any whole numbers a, b, and q with $b \neq 0$. Thus, for example, $6 \div 3 = 2$ if $6 = 3 \times 2$ and $3 \times 2 = 6$ if $6 \div 3 = 2$. (Recall that a common check for a division problem is to perform a multiplication!) If (1) holds, we say that b is a **divisor** of a. Thus, since $6 \div 2 = 3$ and $6 \div 3 = 2$, 2 and 3 are both divisors of 6.

If a divided by b gives a quotient, q, and remainder $r \neq 0$, we then have:

2. $a = (b \times q) + r \quad$ where $0 < r < b$

By considering the remainder in (1) to be 0, we can combine (1) and (2) as:

3. $a = (b \times q) + r \quad$ where $0 \leq r < b$

and q and r are whole numbers. For example, if $a = 6$ and $b = 2$, we have

$$6 = (2 \times 3) + 0 \quad (q = 3, r = 0)$$

If $a = 7$ and $b = 2$, we have

$$7 = (2 \times 3) + 1 \quad (q = 3, r = 1)$$

If $a = 2$ and $b = 7$, we have

$$2 = (7 \times 0) + 2 \quad (q = 0, r = 2)$$

5.4　　　　　　　　　　　　　　　　　**Division as the Inverse of Multiplication　111**

Terminology here, by the way, tends to get confusing. We sometimes say, for example, "2 is not a divisor of 7" or "2 does not divide 7." On the other hand, we sometimes say, "7 divided by 2 is 3 with a remainder of 1." Here, however, by "divide" we will mean statement (1) when we are talking about whole numbers.

What about division by 0? Can we, for example, find an answer to the division problem, $2 \div 0$? By (1) we know that

$$2 \div 0 = q \qquad \text{if and only if } 2 = 0 \times q.$$

But we know (Section 5.1) that $0 \times q = 0$ for all whole numbers q. Thus $2 \neq 0 \times q$ for any whole number q and hence $2 \div 0 \neq q$ for any whole number q.

Considering division in terms of subtraction (Section 5.3) yields the same conclusion. We have

$$2 - 0 = 2, \quad 2 - 0 - 0 = 2, \quad 2 - 0 - 0 - 0 = 2, \quad \text{etc.,}$$

never getting 0 as a result of successive subtractions.

The same sort of reasoning applies to $a \div 0$ for any whole number $a \neq 0$. What about $0 \div 0$? By the subtractive procedure, we certainly quickly get to 0. Indeed,
$$0 - 0 = 0.$$

But, also,

$$0 - 0 - 0 = 0.$$

Is $0 \div 0$ equal to 1 or 2, or, since

$$0 - 0 - 0 - 0 = 0, \quad 0 - 0 - 0 - 0 - 0 = 0,$$

3, or 4, etc.? Similarly, considering

$$0 \div 0 = q \qquad \text{if and only if } \quad 0 = 0 \times q,$$

our problem is that *any* number q will give $0 \times q = 0$. *Conclusion:* No unique answer is possible, and we agree that $a \div 0$ is *undefined* for any whole number a. (See the article at the end of this chapter for a further discussion of the role of 0 in arithmetic.)

Programed Lesson 5.4

1. $6 - \square - \square = 0$ and so $6 \div \underline{\ \ } = \underline{\ \ }$.

     3, 3, 3, 2

2. $12 \div 3 = 4$ and so $12 = \underline{\ \ } \times 4$.

     3

112   **Basic Concepts of Multiplication and Division of Whole Numbers**   5.5

3. $15 = 3 \times 5$ and so $15 \div 3 =$ __.

   5

4. If $a = 8$ and $b = 3$, then $a = (b \times q) + r$, where $0 \le r < b$, becomes __ $=$ $(\_ \times \_) + \_.$

   $8 = (3 \times 2) + 2$

5. If $a = 12$ and $b = 3$, then $a = (b \times q) + r$, where $0 \le r < b$, becomes __ $=$ $(\_ \times \_) + \_.$

   $12 = (3 \times 4) + 0$

6. If $a = 4$ and $b = 9$, then $a = (b \times q) + r$, where $0 \le r < b$, becomes __ $=$ $(\_ \times \_) + \_.$

   $4 = (9 \times 0) + 4$

5.5 PROPERTIES OF MULTIPLICATION AND DIVISION

The two basic properties of multiplication: For all whole numbers a, b, and c,

1. $a \times b = b \times a$ **(Commutative property of multiplication)**

and

2. $a \times (b \times c) = (a \times b) \times c$ **(Associative property of multiplication)**

are easily understood by children by considering examples using particular numbers and by using various visual aids. For example,

$$2 \times 3 = 3 + 3 = 6 \quad \text{and} \quad 3 \times 2 = 2 + 2 + 2 = 6$$

illustrates the commutative property and

$$2 \times (3 \times 4) = 2 \times 12 = 24 \quad \text{and} \quad (2 \times 3) \times 4 = 6 \times 4 = 24$$

illustrates the associative property.

A visual approach to the commutative property is shown in Fig. 5.8. The figure on the left shows six blocks glued to a sheet of cardboard, to illustrate 3×2 (3 sets of 2 objects each), and the figure on the right, obtained by rotating the cardboard 90°, illustrates 2×3 (2 sets of 3 objects each).

To illustrate the associative property, a set of blocks, as shown in Fig. 5.9, is useful. To determine the total number of blocks, we can first calculate the number of blocks in the layer indicated in the figure as 3×4 and then, because there are 2 such layers, get $2 \times (3 \times 4)$ for the total number of blocks. Or we can calculate

5.5 Properties of Multiplication and Division   113

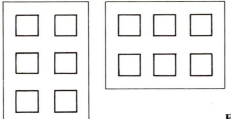

Figure 5.8

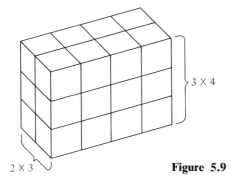

Figure 5.9

the number of blocks in the other layer as 2 × 3 and then, because there are four such layers, get (2 × 3) × 4 as the total number of blocks. (Actually, by our formal definition of multiplication in terms of addition, we get 4 × (2 × 3), but by the commutative property, 4 × (2 × 3) = (2 × 3) × 4.)

In addition to the commutative and associative properties we have, for any whole number a,

3. $a \times 1 = 1 \times a = a$ (**Multiplicative property of 1**)

We say that 1 is the **multiplicative identity.** (Recall that 0 is the additive identity.) Also, for any whole number a, it is true that

4. $a \times 0 = 0 \times a = 0$ (**Multiplicative property of 0**)

Finally, we have a very useful property that serves as a connecting link between the operation of addition and the operation of multiplication: For all whole numbers a, b, and c,

5. $a \times (b + c) = (a \times b) + (a \times c)$

(**Distributive property of multiplication over addition**)

This property is usually written more simply as

   $a \times (b + c) = a \times b + a \times c$,

following the convention that, in a sequence of operations involving both addition and multiplication, the multiplications are to be performed first unless the contrary is indicated. Thus

$$2 \times (3 + 4) = 2 \times 7 = 14,$$

because the parentheses are there to instruct us to do the addition first. But

$$2 \times 3 + 2 \times 4 = 6 + 8 = 14$$

without parentheses being needed. So

$$2 \times (3 + 4) = 2 \times 3 + 2 \times 4 = 14$$

On the other hand,

$$2 \times (3 + 2) \times 4 = 2 \times 5 \times 4 = 10 \times 4 = 40$$

Note that the distributive property of multiplication over addition (commonly referred to simply as "distributive property") can also be written as

6. $\quad (a \times b) + (a \times c) = a \times (b + c)$

and, in this form, is often referred to in algebra as the process of "factoring out the a". (The programed lesson below investigates the existence of other kinds of distributive properties.)

Also note that, because of the commutative property of multiplication, the distributive property can be written as

7. $\quad (b + c) \times a = (b \times a) + (c \times a)$

We sometimes refer to (7) as the **righthand** distributive property and to (5) as the **lefthand** distributive property. But when we speak of *the* distributive property we normally have reference to the lefthand version.

For children, the distributive property (sometimes called for them the **multiplication–addition principle**) can be illustrated by numerical examples as we have done above in showing that $2 \times (3 + 4) = 2 \times 3 + 2 \times 4$, and also by blocks as indicated in Fig. 5.10.

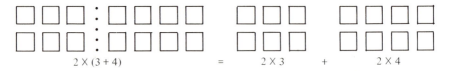

Figure 5.10

The various properties of division are, of course, related to properties of multiplication and will be considered in the following programed lesson.

Programed Lesson 5.5

1. The property that $a \times b = b \times a$ for all whole numbers a and b, is called the _____ property of multiplication.

   commutative

2. The property that $a \times (b \times c) = (a \times b) \times c$ for all whole numbers a, b, and c is called the _____ property of multiplication.

   associative

3. The multiplicative property of 1 states that, for any whole number a, ___ = ___ = ___.

   $a \times 1$, $1 \times a$ (in either order), a

4. 1 is called the _____ _____ for multiplication.

   multiplicative identity

5. The multiplicative property of 0 states that for any whole number a, ___ = ___ = ___.

   $a \times 0$, $0 \times a$ (in either order), 0

6. The following illustration is from a third-grade book. What property of the operations on whole numbers does it illustrate?

   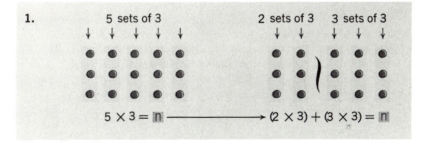

   The distributive property, $(2 + 3) \times 3 = (2 \times 3) + (3 \times 3)$

7. The property that $a \times (b + c) = (a \times b) + (a \times c)$ for all whole numbers a, b, and c is called the _____ property of _____ over _____.

   distributive, multiplication, addition

8. By the commutative property of multiplication, we can also write the distributive property as $(b + c) \times a = (\text{_____}) + (\text{_____})$.

116   Basic Concepts of Multiplication and Division of Whole Numbers   **5.5**

$b \times a, \quad c \times a$

9. Use the commutative and associative properties of multiplication to perform the following computations with little effort.
   a) $(25 \times 37) \times 4$

   $(25 \times 4) \times 37 = 100 \times 37 = 3700$

   b) $(4 \times 138) \times 25$

   $(4 \times 25) \times 138 = 100 \times 138 = 13{,}800$

10. Use the distributive property to perform the following computations with little effort:
    a) $(12 \times 58) + (12 \times 42)$

    $(12 \times 58) + (12 \times 42) = 12 \times (58 + 42) = 12 \times 100 = 1200$

    b) $(18 \times 37) + (18 \times 63)$

    $(18 \times 37) + (18 \times 63) = 18 \times (37 + 63) = 18 \times 100 = 1800$

11. State the distributive property of multiplication over subtraction for whole numbers.

    For all whole numbers a, b, and c, with $b \geq c$, $a \times (b - c) = (a \times b) - (a \times c)$. (*Note*. The requirement that $b \geq c$ is to ensure that $b - c$ is a whole number.)

12. Is it true that $(b + c) \div a = (b \div a) + (c \div a)$ for all whole numbers a, b, and c such that a divides b and a divides c?

    Yes. (*Note*. The requirement that a divides b and a divides c is to ensure that $b \div a$ and $c \div a$ are whole numbers.)

13. Is it true that $a \div (b + c) = (a \div b) + (a \div c)$ for all whole numbers a, b, and c such that b divides a and c divides a?

    No. Counterexample: If $a = 6$, $b = 2$, and $c = 3$, $a \div b = 6 \div 2 = 3$, $a \div c = 6 \div 3 = 2$, so that $(a \div b) + (a \div c) = 3 + 2 = 5$. But $a \div (b + c) = 6 \div 5$ is not even a whole number.

14. Is it true that addition distributes over multiplication? That is, is it true that $a + (b \times c) = (a + b) \times (a + c)$ for all whole numbers a, b, and c?

    No. Counterexample: $2 + (3 \times 4) = 2 + 12 = 14$ but $(2 + 3) \times (2 + 4) = 5 \times 6 = 30$.

5.5 **Properties of Multiplication and Division** **117**

15. True or false?
   a) For all whole numbers a, $a \div 1 = a$.
   b) For all whole numbers a with $a \neq 0$, $1 \div a = a$.
   c) For all whole numbers a and b with $a \neq 0$ and $b \neq 0$, $a \div b = b \div a$.
   d) For all whole numbers a, b, and c with $b \neq 0$ and $c \neq 0$, $a \div (b \div c) = (a \div b) \div c$.
   e) For all whole numbers a and b with $b \neq 0$, $(a \div b) \times b = a$.
   f) For all whole numbers a and b with $b \neq 0$, $(a \times b) \div b = a$.

   (a), (e), and (f) are true; (b), (c), and (d) are false

16. Use the properties given in (e) and (f) of Exercise 15 to perform the following computations with little effort.
   a) $(308 \div 24) \times 24$

   $(308 \div 24) \times 24 = 308$, by 15(e)

   b) $(16{,}124 \times 387) \div 387$

   $(16{,}124 \times 387) \div 387 = 16{,}124$, by 15(f)

5.6 NUMBER ACTIVITIES INVOLVING MULTIPLICATION AND DIVISION

With the operations of multiplication and division available, the range of possible number activities is greatly extended. Here are three possibilities—others are suggested in the exercises.

1. The use of frames can be made much more challenging. For example, consider the problem of putting $+$, $-$, $\times$, or $\div$ in the various frames of

$$\{6 \,\square\, [(2 \,\triangle\, 3) \,\bigcirc\, 2]\} \,\diamondsuit\, 2 = 5 *$$

It will surely take a child considerable experimentation (and hence will entail considerable practice in arithmetic) to conclude that a correct answer (others *may* exist) is

$$\{6 + [(2 \times 3) - 2]\} \div 2 = 5$$

2. Multiplication "rabbits" exist! Consider Fig. 5.11. Note that, in asking children for the completion of a rabbit such as the one shown in Fig. 5.12, you must be careful (before fractions are introduced) to avoid "impossible" divisions. Thus, changing the "15" to "14" would produce a problem with no solution in whole numbers.

---

* Recall the agreement that the same symbol *may* be used in different-shaped frames, whereas the same symbol *must* be used in frames of the same shape.

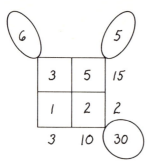
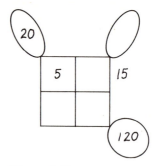

Figure 5.11   **Figure 5.12**

3. Many kinds of number "tricks" involve the use of multiplication or division. Here is one involving division: Ask someone to write down a number with three or more digits, to divide it by 9, and then to give you the remainder. (For example, suppose this person writes down 2453. Then he would give you 5, since 2453 divided by 9 gives a remainder of 5.) Now tell him to cross out one digit in the number he wrote down, again divide by 9, and tell you the new remainder. (For example, suppose he crosses out "4". Then he would have 253 which, when divided by 9, gives a remainder of 1.) You will now be able to tell him the digit that he crossed out. How? If the second remainder is less than the first, simply subtract it from the first remainder to get the number crossed out. (5 − 1 = 4, in the example); if the second remainder is greater than the first, add 9 to the first remainder before subtracting the second. (For example, if 758 was the number chosen and 5 crossed out, you would have (2 + 9) − 6 = 5.)

Exercises 5.6

In Exercises 1 through 5, fill in the frames with +, −, ×, or ÷, to produce a true statement.

1. [(7 □ 8) △ 3] ○ 2 = 3

2. [(4 □ 3) △ 2] ○ 5 = 2

3. [(9 □ 3) □ 3] △ [(7 ○ 2) △ 5] = 11

4. [(6 □ 3) △ 6] ○ (8 ○ 2) = 22

5. [(7 □ 3) △ 3] ○ (4 ○ 1) = 4

6. Complete the multiplication rabbit given in the text.

7. Test the instructions for the trick given in the text, for the number 3572 and the number 3574, striking out the "7" in each case.

8. Pick a number from 1 to 9. Multiply it by 3 and then by 37,037. What do you observe that is interesting about the answer?

Chapter Test **119**

9. Have someone pick any number with two or more digits with not all its digits the same, reverse the digits, and subtract the smaller of the two numbers from the larger (e.g., if 249 were chosen, 249 would be subtracted from 942). Now have him multiply this difference by any nonzero number and then cross out one digit other than 0 from the result. Finally, he is to give you the sum of the remaining digits. You will then be able to tell him the digit that he crossed out. To do this, you simply subtract the sum given you from the multiple of 9 greater than this sum. (Try this yourself first with 249, using as a multiplier the number 24.)

10. Have someone pick a number less than or equal to 105 and divide it in turn by 3, 5, and 7, giving you each of the remainders in order. You will then be able to determine the number chosen. To do this let r_3, r_5, and r_7 be the remainders given to you upon division by 3, 5, and 7, respectively. Then the number chosen was $N = 70r_3 + 21r_5 + 15r_7$ if $N \leq 105$, whereas if $N > 105$, it was the remainder when N is divided by 105. (Try this yourself first using 15 and then 101.)

Chapter Test

In preparation for this test you should review the meaning of the following terms and symbols:

Cartesian product (104)

Ordered pair (105)

$A \times B$ for sets A and B (105)

Multiple (106)

Divisor (110)

Partition of a set (108)

Remainder (109)

Commutative and associative properties of multiplication (112)

Multiplicative property of 1 (113)

Multiplicative property of 0 (113)

Multiplicative identity (113)

Distributive property of multiplication over addition (113)

You should also review how to:

1. Find the Cartesian product of two sets (105);

2. Show multiplication of whole numbers in terms of the Cartesian product of two sets (105);

3. Fill in frames with the correct numerals or symbols of operations (117);

4. Complete multiplication "rabbits" (117).

1. If (a, b) and (x, y) are ordered pairs, then $(a, b) = (x, y)$ if and only if __ = __ and __ = __.

2. If A and B are any sets, then $A \times B = \{(x, y): $ _____ $\}$.

3. If $A = \{1, 2\}$ and $B = \{1, 2, 3\}$, then $A \times B = \{$ _____ $\}$.

4. If A and B are sets and $A \times B = B \times A$, then either $A = $ __, or $A = $ __, or $B = $ __.

120 Basic Concepts of Multiplication and Division of Whole Numbers

5. The product of two whole numbers a and b, where $n(A) = a$ and $n(B) = b$, can be defined by $n(A) \cdot n(B) = n(\underline{\hspace{1em}})$.

6. If a and b are any whole numbers such that $b \neq 0$, then $a \div b = c$ if and only if $a = \underline{\hspace{1em}}$.

7. If a and b are any whole numbers with $b \neq 0$, then $a = b \times q + r$ where $0 \leq r <$ \underline{\hspace{1em}} and \underline{\hspace{1em}} and \underline{\hspace{1em}} are whole numbers.

8. If $2 \div 0 = q$, then $2 = 0 \times \underline{\hspace{1em}}$. Since $0 \times \underline{\hspace{1em}} = \underline{\hspace{1em}} \neq 2$, we conclude that $2 \div 0$ is \underline{\hspace{1em}}.

9. The fact that $2 \times 3 = 3 \times 2$ illustrates the \underline{\hspace{2em}} property of \underline{\hspace{2em}}.

10. The fact that $(2 \times 3) \times 4 = 4 \times (2 \times 3)$ illustrates the \underline{\hspace{2em}} property of \underline{\hspace{2em}}.

11. The fact that $(2 \times 4) + (2 \times 5) = 2 \times (4 + 5)$ illustrates the \underline{\hspace{2em}} property of \underline{\hspace{2em}} over \underline{\hspace{2em}}.

12. The fact that $2 \times 0 = 0$ illustrates the \underline{\hspace{2em}} \underline{\hspace{2em}} of 0.

13. Complete the following multiplication "rabbit".

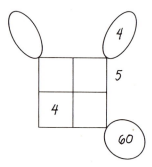

14. Fill in the frames with $+$, $-$, $\times$, or $\div$, to produce a true statement: $[(9 \square 3) \triangle 2] \bigcirc (4 \triangle 1) = 2$.

15. For all sets A and B it is true that $A \times B \underline{\hspace{1em}} B \times A$.

16. Since $A \times B \sim B \times A$ for all sets A and B, it follows that $n(A \times B) \underline{\hspace{1em}} n(B \times A)$.

17. $n(A) \cdot n(B) = n(\underline{\hspace{1em}})$ and $n(B) \cdot n(A) = n(\underline{\hspace{1em}})$.

18. Thus $n(A) \cdot n(B) \underline{\hspace{1em}} n(B) \cdot n(A)$.

19. $n(\emptyset) = \underline{\hspace{1em}}$ and, for all sets A, $A \times \emptyset = \underline{\hspace{1em}}$.

20. Hence $n(A) \cdot 0 = n(A) \cdot n(\underline{\hspace{1em}}) = n(A \times \underline{\hspace{1em}}) = n(\underline{\hspace{1em}}) = 0$.

TEST ANSWERS

1. a, x, b, y
2. $x \in A$ and $y \in B$
3. $(1, 1), (1, 2), (1, 3), (2, 1), (2, 2), (2, 3)$
4. $\emptyset, B$ (in either order), $\emptyset$
5. $A \times B$
5. $b \times c$ (or $c \times b$)
7. b, q, r
8. $q, q, 0$, undefined
9. Commutative, multiplication
10. Commutative, multiplication
11. Distributive, multiplication, addition
12. Multiplicative property
13.

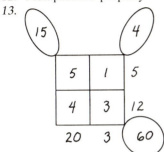

14. $\div, \times, -, \times$
15. $\sim$
16. $=$
17. $A \times B, \quad B \times A$
18. $=$
19. $0, \emptyset$
20. $\emptyset, \emptyset, \emptyset$

Zero, the troublemaker

by Boyd Henry

Reprinted from *The Arithmetic Teacher*, Vol. 16 (1969), pp. 365–367.

Every elementary school teacher will agree that the concept of zero is difficult for many children to grasp. In fact, many teachers themselves are uncomfortable when they must work with numbers involving zero. One veteran fifth-grade teacher was observed drilling her students to repeat that $6 \times 0 = 0$, but that $0 \times 6 = 6$. Apparently she didn't really believe the commutative law which she had previously "taught" her students. Moreover, the role of zero in the structure of the number system was at best vague and certainly confusing to her. We will never know how many hundreds of students she had confused about zero over the long years of her teaching career. The fault does not lie entirely with the teacher alone, of course. At least one textbook she had used in years past made quite a point of stating that zero is not a number. Such misinformation as this could only confuse both teacher and students. After all, she might reason, if zero is not a number, it is not obliged to follow the laws of numbers such as the commutative law for multiplication.

Let us take a look at the role of zero in the number system. First of all, it should be noted that zero differs in several ways from each of the other numbers. One important difference is that we use the numeral for zero, 0, in two distinctly different roles. Sometimes we use the numeral 0 when we are communicating how many of something we might have. For example, in writing the numeral 307, just as the 3 tells us that we have three hundreds and the 7 tells us that we have seven units, the 0 tells us that we have no tens. Note that in this case the 0 is used to represent a counting number in that it states how many tens we have—namely, not any. This is a role similar to that played by each of the other digits. It is used to tell how many of each grouping we have in a number such as 307.

On the other hand, the numeral 0 is also used as a placeholder when we don't really mean zero in its role as a counting number, but when we must complete the numeral with something that uses up space. In such a case, 0 acts solely as a spacefiller. For example, the distance to the sun is given as 93,000,000 miles. Here we use 0 in a totally different role than we used it in a numeral such as 307. We don't really mean to imply that the distance to the sun is 93,000,000 miles correct to the nearest mile and that each of the 0's appearing in the numeral means zero. For example, there exists an instant during the year when the distance is 93,127,904 miles. Clearly, only one of the 0's in the approximation 93,000,000 is the correct numeral. In this example, 0 was also used to replace 1, 2, 7, 9, and 4. We have used the 0's to fill up space so that we might distinguish 93 million from 93 thousand or from 93.

Perhaps it would have been clearer if, in the history of the development of notation, man had decided to use a symbol such as — when he didn't know which digit properly belonged in a given position. Thus, 0 could have been retained to tell how many, and — could have been used as a spaceholder. In such notation, 93 million, merely correct to the nearest million miles, would have been written 93,———,———. This might better

Zero, the Troublemaker 123

imply that we really don't know which numerals belong in the blanks, but realize that something belongs there to distinguish it from 93 billion or 93 thousand. On the other hand, if the exact count of a large number of items happened to be exactly 93 million, it could then have been written 93,000,000.

Of course, we are not advocating the addition of a new symbol in arithmetic to take care of the dual role of the numeral 0. Rather, we are advocating that the teacher be thoroughly aware of this dual role and work to help her students understand that, while the numeral 0 certainly does represent a number, it is also used in a second role as a space-filler.

The number zero is unique in yet another way. It is the only number which may never be used as a divisor in division. All too often the teacher tells the students that division by zero is not permitted and ends the discussion there. Of course the statement is correct. Division by zero is not permitted. But why make it sound like a teacher-made rule? (The beleaguered student may believe that he must add division by zero to the list of things that he is not permitted to do—such as run in the halls, chew gum in class, or study spelling during arithmetic period.) How much better it would be to let him discover for himself that an attempt to divide by zero leads to no result.

First, however, the teacher should stress that zero can be used as the dividend in dividing by a nonzero divisor. For example, $0 \div 6$ is perfectly clear and well defined. The student can be led to discover the correct answer to $0 \div 6$ in a variety of ways. To do this, we might use a comparable example involving nonzero numbers. What do we mean by $15 \div 3$, for example? There are various ways that we might arrive at the correct answer. If I have 15 marbles, to how many boys could I give 3 marbles each? The child can experiment with various models to confirm that the answer is 5. He can see from a model such as XXX XXX XXX XXX XXX that there are indeed 5 threes in fifteen. Or we might confirm that $15 \div 3 = 5$ because division is the inverse operation of multiplication, and since $5 \times 3 = 15$, then $15 \div 3$ must equal 5. In still another approach, we might say that the problem $15 \div 3$ asks how many threes will add to 15? That is, since $3 + 3 + 3 + 3 + 3 = 15$, by counting we can determine that 5 threes add to 15. That is, there are 5 threes in 15. With several examples of this sort before the student, we can proceed with the development of dividing zero by a nonzero number, such as $0 \div 3$.

First, if I have no marbles to divide equally among 3 boys, how many marbles does each boy get? Nothing divided three ways is clearly just that—nothing. Yet many students fail to grasp this rather obvious point, perhaps because it has never been explained to them. Or we can confirm that $0 \div 3 = 0$ because division is the inverse operation of multiplication and $3 \times 0 = 0$. Therefore $0 \div 3$ must equal 0 for the same reason that, because $3 \times 5 = 15$, then $15 \div 3$ must equal 5. Finally, if we ask how many 3's I must use to add to 0, we quickly see that if I write any 3's at all, I have written too many. Thus, there are no threes in zero. That is, $0 \div 3 = 0$.

By whatever means the teacher chooses to teach this division fact, the important point is that she should teach it and not simply ignore it. Zero divided by a nonzero number is perfectly well-defined and is permitted. The answer is always zero.

What about division by zero? What might we mean by $6 \div 0$? Instead of completely avoiding the issue by merely telling the student that he must never ask about $6 \div 0$, guide him to discover for himself that this operation is without meaning. Let us use examples similar to those used when we considered the meaning of $15 \div 3$. First, if I have 6 marbles to divide equally among 0 boys, how many can I give to each boy? Could I not promise each boy 100 marbles from my stock of 6 marbles—as long as no one shows up to collect? Could I not also promise each boy a thousand or a million or a trillion marbles? There is no limit to the number that I might give to each of 0 boys from my stock of 6 marbles—and

124   Basic Concepts of Multiplication and Division of Whole Numbers

still have 6 marbles left over. Here the child is introduced to a new and difficult concept. However, there should be no difficulty for the child to see that division by zero does indeed lead to a rather odd situation—different from any he has ever previously encountered when dividing by nonzero numbers. As a second approach, let us look at the inverse operation property of division. Using "box" arithmetic, we argue that we can find the answer to $6 \div 0$ if we can find something to multiply by 0 to give 6. That is, if $6 \div 0 = \square$, then $\square \times 0 = 6$. Now all we have to do is search for something to put in the box. If the student really believes that $n \times 0 = 0$ for all n, he will soon decide that it is a useless endeavor to try to find a multiplier for 0 that will yield a product of 6. It is impossible.

Let us next attack the problem from the third approach. If $15 \div 3 = 5$ because $3 + 3 + 3 + 3 + 3 = 15$, then in a similar manner let us try to find the answer to $6 \div 0$. How many zeros are required to add to 6? Let's start. $0 + 0 + 0 + 0 + 0 + 0 = 0$. Here we see that six zeros only add to 0, so perhaps we had better use a few more zeros to see if we can't increase the sum to 6. $0 + 0 + 0 + 0 + 0 + 0 + 0 + 0 + 0 + 0 + 0 + 0 + 0 + 0 + 0 + 0$. This time we used 16 zeros. Alas! We still haven't used enough zeros. The sum is still only 0 and not 6. Perhaps we had better plan to use many, many more zeros. How many are we likely to need? Will a hundred zeros be enough? Hardly. What about a thousand—or a million—or a trillion? No matter how many zeros we add to the string of zeros in the sum, $0 + 0 + 0 + 0 + 0 + 0 + 0 + 0 + 0 + 0 + 0$, the sum will never exceed zero and certainly will never get to 6.

No matter how we approach division by zero, we are led to an impossible situation. It may be more nearly correct to say that we must not try to divide by zero because the operation is impossible, rather than to state darkly that it is not permitted.

We now turn our attention to another potential division fact. What might we mean by $0 \div 0$? Isn't any number divided by itself equal to 1? Here we have a question with yet new implications. We can argue that since 5 zeros add to 0, that is, since $0 + 0 + 0 + 0 + 0 = 0$, then $0 \div 0 = 5$; just as the fact that $3 + 3 + 3 + 3 + 3 = 15$ implies that $15 \div 3 = 5$. We can also verify this through the inverse relationship of division with multiplication. Since $5 \times 0 = 0$, then it immediately follows that $0 \div 0 = 5$, just as $5 \times 3 = 15$ implies that $15 \div 3 = 5$. Clearly, however, we can in a similar manner establish the fact that $0 \div 0$ also equals 7, or 11, or 99, or 173,237,868—or any other number in the entire number system including 0 itself. Thus we see that while a division such as $6 \div 0$ is impossible; $0 \div 0$ certainly isn't impossible, but rather it leads to a whole multitude of equally logical and correct answers. There is no limit to the number of correct quotients that one might report for $0 \div 0$. Here again, the results of dividing 0 by 0 are totally unlike the results we get if we divide any other number by itself, such as $7 \div 7$. For the first time we are faced with a fundamental operation involving a rational number that yields more than one correct answer, and in fact yields a limitless supply of correct answers. Since all numbers are correct answers for $0 \div 0$, and since we cannot possibly list all of the numbers in the entire number system, once again we are faced with an impossible situation. This is why we say that division by zero is impossible.

If the teacher will lead the child to discover these concepts for himself, perhaps some of the mystery with which we have surrounded zero can be eliminated.

REFERENCES FOR FURTHER READING

1. Allendoerfer, C. B., *Principles of Arithmetic and Geometry for Elementary School Teachers*, New York: The Macmillan Co., 1971. Chapter 7 and Sections 12.2 and 12.3 deal with the relation between Cartesian products of sets and multiplication of whole numbers.

References for Further Reading   125

2. Bender, M. L., "Order of operations in elementary arithmetic," *AT*, **9** (1962), pp. 263–267. A detailed discussion of the conventions regarding order of operations mentioned briefly in Section 5.5 in our discussion of the distributive property.

The next two references deal with the question of division by zero (Section 5.4).

3. Bender, M. L., "Dividing by zero," *AT*, **8** (1961), pp. 176–179.

4. Duncan, H. F., "Division by zero," *AT*, **18** (1971), pp. 381–387.

There are many books of number tricks and puzzles. The following are all available in paperback editions.

7. Dudeney, H. E., *Amusements in Mathematics*, New York: Dover Publications, 1958.

8. Fröhlichstein, J., *Mathematical Fun, Games, and Puzzles.* New York: Dover Publications, 1962.

9. Gardner, M. (editor), *Mathematical Puzzles of Sam Loyd.* Vols. 1 and 2. New York: Dover Publications, 1959 and 1960.

10. Hunter, J. A. H., *Fun with Figures.* New York: Dover Publications, 1965.

11. Hunter, J. A. H., *More Fun with Figures.* New York: Dover Publications, 1966.

12. Kraitchek, M., *Mathematical Recreations.* New York: Dover Publications, 1942.

13. Merrill, H., *Mathematical Excursions.* New York: Dover Publications, 1958.

14. Shuh, F., *The Master Book of Mathematical Recreations.* New York: Dover Publications, 1968.

Chapter 6

Multiplication and Division Algorithms for Whole Numbers

6.1 INTRODUCTION

Examine the following multiplication problem:

```
  372
×164
  372
 2232
 1488
61008
```

Is the answer correct? What about the format? Looking closely at the work, we note that it seems as if multiplication began on the left with 1 and then continued with 6 and 4 (the "lefthand" method). You are most likely to have become accustomed to performing the multiplication like this (the "righthand" method):

```
  372
×164
 1488
 2232
  372
61008
```

Both procedures are correct and, with practice, equally easy. Remember, however, that if you find the first method somewhat difficult to perform at first, your pupils, in the beginning, would probably find either method equally difficult!

Exercises 6.1

In Exercises 1 through 4, perform the given multiplication by the "lefthand" method.

1. 232×574 **2.** 516×173

128   Multiplication and Division Algorithms for Whole Numbers   6.2

3. 218×607 **4.** 2134×8032

5. Recall the "shockingly" different method of performing subtraction given in Section 4.1 (the method of complements). Here is an equally "shocking" way of performing multiplication, commonly known as the **Russian peasant** method. We present three examples of it here, and you are to figure out what the method is from these examples.

$19 \times 26 = 494$		$22 \times 31 = 682$		$51 \times 12 = 612$	
19	26	22	~~31~~	51	12
9	52	11	62	25	24
4	~~104~~	5	124	12	~~48~~
2	~~208~~	2	~~248~~	6	~~96~~
1	416	1	496	3	192
	494		682	1	384
					612

6. Perform the following multiplications by the Russian peasant method.
a) 27×13 **b)** 36×13

7. We can write $19 = 1 \times 2^4 + 0 \times 2^3 + 0 \times 2^2 + 1 \times 2 + 1$. (Note that this says that $19_{ten} = (10011)_{two}$. Then,

$$19 \times 26 = (1 \times 2^4 + 0 \times 2^3 + 0 \times 2^2 + 1 \times 2 + 1) \times 26$$
$$= (1 \times 2^4 \times 26) + (0 \times 2^3 \times 26) + (0 \times 2^2 \times 26) + (1 \times 2 \times 26) + (1 \times 26)$$
$$= (16 \times 26) + 0 + 0 + (2 \times 26) + (1 \times 26)$$
$$= 416 + 0 + 0 + 52 + 26 = 494$$

Use a similar procedure to explain why $22 \times 31 = 682$ and $51 \times 12 = 612$ (the other two examples in Exercise 5) are obtained by the Russian peasant method of multiplication.

6.2 MULTIPLICATION ALGORITHMS

Both the standard (for the U.S.A.!) "righthand" multiplication algorithm and the "lefthand" multiplication algorithm are founded on three basic ideas:

1. The meaning of our decimal (base ten) notation in terms of powers of ten (expanded notation—see Section 3.7), e.g.,

$$1234 = 1 \times 10^3 + 2 \times 10^2 + 3 \times 10 + 4$$

2. A simple rule for multiplying any whole number by a power of ten.

3. The distributive property plus, of course, a knowledge of basic multiplication facts, i.e., the products $a \times b$ for a and b in $\{0, 1, 2, 3, 4, 5, 6, 7, 8, 9\}$.

It is, of course, easy to develop by examples the idea that multiplying a whole number n by a power of ten can be indicated by simply "adding" 0's to the numeral for n. Thus

$$7 \times 100 = 7 \times 10^2 = 700$$
$$82 \times 10 = 820$$
$$1023 \times 1000 = 1023 \times 10^3 = 1,023,000$$

etc. (Note that this procedure applies equally well in bases other than ten. Thus, for example, $42_{\text{five}} \times 100_{\text{five}} = 4200_{\text{five}}$.)

Now we consider such problems as

$$7 \times 14 = 7 \times (10 + 4) = (7 \times 10) + (7 \times 4) = 70 + 28 = 98$$

and

$$6 \times 324 = 6 \times (3 \times 100 + 2 \times 10 + 4)$$
$$= 6 \times (3 \times 100) + 6 \times (2 \times 10) + (6 \times 4)$$
$$= (6 \times 3) \times 100 + (6 \times 2) \times 10 + (6 \times 4)$$
$$= (18 \times 100) + (12 \times 10) + 24$$
$$= 1800 + 120 + 24 = 1944$$

In such problems, of course, we are using, in addition to the distributive property, the associative property of addition (otherwise $1800 + 120 + 24$, for example, would have no meaning without signs of grouping), and the associative property of multiplication (in, for example, writing $6 \times (3 \times 100)$ as $(6 \times 3) \times 100$). It is the distributive property, however, which is conspicuously used.

In our second example the process can be indicated in a more concise form as follows:

$$
\begin{array}{rl}
324 & \\
\underline{\times 6} & \\
24 & (6 \times 4) \\
120 & (6 \times 20) \\
\underline{1800} & (6 \times 300) \\
1944 &
\end{array}
$$

So far, matters are fairly simple (although, of course, much more time must be spent with primary school children in order for them to achieve competence). Complications arise, however, when both factors in a multiplication problem have two or more digits. Thus, suppose we need to multiply 32 by 47. We can write

$$32 \times 47 = 32 \times (40 + 7)$$
$$= (32 \times 40) + (32 \times 7)$$
$$= (30 + 2) \times 40 + (30 + 2) \times 7$$
$$= [(30 \times 40) + (2 \times 40)] + [(30 \times 7) + (2 \times 7)]$$
$$= 1200 + 80 + 210 + 14 = 1504$$

(We have left out details such as

$$30 \times 40 = (3 \times 10) \times (4 \times 10) = (3 \times 4) \times (10 \times 10) = 12 \times 100 = 1200$$

130 **Multiplication and Division Algorithms for Whole Numbers**

which it would be necessary to discuss with primary school children. Note, by the way, that this step involves the use of both the associative and the commutative properties of multiplication.)

In a more condensed form, this multiplication can be written as follows:

$$
\begin{array}{rl}
32 & \\
\times 47 & \\
\hline
14 & (7 \times 2) \\
210 & (7 \times 30) \\
80 & (40 \times 2) \\
1200 & (40 \times 30) \\
\hline
1504 &
\end{array}
$$

After a large number of problems like this have been worked in this way (which we call the "long" method), the general pattern should become evident to children. Thus, for example, it should appear natural to them to proceed as follows to solve the multiplication problem introduced at the beginning of this chapter:

$$
\begin{array}{rl}
372 & \\
\times 164 & \\
\hline
\end{array}
$$

$$
\text{(multiplying by 4)} \left\{ \begin{array}{rl}
8 & (4 \times 2) \\
280 & (4 \times 70) \\
1200 & (4 \times 300)
\end{array} \right.
$$

$$
\text{(multiplying by 60)} \left\{ \begin{array}{rl}
120 & (60 \times 2) \\
4200 & (60 \times 70) \\
18000 & (60 \times 300)
\end{array} \right.
$$

$$
\text{(multiplying by 100)} \left\{ \begin{array}{rl}
200 & (100 \times 2) \\
7000 & (100 \times 70) \\
30000 & (100 \times 300)
\end{array} \right.
$$

$$
\overline{61008}
$$

or, in left-to-right order,

$$
\begin{array}{r}
372 \\
\times 164 \\
\hline
30000 \\
7000 \\
200 \\
18000 \\
4200 \\
120 \\
1200 \\
280 \\
8 \\
\hline
61008
\end{array}
$$

6.2 **Multiplication Algorithms** **131**

This of course, is certainly not nearly as concise as the solution:

```
    372
   ×164
   ─────
   1488
   2232
   372
   ─────
  61008
```

given in the introduction. Let us begin to condense the longer format. First of all, what is 1488? It is simply 8 + 280 + 1200, the sum of the first 3 numbers listed in the long form. It is usually found by saying something like this: "Four times two is eight; put down the eight; four times seven is twenty-eight [but note that it's really 4×70 we are considering]; put down the eight and carry the two; four times three is twelve [but note that it's really 4×300 we are considering] plus two is fourteen; put down the fourteen." Then we say "Indent [or move to the left]; six times two is twelve; put down the two and carry one," etc.

Effective as this kind of incantation may be to those who know how to use it, it should be clear that it is mathematical gibberish that should not be presented to children at an early stage in their study of multiplication (if, indeed, it should ever be presented). On the other hand, if a child can find out for himself that 8, 280, and 1200, etc., can be added mentally (and this is all that the above "gibberish" says!) so that his work can be simplified, this is all to the good. When difficulties arise, however, a child should always be able to go back to the more understandable procedure.

There is one additional point: Even with the more condensed method (which we will call the "short" method), it still may be a good idea to have children retain zeros in their work, as in

```
    372                    372
   ×164                   ×164
   ─────                  ─────
   1488        or         37200
  22320                   22320
  37200                    1488
  ─────                   ─────
  61008                   61008
```

After all, we are *not* adding 1488, 2232, and 372 to get 61,008!

Exercises 6.2

Perform the following multiplications, first by the long method and then by the short method (but retaining zeros).

1. 262×37 **2.** 365×42 **3.** 402×67

4. 314×142 **5.** 655×278 **6.** 605×280

132   Multiplication and Division Algorithms for Whole Numbers   6.3

6.3 DIVISION ALGORITHMS

The standard division algorithm is based squarely on viewing division in terms of successive subtraction. (See Section 5.3.) Thus, one way of finding $36 \div 12$ is to write

$$
\begin{array}{r}
12 \overline{\smash{)}\,36} \\
-12 \\
\hline
24 \\
-12 \\
\hline
12 \\
-12 \\
\hline
0
\end{array}
$$

Since we had to subtract three twelve's to get a remainder of 0, we can conclude that $36 \div 12 = 3$. The work can be shortened, of course, to

$$
\begin{array}{r}
12 \overline{\smash{)}\,36} \\
-24 \quad (2 \times 12) \\
\hline
-12 \\
-12 \quad (1 \times 12) \\
\hline
0
\end{array}
\qquad \text{or even to} \qquad
\begin{array}{r}
12 \overline{\smash{)}\,36} \\
-36 \quad (3 \times 12) \\
\hline
0
\end{array}
$$

The important thing to observe is that, desirable as it may be for efficiency to shorten the work, this shortening is not vital and, if pushed too hard, may easily interfere with understanding.

Now let us do a harder division problem without worrying too much about efficiency. Here the numerals on the right indicate what multiple of the divisor we are considering, and we've left out the minus signs. So our answer is 574 with remainder 210.

$$
\begin{array}{r}
234 \overline{\smash{)}\,134526} \\
46800 \quad 200 \\
\hline
87726 \\
46800 \quad 200 \\
\hline
40926 \\
23400 \quad 100 \\
\hline
17526 \\
9360 \quad 40 \\
\hline
8166 \\
4680 \quad 20 \\
\hline
3486 \\
2340 \quad 10 \\
\hline
1146 \\
936 \quad 4 \\
\hline
210 \quad 574
\end{array}
$$

6.3   **Division Algorithms**   **133**

(In a problem such as this, of course, *some* concern with efficiency is needed. It would certainly take far too long to subtract 234's one at a time for a total of 574 subtractions!)

Now with experience (and luck!), we can greatly condense this work as follows:

$$
\begin{array}{r}
234\overline{)134526} \\
\underline{117000} \quad 500 \\
17526 \\
\underline{16380} \quad \ 70 \\
1146 \\
\underline{\quad 936} \quad \ \ \underline{\ 4} \\
210 \quad \ \, 574
\end{array}
$$

Although the work has been very considerably abbreviated, you should note that it is based on the same basic procedure and that we have written down 500, 70, and 4, as well as retaining the 0's in 117000 and 16380. Contrast this with the completely abbreviated procedure in common use, which begins with

$$
\begin{array}{r}
5 \\
234\overline{)134526} \\
\underline{1170} \\
175
\end{array}
$$

Note that the "5" appearing at the top is not clearly indicated at this stage as really standing for 500, and that it appears as if we are subtracting 1170 from 134526 in a very peculiar fashion. (Certainly $134526 - 1170 \neq 175$!). The next stage might look like this:

$$
\begin{array}{r}
5 \\
234\overline{)134526} \\
\underline{1170} \\
17526
\end{array}
$$

Now we know that the "26" is there because $134526 - 117000 = 17526$ but the customary "explanation" is that we "bring down" the "26"—surely not a very meaningful mathematical operation!

While abbreviation from our first lengthy format to the format given at the top of this page is certainly desirable (although by no means essential), there is no really good reason to encourage children to go to the format:

$$
\begin{array}{r}
574 \\
234\overline{)134526} \\
\underline{1170} \\
17526 \\
\underline{1638} \\
1146 \\
\underline{\ 936} \\
210
\end{array}
$$

134   **Multiplication and Division Algorithms for Whole Numbers**   **6.4**

and there is good reason to discourage them until you are positive that they understand the process thoroughly. Writing a few extra 0's and the "side" numbers 500, 70, and 4 takes very little extra time and helps greatly to make the work meaningful at each stage.

Exercises 6.3

Perform the following divisions by first making some "wrong" guesses of the proper multiples, as in our first computation of $134{,}526 \div 234$. Then show a condensed version of your work, retaining, however, 0's and the "side" numbers.

1. $492 \div 8$ **2.** $843 \div 7$

3. $23{,}969 \div 42$ **4.** $43{,}169 \div 67$

5. $423{,}987 \div 259$ **6.** $837{,}426 \div 628$

6.4 MULTIPLICATION AND DIVISION IN OTHER BASES

A good test of your understanding of the reasoning behind the multiplication and division algorithms using base ten numerals is to try to use such algorithms with numbers written in other bases. In this way, too, you will better appreciate the necessity for the intermediate steps we introduced before presenting the final forms of the algorithms. We will consider here multiplication and division in base five, with work in other bases to be considered in the exercises.

Since you already know how to multiply and divide in base ten, much of your work in base five can be done by first doing it in base ten and then "translating" it into base five. For example, $4 \times 3 = 12_{\text{ten}} = 22_{\text{five}}$. If, however, you were learning about multiplication for the first time in base five, you would first need to construct and memorize a table of basic multiplication facts in base five. (At least it would be useful to do so!) Such a table is given in Fig. 6.1.

$\times$	0	1	2	3	4
0	0	0	0	0	0
1	0	1	2	3	4
2	0	2	4	11	13
3	0	3	11	14	22
4	0	4	13	22	31

Figure 6.1

How could we get such a table if we knew no base ten arithmetic? One way would be to use our knowledge of addition in base five, together with our definition of multiplication in terms of addition. Thus to find $4_{\text{five}} \times 3_{\text{five}}$ we could write:

$$4_{\text{five}} \times 3_{\text{five}} = 3_{\text{five}} + 3_{\text{five}} + 3_{\text{five}} + 3_{\text{five}}$$

$$= 11_{\text{five}} + 11_{\text{five}} = 22_{\text{five}}$$

Our eventual goal in terms of multiplication and division would be to develop the ability to apply, in base five, the standard algorithms for multiplication

6.4 **Multiplication and Division in Other Bases** **135**

and division to perform multiplications in base five such as 343×134 and divisions such as $12{,}342 \div 132$.

In the abbreviated standard form the computation for 343×134 would look like this (keep in mind that all computation is in base five):

```
    343
   ×134
   3032
   2134
    343
 114222
```

Did you have difficulty in checking this computation and seeing exactly what is going on? I rather hope so, because, if you did, you will better appreciate children's difficulties with the abbreviated form of multiplication in base ten, and hence will see the value of first introducing the long method. Doing the problem by this long method, we have

```
        343
       ×134
      22   (4 × 3)
     310   (4 × 40)    3032
    2200   (4 × 300)
     140   (30 × 3)
    2200   (30 × 40)   21340
   14000   (30 × 300)
     300   (100 × 3)
    4000   (100 × 40)  34300
   30000   (100 × 300)
  114222
```

Now let us consider the "long" division problem, $12{,}342 \div 132$ in base five. Note here how difficult it is to guess the "right" divisor at each stage because of unfamiliarity with the multiplication table for base five. No matter. If we do the division "slowly," as in our discussion of the first stage in teaching "long" division in base ten, we can handle the problem fairly easily. And, once again, you will better appreciate the problems of teaching the corresponding operation in base ten.

A "slow" method

```
132 ⟌ 12342
        3140   20
        4202
        3140   20
        1012
         314    2
         143
         132    1
          11   ‾43
```

A "fast" method

```
132 ⟌ 12342
       11330   40
        1012
        1001    3
          11   43
```

136   Multiplication and Division Algorithms for Whole Numbers   **6.5**

Both methods give us $12{,}342 \div 132$ is 43, with a remainder of 11 (all in base five). As a check, we can proceed as follows:

$$
\begin{array}{r}
132 \\
\times 43 \\
\hline
1001 \\
11330 \\
\hline
12331 \\
+11 \\
\hline
12342
\end{array}
$$

We have introduced the topic of computation in bases other than ten primarily to focus your attention on the need for an understanding of computation in base ten. It should be noted, however, that some work in bases other than ten is frequently included as early as the fifth grade and almost certainly in the seventh or eighth grades.

Exercises 6.4

In Exercises 1 through 8, perform the indicated computations in base five.

1. 232×24 **2.** 322×43

3. 1231×44 **4.** 4312×32

5. $12{,}344 \div 43$ **6.** $43{,}142 \div 34$

7. $30{,}103 \div 24$ **8.** $40{,}203 \div 32$

In Exercises 9 through 12, perform the indicated computations in base four.

9. 312×23 **10.** 213×32

11. $12{,}322 \div 32$ **12.** $13{,}233 \div 23$

In Exercises 13 through 16, perform the indicated computations in base seven.

13. 564×43 **14.** 465×34

15. $316{,}114 \div 36$ **16.** $215{,}614 \div 63$

6.5 THE G.C.D. AND THE L.C.M.

In Section 2.4 we defined the intersection, $A \cap B$, of sets A and B by $A \cap B = \{x : x \in A \quad \text{and} \quad x \in B\}$. We now apply this concept to the finding of the greatest common divisor (G.C.D.) and the least common multiple (L.C.M.) of a set of natural numbers. These concepts are, as we shall see in Chapter 10, of considerable use in calculation with fractions.

First, the G.C.D.: Recall (Section 5.4) that we say that if a and b are natural numbers, then b is a divisor of a if there exists a natural number d such that $a \div b = d$. Thus, for example, 3 is a divisor of 6 because $6 \div 3 = 2$. (Note that any natural number a has the divisors a and 1, since $a \div a = 1$ and $a \div 1 = a$.)

Suppose now that we have a set of natural numbers, $\{a_1, a_2, \ldots, a_n\}$. (We read a_1 as "a sub one", a_2 as "a sub two", $\ldots$, a_n as "a sub n".) Then b is said to be a **common divisor** of $a_1, a_2, \ldots, a_n$ if it is a divisor of each of $a_1, a_2, \ldots, a_n$. Thus, for example, 3 is a common divisor of 6, 12, and 18 since $6 \div 3 = 2$, $12 \div 3 = 4$, and $18 \div 3 = 6$. Now the members of any set of natural numbers may have more than one common divisor. Thus 1 and 6, as well as 3, are common divisors of 6, 12, and 18. The **greatest common divisor** is, as the name implies, simply the largest number in the set of common divisors.

To find the greatest common divisor of, for example, 6, 12, and 18, we can first write the sets D_6, D_{12}, and D_{18} of the divisors of 6, 12, and 18, respectively. They are:

$$D_6 = \{1, 2, 3, 6\}, \qquad D_{12} = \{1, 2, 3, 4, 6, 12\},$$

and

$$D_{18} = \{1, 2, 3, 6, 9, 18\}$$

Then the set of common divisors of 6, 12, and 18 is

$$D_6 \cap D_{12} \cap D_{18} = \{1, 2, 3, 6\},$$

and we see that the greatest common divisor of 6, 12, and 18 is 6.

Similarly, for the least common multiple we first recall (Section 5.2) that, if a and b are natural numbers, then b is a multiple of a if there exists a natural number c such that $b = c \times a$. Thus, for example, 6 is a multiple of 3 since $6 = 2 \times 3$. Then if $\{a_1, a_2, \ldots, a_n\}$ is a set of natural numbers we say that b is a **common multiple** of $a_1, a_2, \ldots, a_n$ if it is a multiple of each of $a_1, a_2, \ldots, a_n$. Thus, for example, 12 is a common multiple of 2, 3, and 4 since $12 = 6 \times 2 = 4 \times 3 = 3 \times 4$. Other common multiples of 2, 3, and 4 are 24, 36, 48, $\ldots$ The **least common multiple,** then, is the smallest number of the set of common multiples.

To find the least common multiple of, for example, 2, 3, and 4, we can first write the sets M_2, M_3, and M_4 of the multiples of 2, 3, and 4, respectively. They are

$$M_2 = \{2, 4, 6, 8, 10, 12, 14, 16, \ldots\},$$
$$M_3 = \{3, 6, 9, 12, 15, 18, \ldots\},$$

and

$$M_4 = \{4, 8, 12, 16, 20, 24, \ldots\}$$

Then the set of common multiples of 2, 3, and 4 is

$$M_2 \cap M_3 \cap M_4 = \{12, 24, 36, \ldots\}$$

and we see that the least common multiple of 2, 3, and 4 is 12.

138 Multiplication and Division Algorithms for Whole Numbers 6.5

Although the calculation of G.C.D.'s and L.C.M.'s via intersection of sets is certainly the most meaningful way, there are shorter methods. The ones we will discuss involve the **factorization** of natural numbers into products of **prime** numbers: numbers with exactly two divisors. Thus 2, 3, 5, 7, 11, and 13 are the first six prime numbers. The number 4, for example, has the three divisors 1, 2, and 4 and so is not a prime number. In particular, note that 1 is not a prime number, because it has only one divisor, namely, 1. Any number other than 1 can be written as a power of a prime number or as the product of powers of prime numbers, if we consider a prime number itself to be a power—namely, a first power. Thus, for example, $2 = 2^1$, $3 = 3^1$, etc.

To write a number other than 1 as a power of a prime number or as the product of powers of prime numbers, we simply test the number for divisibility by the various prime numbers in order. (See Chapter 18 for some tests for divisibility.) For example, 360 is divisible by 2 and $360 = 2 \times 180$. Since 180 is also divisible by 2 we have $360 = 2 \times (2 \times 90)$, where 90 is still divisible by 2. Hence $360 = 2^2 \times 90 = 2^2 \times (2 \times 45) = 2^3 \times 45$. Now 45 is not divisible by 2 but it is divisible by 3 and we have $360 = 2^3 \times (3 \times 15)$. Since 15 is divisible by 3 we have

$$360 = 2^3 \times 3 \times (3 \times 5) = 2^3 \times 3^2 \times 5.$$

Finally, since 5 is a prime number, we can write, for the **prime factorization** of 360,

$$360 = 2^3 \times 3^2 \times 5^1$$

To apply this prime factorization to finding the L.C.M. of 90, 12, and 24, we write each number in this way: Thus we write

$$90 = 2^1 \times 3^2 \times 5^1, \qquad 12 = 2^2 \times 3^1, \qquad \text{and} \qquad 24 = 2^3 \times 3^1$$

Since the L.C.M. of 90, 12, 24 must be divisible by 90, 12, and 24, it must be divisible by 2^3 (because 2^3 is a factor of 24), by 3^2 (because 3^2 is a factor of 90), and by 5^1 (because 5^1 is a factor of 90). Thus the L.C.M. of 90, 12, and 24 is the product of all the prime numbers occurring in the factorizations, each to the *highest* power it occurs in *any* of the factorizations. Thus

$$\text{L.C.M. of 90, 12, and 24} = 2^3 \times 3^2 \times 5^1 = 8 \times 9 \times 5 = 360$$

On the other hand, the G.C.D. of 90, 12, and 24 must divide 90, 12, and 24. Now any power of 2 higher than 2^1 does not divide 90; any power of 3 higher than 3^1 does not divide 12; and no power of 5 divides 12 (and also 24). Hence to find the G.C.D. of 90, 12, and 24, we take the product of those prime numbers that occur in *all* of the factorizations, each to the *lowest* power it occurs. Thus

$$\text{G.C.D. of 90, 12, and 24} = 2^1 \times 3^1 = 6$$

If there are no prime factors in common as in $15 = 3 \times 5$, and $26 = 2 \times 13$, the G.C.D. is 1. If the G.C.D. of a and b is 1, we say that a and b are **relatively prime.**

6.5   **The G.C.D and the L.C.M.**   **139**

If any one of the numbers we are considering is a prime, no prime factorization is possible, of course. For example, to find the L.C.M. and G.C.D. of 7, 42, and 48 we write

$$7 = 7^1, \qquad 42 = 2^1 \times 3^1 \times 7, \qquad \text{and} \qquad 48 = 2^4 \times 3^1$$

Then

$$\text{L.C.M. of } 7, 42, \text{ and } 48 = 7^1 \times 2^4 \times 3^1 = 336$$

and

$$\text{G.C.D. of } 7, 42, \text{ and } 48 = 1$$

(See the article at the end of this chapter for still another way of finding G.C.D.'s and L.C.M.'s.)

A commonly used symbol for the G.C.D. of a and b is (a, b) and, for the L.C.M., $[a, b]$. Thus $(8, 12) = 4$ and $[8, 12] = 24$. (Note that we have previously used, in Section 5.2, the symbol (a, b) for an ordered pair, where, for example, $(8, 12) \neq (12, 8)$. Here, of course, where (a, b) stands for the G.C.D. of a and b we have $(8, 12) = (12, 8) = 4$. Such multiple use of symbols for different things in mathematics is rather unfortunate, but, nevertheless, common.)

The G.C.D. is also sometimes called the **G.C.F. (greatest common factor)** and sometimes the **H.C.F. (highest common factor)**. This is because if a is a divisor of b, then $b \div a = d$ and so $b = a \times d$. Hence a is also a factor of b.

Exercises 6.5

In Exercises 1 through 9, find the G.C.D and the L.C.M. of the given set of numbers. For Exercises 1 and 2, use both the set intersection method and the prime factorization method; for the remaining exercises use only the prime factorization method.

1. $\{15, 25\}$ **2.** $\{4, 6, 9\}$ **3.** $\{2, 24, 15\}$

4. $\{9, 18, 15\}$ **5.** $\{6, 7, 8\}$ **6.** $\{1, 3, 5\}$

7. $\{3, 18, 24\}$ **8.** $\{5, 15, 25, 50\}$ **9.** $\{8, 12, 16, 20\}$

10. What can you say about $(a, 1)$ for any natural number a? About $[a, 1]$?

11. If $(a, b) = d$, what can you say about $(k \times a, k \times b)$ for any natural number k?

12. If $[a, b] = d$, what can you say about $[k \times a, k \times b]$ for any natural number k?

13. Complete the entries in the following table.

a	b	(a, b)	$[a, b]$	$a \times b$	$(a, b) \times [a, b]$
4	6				
8	9				
1	7				
12	18				

140 Multiplication and Division Algorithms for Whole Numbers 6.5

14. What does the table that you constructed in Exercise 13 seem to imply about $(a, b) \times [a, b]$?

15. A certain number of Gizmos can be packed into boxes holding 8 Gizmos with each box completely filled. It is also possible for the same number of Gizmos to be packed into boxes holding 12 Gizmos with each box completely filled. What is the smallest number of Gizmos that we could have that would satisfy both these conditions?

16. A useful device for finding the prime factors of a natural number is a factor tree, described below in material taken from a fifth-grade text. Do the exercises given in this material.

The diagrams below are called "factor trees." Factor trees give information about factors of a number. Since every number has itself and 1 as factors, a factor tree using the number 1 would give little extra information. We do not use the number 1 in a factor tree.

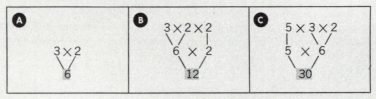

DISCUSSION EXERCISES

1. Does 6 have a factor tree that is different from the one shown in example A? Explain.

2. Do you think 12 has another factor tree? Explain.

3. The number 30 has two more factor trees. Copy each one on your paper and give the missing factors. Explain how you completed the trees.

4. Show that each row of each factor tree for 30 contains numbers that are factors of 30.

5. Which row of the three trees contains the same factors of 30?

EXERCISES

1. Copy each factor tree and give the missing factors.

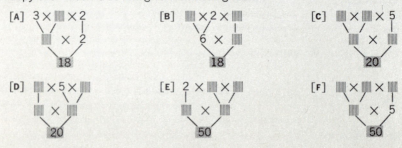

Chapter Test 141

Chapter Test

In preparation for this test, you should review the meaning of the following terms:

Greatest common divisor (G.C.D.) (137) Factorization (138)

(a, b) for the G.C.D. (139) Prime number (138)

Least common multiple (L.C.M.) (137) Prime factorization (138)

$[a, b]$ for the L.C.M. (139) Relatively prime (138)

You should also review how to:

1. Multiply whole numbers by both the "righthand" and "lefthand" method (127):

2. Do multiplication of whole numbers by the "long" method (130);

3. Multiply whole numbers by the Russian peasant method (128);

4. Divide whole numbers using the method given in Section 6.3 (132–133);

5. Perform multiplications and divisions of whole numbers in bases other than ten (any method) (134–136);

6. Find G.C.D.'s and L.C.M.'s both by the set intersection method and by factorization into products of primes (137–139).

1. Multiply 635 by 246 by the "lefthand" method.

2. Multiply 26 by 17 by the Russian peasant method.

3. Multiply 628 by 125, first by the long method and then by the short method (but retaining zeros).

4. Divide 143,521 by 314 by first making some "wrong" guesses of the proper multiples. Then show a condensed version of your work, retaining, however, the 0's.

5. Multiply 324_{five} by 43_{five}.

6. Divide $13,321_{\text{five}}$ by 43_{five}.

7. Find the G.C.D. of 16, 24, and 96 by the use of set intersection and also by factorization into products of prime numbers.

8. Find the L.C.M. of 6, 8, and 12 both by the use of set intersection and also by factorization into products of prime numbers.

9. Two natural numbers are said to be relatively prime if their G.C.D. is ____.

10. For any natural number a, $(a, 1) =$ ____ and $[a, 1] =$ ____.

142   **Multiplication and Division Algorithms for Whole Numbers**

TEST ANSWERS

1.   635
　×246
　127000
　25400
　3810
　156210

2. 26   1̷7̷
　13   34
　6   6̷8̷
　3   136
　1   272
　 442

3.   628   628
　×125   ×125
　40   3140
　100   12560
　3000   62800
　160   78500
　400
　12000
　800
　2000
　60000
　78500

4. (with "wrong" guesses)

$314\,\overline{)143521}$
　31400   100
　112121
　62800   200
　49321
　31400   100
　17921
　9420   30
　8501
　6280   20
　2221
　942   3
　1279
　942   3
　337
　314   1
　23   457

(short method)

$314\,\overline{)143521}$　457
　125600
　17921
　15700
　2221
　2198
　23

(Other arrangements are possible, of course.)

5.   324
　×43
　22
　110
　1400
　310
　1300
　22000
　31142

6.   142
$43\,\overline{)13321}$
　4300   100
　4021
　1410   20
　2111
　1410   20
　210
　141   2
　10   142

(Other arrangements are
possible, of course.)

7. $D_{16} = \{1, 2, 4, 8, 16\}$, $D_{24} = \{1, 2, 3, 4, 6, 8, 12, 24\}$,
$D_{96} = \{1, 2, 3, 4, 6, 8, 12, 16, 24, 32, 48, 96\}$
$D_{16} \cap D_{24} \cap D_{96} = \{1, 2, 4, 8\}$

G.C.D. of 16, 24, and 96 = 8

$16 = 2^4$, $24 = 2^3 \times 3$, $96 = 2^5 \times 3$

G.C.D. of 16, 24, and 96 = 2^3 = 8

8. $M_6 = \{6, 12, 18, 24, 30, 36, \ldots\}$,
$M_8 = \{8, 16, 24, 32, 40, 48, \ldots\}$
$M_{12} = \{12, 24, 36, 48, 60, 72, \ldots\}$
$M_6 \cap M_8 \cap M_{12} = \{24, 48, \ldots\}$

L.C.M. of 6, 8, and 12 = 24

$6 = 2 \times 3$, $8 = 2^3$, $12 = 2^2 \times 3$

L.C.M. of 6, 8, and 12 = $2^3 \times 3$ = 24

9. 1

10. 1, a

A simplified presentation for finding the L.C.M. and the G.C.F.

by Laurence Sherzer

Reprinted from *The Arithmetic Teacher*, Vol. 21 (1974), pp. 415–416.

Given the prime factors of two positive integers, the least common multiple (L.C.M.) of these two numbers is the product of the union of these prime factors, and the greatest common factor (G.C.F.) is the product of the intersection of these prime factors. If we could just state this fact to our students and be understood, our job of teaching them to find the L.C.M. or the G.C.F. of two numbers would be greatly simplified. Unfortunately, as in most teaching, simple verbal statements do not suffice.

This is especially true in the case of finding the L.C.M. and G.C.F. It requires a certain sophistication on the part of the student to be able to think of a set of prime factors as a set at all, for he has been taught that an element cannot appear in a set more than once. He has been taught, for example, that $\{2, 2\} = \{2\}$ and $\{2, 3, 3\} = \{2, 3\}$. The student is not immediately aware, when we speak of the set of prime factors of a number, that though some of these factors may be the same number they are different factors. To clarify this, we would probably have to use some subnotation to distinguish different factors that are the same number. That is, we might write the set of prime factors of 12, $\{2, 2, 3\}$, as $\{2_1, 2_2, 3\}$ to distinguish the first factor 2 from the second factor 2.

The usual alternative for finding the L.C.M., given the prime factors of the numbers concerned, is to instruct the student to choose the number with the greatest power that appears in either set of prime factors. Therefore, to find the L.C.M. of 6 and 9 we establish that $6 = 2 \cdot 3$ and $9 = 3^2$, and choose 2 and 3^2, and derive $2 \cdot 3^2 = 18$. But this requires a knowledge of exponents and a certain dexterity in their manipulation, which may seem formidable to very young grade-school students.

The following approach, which can be presented with a minimum of verbal directions, should bypass the problems of either of the aforementioned procedures. Assume we wish to find the L.C.M. of 6 and 9. We write the factors as shown in Figure 1.

$$6 = 2 \cdot 3$$
$$9 = \quad 3 \cdot 3$$
$$\quad 2 \cdot 3 \cdot 3 = 18$$

Figure 1

Again, to find the L.C.M. of 18 and 30, we write factors as shown in Figure 2. Once more, to find the L.C.M. of 4 and 12, we write factors as shown in Figure 3.

$$18 = 2 \cdot 3 \cdot 3$$
$$30 = 2 \cdot 3 \quad \cdot 5$$
$$\quad 2 \cdot 3 \cdot 3 \cdot 5 = 90$$

Figure 2

$$4 = 2 \cdot 2$$
$$12 = 2 \cdot 2 \cdot 3$$
$$\quad 2 \cdot 2 \cdot 3 = 12$$

Figure 3

What we are doing is setting up a correspondence between those factors that appear in both factorizations, while leaving unmatched those factors that do not have a corresponding factor. Then under each column we write the corresponding prime factor whether it appears once or twice.

Let us examine the model for finding the G.C.F. of two integers. To find the G.C.F. of 6 and 9, we write the factors as shown in Figure 4. To find the G.C.F. of 18 and 30, we write as shown in Figure 5. To find the G.C.F. of 4 and 12, we write as shown in Figure 6.

$$6 = 2 \cdot 3$$
$$9 = \quad 3 \cdot 3$$
$$3 \quad = 3$$

Figure 4

$$18 = 2 \cdot 3 \cdot 3$$
$$30 = 2 \cdot 3 \quad \cdot 5$$
$$2 \cdot 3 \quad = 6$$

Figure 5

$$4 = 2 \cdot 2$$
$$12 = 2 \cdot 2 \cdot 3$$
$$2 \cdot 2 \quad = 4$$

Figure 6

We see that this pictorial representation or model gives the student a simple method for finding the L.C.M. or G.C.F. of two numbers without recourse to complex verbal explanations. A number of illustrative examples will suffice and the student will develop an operational conceptualization. This conceptualization may in turn help the student in his understanding of the concepts of union and intersection of sets.

REFERENCES FOR FURTHER READING

The first two articles discuss what I have called the Russian peasant method of multiplication.

1. Adkins, B. E., "A rationale for duplication–mediation multiplying," *AT*, **11** (1964), pp. 251–253.

2. Stern, C., and M. B. Stern, "Comments on Ancient Egyptian Multiplication," *AT*, **11** (1964), pp. 254–257.

3. Dubisch, R., "The sieve of Eratosthenes," *AT*, **18** (1971), pp. 236–237. Discusses an ancient procedure for finding prime numbers.

4. Tucker, B. F., "The division algorithm," *AT*, **20** (1973), pp. 639–641. Describes the use of blocks to develop the procedures for division of whole numbers.

5. Johnson, P. B., "Modern math in a toga," *AT*, **12** (1965), pp. 343–347. Shows how to do multiplication and division using Roman numerals.

6. Allendoerfer, C. B., *Principles of Arithmetic and Geometry for Elementary Teachers.* New York: The Macmillan Co., 1971. Section 16.5 describes an ancient way of multiplying whole numbers.

The following two articles deal with various shortcuts for computation with whole numbers.

7. Brumfiel, C., and I. Vance, "On whole number computation," *AT*, **16** (1969), pp. 253–257.

8. Freitag, H., and A. H. Freitag, "Shortcuts for the human computer," *AT*, **13** (1966), pp. 671–676.

Summary Test for Chapters 1 through 6

1. State which of the following curves are (i) closed curves; (ii) simple curves; and (iii) simple closed curves.

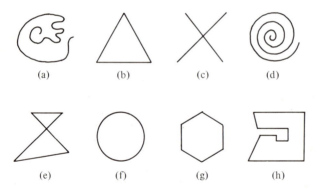

2. True or false?
   a) $0 \in \emptyset$
   b) $x \subseteq \{x\}$
   c) $5 \in \{1, 2, 3, 4\}$
   d) $5 \in \{1, 2, 3, 4, \ldots\}$
   e) $\emptyset \subseteq \emptyset$
   f) $(1, 2) \in \{1\} \times \{2\}$

3. Which of the following sets are equivalent?
   a) $\{\square \triangle\}$
   b) $\{x: x \text{ is a whole number}\}$
   c) $\{x: x = n^2, \text{ where } n \text{ is a whole number}\}$
   d) $\{x: x \text{ is a whole number and } x < 2\}$

4. If A is an infinite set, then there exists a subset B of A such that $B \neq A$ and B___A.

5. $A \cup B = \{x:$ _____$\}$

Summary Test for Chapters 1 through 6   147

6. $A \cap B = \{x: \underline{\qquad\qquad}\}$.

7. State the commutative and associative properties of set union and intersection.

8. State the distributive property of multiplication for whole numbers.

9. For all sets A, $A \cup \emptyset = \underline{\quad}$; $A \cap \emptyset = \underline{\quad}$; $A \cup A = \underline{\quad}$; $A \cap A = \underline{\quad}$.

10. For all sets A and B, if $B \underline{\quad} A$, we define $A \backslash B = \{x: x \underline{\quad} A$ and $x \underline{\quad} B\}$.

11. For all sets A, $A \backslash A = \underline{\quad}$ and $A \backslash \emptyset = \underline{\quad}$.

12. 0 is the $\underline{\qquad\qquad}$ identity.

13. 1 is the multiplicative $\underline{\qquad\qquad}$.

14. Show the addition $14 + 38$ with (i) stick bundles and (ii) an abacus.

15. Show the subtraction $43 - 24$ with (i) stick bundles and (ii) an abacus.

16. Do the following addition by the method of partial sums:

$$\begin{array}{r} 234 \\ 578 \\ +385 \\ \hline \end{array}$$

17. Do the following subtraction (i) by the "borrowing" method, (ii) by the equal additions method, and (iii) by the method of complements. Show why the method works in (i) and (ii).

$$\begin{array}{r} 342 \\ -156 \\ \hline \end{array}$$

18. Do the following addition in base five:

$$\begin{array}{r} 232 \\ 144 \\ +341 \\ \hline \end{array}$$

19. Do the following multiplication in base five by the long method.

$$\begin{array}{r} 232 \\ \times 43 \\ \hline \end{array}$$

20. Do the following subtraction in base five (any method).

$$\begin{array}{r} 3142 \\ -1343 \\ \hline \end{array}$$

21. Do the division $342 \div 22$ in base five.

22. If $A = \{a, b\}$ and $B = \{x, y, z\}$, then $A \times B = \{\text{_____}\}$.

23. Since, in Problem 22, $n(A) = \_\_$, $n(B) = \_\_$, and $n(A \times B) = \_\_$, we conclude that $2 \cdot \_\_ = \_\_$ because $n(A) \cdot n(B) = n(\_\_)$.

24. Since $12 - 6 - 6 = 0$, we conclude that $\_\_\_\_ = 2$.

25. $14 = (\_\_ \times 4) + 2$.

26. The fact that $2 \times (5 \times 4) = 2 \times (4 \times 5)$ illustrates the _____ property of multiplication.

27. If a, b, and c are whole numbers and $a \div b = c$, then $a = \_\_ \times \_\_$.

28. The distributive property of multiplication states that, for all whole numbers a, b, and c, _____ = _____.

29. Corresponding to $A \times \emptyset = \_\_$ for all sets A is the fact that $a \times \_\_ = \_\_$ for all whole numbers a.

30. Insert "+", "−", "×", or ÷, in the frames so as to make a true statement.

    $(6 \square 2) \triangle (2 \bigcirc 1) = 1$

31. Complete the following multiplication rabbit.

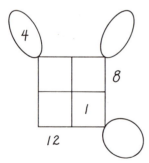

32. The fact that, for all sets A and B, $A \cap (B \cap C) = (A \cap B) \cap C$ is called the _____ property of set _____.

33. Find the G.C.D. of 15, 18, and 30 by using intersection of sets and also by factorization into products of primes.

34. Find the L.C.M. of 9, 18, and 24, by two methods. (See Problem 33.)

Chapter 7

The Integers

7.1 INTRODUCTION

It is a curious fact that, in terms of difficulty, traditional primary and secondary mathematics programs are not well ordered. Thus, for example, it has been customary in the past to reserve any work on algebra (including negative numbers) for the ninth grade. But is there any doubt in the world that it is far easier to understand how to solve such a simple algebraic equation as $x + 1 = 3$ than to learn how to add fractions or to do long division? Similarly, the basic idea of negative numbers is a very natural and useful one. Negative numbers practically beg to be put on the left of zero on a number line, and there are plenty of illustrations involving temperatures below zero, depths below sea level, and rocket countdowns.

So, today, we find negative numbers introduced as early as the third grade and certainly by the sixth grade. These introductions are, of course, just introductions. At first only the existence of negative numbers is considered by means of a number line and such illustrations as we have mentioned. Later, some work on addition is presented, followed by subtraction, multiplication, and division in the upper grades. Thus, before entering ninth grade, children today should be quite familiar with operations with negative numbers.

Since we have not yet considered fractions, we will confine ourselves in this chapter to whole numbers and their negatives. Now there is, as we shall see, a connection between negative numbers and the operation of subtraction. Initially, however, the concept of negative numbers should be considered independently of the operation of subtraction. From this point of view, then, it is unfortunate that the same symbol ($-$) is used to indicate both subtraction and negative numbers. To minimize the possibility of confusion between the two, most texts today initially use a raised minus sign ($^-$) to indicate negative numbers, as in $^-2$, in contrast to $3 - 2$. We read $^-2$ as "**negative** two" or, sometimes, as "the **opposite** of two" or, in more advanced work, for reasons discussed later, as "the **additive**

149

inverse of two". (We also find ⁻2 read as "minus two"—a terminology sanctioned by long usage. It is, however, important not to use this terminology in introducing negative numbers to children, in order to avoid confusion with the concept of subtraction.)

Let us consider a number line, then, as shown in Fig. 7.1, which is taken from a fourth-grade book. Looking at this number line we can see the reason for the term "opposite"; "⁻2", for example, is opposite "2" and is at the same distance from "0" as "2" is. Also, the opposite of ⁻2 is 2 and this observation quickly yields the fact that ⁻(⁻2) = 2, ⁻(⁻3) = 3, etc.

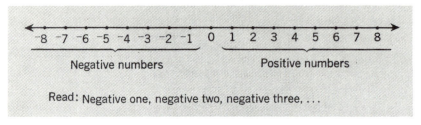

Figure 7.1

The numbers ⁻1, ⁻2, ⁻3, ⁻4, ... form the set of **negative whole numbers**; 1, 2, 3, 4, ... form the set of **positive whole numbers**; 0 is neither positive nor negative. The set

$$\{\ldots {}^-4, {}^-3, {}^-2, {}^-1, 0, 1, 2, 3, 4, \ldots\}$$

is called the set of **integers**, so that we also call {⁻1, ⁻2, ⁻3, ⁻4, ...} the set of **negative integers** and {1, 2, 3, 4, ...} the set of **positive integers**. Again, 0 is neither a positive nor a negative integer. 0, however, is a member of the set {0, 1, 2, 3, 4, ...}, the set of whole numbers or **nonnegative integers**. We define ⁻0 to be 0; i.e., 0 is its own opposite.

The ideas of inequality carry over easily from whole numbers to integers. Just as we say that 2 < 3 because 2 is to the left of 3 on a number line, so we say that ⁻3 < ⁻2 because ⁻3 is to the left of ⁻2 on a number line. Similarly, ⁻2 > ⁻3. Notice that any negative integer is less than any positive integer and, indeed, less than 0. These ideas can be made more vivid by such examples as "three degrees below zero is colder than two degrees above zero" and "a man with a $3 debt is worse off financially than a man with a $2 debt".

It is interesting to note that there was at one time considerable resistance to the use of negative numbers (once called "fictitious" numbers!). This is indicated by the following quotation from *Principles of Algebra* by William Frend, written in 1796:

> After seeing that so able a mathematician as Monsieur Clairaut could be so far blinded and puzzled by the strange doctrine of negative quantities (unnecessarily introduced into the otherwise clear and simple science of Algebra, or Universal Arithmetick [*sic*]) as to reason so very weakly in support of it as we have seen in the foregoing passage, it is surely high time for every true lover of this science, who is

7.1   **Introduction**   **151**

zealous for the honour of its purity and perspicuity, to exclaim, as the good Archbishop Tillotsen did with respect to the Athanasian Creed, "I wish we were fairly rid of it!".

Programed Lesson 7.1

1. $^-2$ should be read as _____ two or as the _____ of two.

   negative, opposite

2. $^-2$ is also sometimes read as _____ two, although such terminology should be avoided with young children.

   minus

3. Show the numbers $^-5$, $^-2$, 0, 3, and 4 on a number line.

4. The set $\{\ldots \,^-4, \,^-3, \,^-2, \,^-1, 0, 1, 2, 3, 4, \ldots\}$ is called the set of _____.

   integers

5. The set $\{^-1, \,^-2, \,^-3, \,^-4, \ldots\}$ is called the set of _____ _____ numbers or _____ _____.

   negative whole, negative integers

6. We have previously referred to the set $\{1, 2, 3, 4, \ldots\}$ as the set of _____ or _____ numbers.

   counting, natural (either order)

7. We also refer to the set $\{1, 2, 3, 4, \ldots\}$ as the set of _____ _____ numbers or _____ _____.

   positive whole, positive integers

8. $^-(^-5) =$ __.

   5

9. If a is a positive integer, then ^-a is a _____ integer.

   negative

10. If a is a negative integer, then ^-a is a _____ integer.

    positive

152   The Integers   **7.2**

11. $4 > 2$ but $^-4$ __ $^-2$.

   $<$

12. $^-3 > ^-4$ but 3 __ 4.

   $<$

13. If a and b are any integers such that $a > b$, then ^-a __ ^-b.

   $<$

14. 0 is neither a _____ integer nor a _____ integer.

   positive, negative (either order)

15. The opposite of 0 is __; that is, $^-0$ = __.

   $0, 0$

16. Insert $>$ or $<$ between each pair of numerals below so as to make a true statement.
   a) 2 __ 0 **b)** $^-3$ __ $^-1$ **c)** $^-5$ __ 0 **d)** $^-3$ __ 2
   e) 2 __ 0 **f)** 0 __ $^-4$ **g)** $^-4$ __ $^-6$

   $a) >$ $b) <$ $c) <$ $d) <$ $e) >$ $f) >$ $g) >$

7.2 ADDITION OF INTEGERS

We have previously discussed procedures for adding two nonnegative integers. The sum of a positive and a negative integer can be easily illustrated by physical examples as, for example, cash on hand and debts. Thus a person with $5 cash (positive) and a debt of $2 (negative) has a net worth of $3. This illustrates $5 + ^-2 = 3$. Similarly, we can illustrate $2 + ^-5 = ^-3$ by a person with $2 cash and a debt of $5; $^-2 + ^-3 = ^-5$ by a person having no cash but debts of $2 and $3; and $2 + ^-2 = 0$ by a person having $2 cash and a debt of $2.

In Section 2.7 we saw how the addition of whole numbers can be pictured on a number line by jumps to the right. Similarly, we can picture addition of negative whole numbers by jumps to the left. Thus our four examples of addition above can be pictured on a number line as shown in Fig. 7.2.

If we assume the associative property of addition for integers, we can also calculate sums of positive and negative integers as follows:

$$5 + ^-2 = (3 + 2) + ^-2 = 3 + (2 + ^-2) = 3 + 0 = 3$$
$$2 + ^-5 = 2 + (^-2 + ^-3) = (2 + ^-2) + ^-3 = 0 + ^-3 = ^-3$$

Here, of course, we are using the fact that 0 is an additive identity for integers.

7.2   **Addition of Integers**

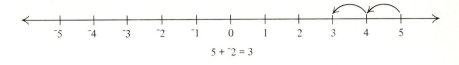

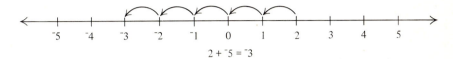

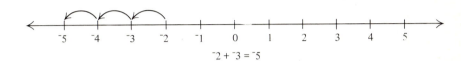

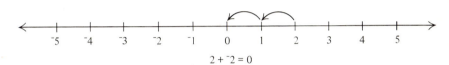

Figure 7.2

Programed Lesson 7.2

1. Illustrate the following additions on number lines:
   a) $^-3 + {^-1}$ b) $^-3 + 4$ c) $4 + {^-3}$
   d) $3 + {^-4}$ e) $^-4 + 3$ f) $^-3 + 3$

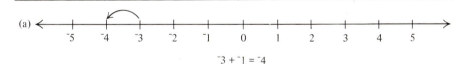

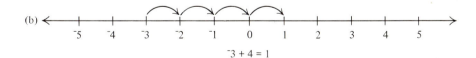

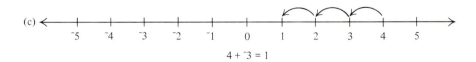

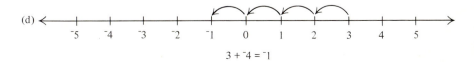

(e)

$^-4 + 3 = ^-1$

(f)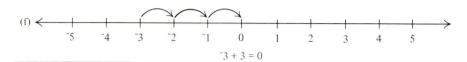

$^-3 + 3 = 0$

2. $6 + {}^-4 = (2 + \_\_) + {}^-4 = 2 + (\_\_ + {}^-4) = 2 + \_\_ = \_\_.$

   4, 4, 0, 2

3. $8 + {}^-5 = (3 + \_\_) + {}^-5 = 3 + (\_\_ + {}^-5) = 3 + \_\_ = \_\_.$

   5, 5, 0, 3

4. $4 + {}^-6 = 4 + (\_\_ + {}^-2) = (4 + \_\_) + {}^-2 = \_\_ + {}^-2 = \_\_.$

   $^-4, {}^-4, 0, {}^-2$

5. $5 + {}^-8 = 5 + (\_\_ + {}^-3) = (5 + \_\_) + {}^-3 = \_\_ + {}^-3 = \_\_.$

   $^-5, {}^-5, 0, {}^-3$

6. $^-5 + 5 = \_\_.$

   0

7. If a is any integer, then $^-a + a = \_\_.$

   0

8. If a, b, and c are any integers such that $a > b$, then $a + c \_\_ b + c$.

   $>$

9. Complete the following addition "rabbits".

(a)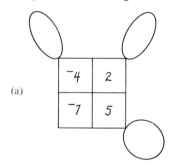

(b)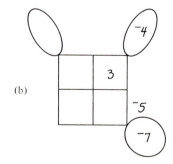

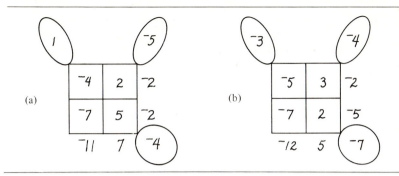

10. A football team gains 7 yards in one play, loses 9 yards on the next, and gains 10 yards on the next. Write an equality involving a negative number to describe the situation.

   7 + ⁻9 + 10 = 8 yards gained

7.3 SUBTRACTION OF INTEGERS

As indicated in Section 2.7, we can show on a number line the subtraction of one positive integer from another positive integer by jumps to the left. It then follows that subtraction of negative integers, like addition of positive integers, corresponds to jumps to the right. All of these rules for "jumping," however, can become quite confusing; and I feel it is best to use a number line only for addition of integers and subtraction of one positive integer from another positive integer.

If we don't use the number line to show subtraction of integers, how can we proceed? The best way, in my opinion, is to connect subtraction with addition and thus to lead children to the rule "to subtract a number, we can add its opposite." (Not that such a rule should be announced by the teacher, or even necessarily verbalized at all.) To begin with, then, we might have the class first note that

   5 − 3 = 2   and also   5 + ⁻3 = 2

Now why is 5 − 3 = 2? Recall the definition (Section 2.6) of subtraction of whole numbers, which is: For any whole numbers a, b, and c,

   $c - b = a$ if and only if $c = a + b$.

Thus

   5 − 3 = 2,   because   5 = 2 + 3.

Now if we want to find what a is if

   5 − ⁻3 = a,

we must have, if our definition of subtraction is to continue to hold for integers as well as for whole numbers,

156 The Integers 7.3

$$5 = a + {}^-3.$$

Then, from our knowledge of addition of integers, we know that $a = 8$; that is, $5 = 8 + {}^-3$ and so $5 - {}^-3 = 8$. Next we note that

$$5 - {}^-3 = 8 \quad \text{and also} \quad 5 + 3 = 8$$

Thus subtracting ${}^-3$ is the same as adding its opposite, 3. Similarly, we have

$$2 - {}^-3 = 5 \quad \text{because} \quad 2 = 5 + {}^-3.$$

Thus

$$2 - {}^-3 = 5 \quad \text{and also} \quad 2 + 3 = 5.$$

Continuing, we have

$${}^-2 - {}^-3 = 1, \quad \text{because} \quad {}^-2 = 1 + {}^-3;$$

so that

$${}^-2 - {}^-3 = 1 \quad \text{and also} \quad {}^-2 + 3 = 1.$$

The above approach may certainly appear too sophisticated for children. Remember, however, that (1) many more examples would be presented to children, and at a slower pace; (2) we are talking about work that almost certainly will not be presented before grade 7; and (3) in a modern mathematics program, children will have had considerable work with the use of letters in equations prior to tackling subtraction of integers. Here (but certainly not in the first introduction to the topic) we can go further and show that, in general,

$$a - b = a + {}^-b \quad (= a + \text{opposite of } b)$$

by simply using the fact that

$$a - b = c \quad \text{if and only if} \quad a = c + b.$$

Thus to say that

1. $\qquad a - b = a + {}^-b = c$

is to assert that

2. $\qquad a = (a + {}^-b) + b.$

But

$$\begin{aligned}
(a + {}^-b) + b &= a + ({}^-b + b) \quad &&\text{(Why?)} \\
&= a + 0 \quad &&\text{(Why?)} \\
&= a \quad &&\text{(Why?)}
\end{aligned}$$

Thus (2) is established and hence (1) is established.

7.3   **Subtraction of Integers   157**

Note that, as a special case of this result, we have

$$0 - a = 0 + {}^-a = {}^-a.$$

This shows the connection between subtraction and "oppositing" and helps explain why we write (except in an introductory treatment) $-a$ rather than ${}^-a$ and even sometimes read ${}^-a$ as "minus a". There are, however, sound pedagogical reasons for beginning with ${}^-a$ and retaining the raised sign for some time. It is even more important to read ${}^-a$ as "negative a" rather than "minus a" in the beginning, and, indeed, it is preferable never to use "minus a".

   We have mentioned two alternatives to "negative a": "the opposite of a" and "the additive inverse of a", and have already seen how the word "opposite" arises (Section 7.1); "additive inverse" is used because

$$(a + b) + {}^-b = a + (b + {}^-b) = a + 0 = .a;$$

that is, ${}^-b$ "undoes" what b "does" to a under addition.

   The commonly used "negative a" suffers from one serious disadvantage: It seems to imply that ${}^-a$ is a negative number. We know, however, that ${}^-a$ can be a positive number; if $a = {}^-5$, for example, ${}^-a = {}^-({}^-5) = 5$. Furthermore, if $a = 0$, ${}^-a = {}^-0 = 0$. Thus it would be better to read ${}^-a$ as "the negative of a."

   In practice, with due regard for alternative notations and possible student misconceptions, you are probably best advised to use the language of the text being used in your school.

   We have noted in Chapter 2 that the sum of two whole numbers is a whole number. However, the difference of two whole numbers need not be a whole number; $2 - 3$, for example, is not a whole number. That is, the set of whole numbers is **closed** under the operation of addition but not under the operation of subtraction. Since, however, if a and b are integers, then both $a + b$ and $a - b$ are integers, the set of integers is closed under both the operation of addition and the operation of subtraction. Also, as we will see in Section 7.4, the set of integers, like the set of whole numbers, is closed under the operation of multiplication. However, it is not closed under the operation of division since $5 \div 2$, for example, is not an integer. Later, with the introduction of rational numbers, we will see that the set of nonzero rational numbers is closed under division.

Programed Lesson 7.3

1. If b and c are integers, then $c - b = a$ if and only if $c = $ _____.

   $a + b$

---

2. Therefore, $6 - {}^-4 = a$ if and only if __ $= a + {}^-4$.

   6

158   The Integers   7.3

3. Since $6 = 10 + {}^-4$, we conclude that $a = $ __ and so $6 - {}^-4 = $ __.

10, 10

4. Thus $6 - {}^-4 = 6 + $ __ $= $ __.

4, 10

5. ${}^-3 - 4 = {}^-3 + $ __ $= $ __.

${}^-4, {}^-7$

6. ${}^-5 - {}^-8 = {}^-5 + $ __ $= $ __.

8, 3

7. For all integers a, $0 - a = 0 + $ __ $= $ __.

${}^-a, {}^-a$

8. If a and b are integers, then $a - b$ is an _____.

integer

9. Use subtraction of integers to answer the question: The temperature was $4°$ below zero one day and $10°$ above the next. How much had the temperature increased?

$10 - {}^-4 = 14$

10. Follow the method of Exercise 9 to answer the question:

A miner is working 1000 ft below sea level while a steel worker is on top of a 100-ft tower. How many feet apart vertically are the two workers?

$100 - {}^-1000 = 1100$

Exercises 7.3

In the following exercises, insert a "+" or a "−" in the frames to make a true statement. Remember that the same symbol must be put in frames of the same shape; frames of different shapes need not (but may have) the same symbol inserted.

1. $4 \square {}^-3 = 7$

2. ${}^-3 \square {}^-5 = {}^-8$

3. ${}^-3 \square {}^-8 = {}^-11$

4. ${}^-5 \square {}^-7 = 2$

5. $(2 \square\ ^-3) \square (5 \triangle\ ^-2) = 6$

6. $[(^-3 \square\ ^-4) \triangle\ ^-5] \triangle\ ^-3 =\ ^-7$

7. $(^-2 \square\ ^-1) \triangle (6 \bigcirc\ ^-3) =\ ^-12$

8. $[(5 \square\ ^-7) \triangle 3] \bigcirc\ ^-4 =\ ^-1$

9. $[(8 \square\ ^-9) \triangle 10] \bigcirc\ ^-1 = 28$

10. $[(^-5 \square\ ^-2) \triangle\ ^-3] \bigcirc\ ^-5 = 5$

7.4 MULTIPLICATION OF INTEGERS

You already know how to multiply nonnegative integers (whole numbers). Using the same additive approach as we did for multiplication of nonnegative integers, it is easy to argue that we should take $3 \times\ ^-2$ to be $^-6$ because

$$3 \times\ ^-2 =\ ^-2 +\ ^-2 +\ ^-2 =\ ^-6$$

Similarly,

$$4 \times\ ^-3 =\ ^-12, \qquad 5 \times\ ^-7 =\ ^-35, \qquad \text{etc.}$$

What, however, should we consider as answers to such products as $^-2 \times 3$ and $^-2 \times\ ^-3$? Certainly it makes no sense to talk about adding 3 or $^-3$ negative two times!

   Probably the most common approach to defining such products when they are first introduced (in grades 6 to 8) is as follows: We first ask the class to supply answers to a sequence of problems such as 4×3, 3×3, 2×3, 1×3, and 0×3. The tabulated results will be

$$4 \times 3 = 12$$
$$3 \times 3 = 9$$
$$2 \times 3 = 6$$
$$1 \times 3 = 3$$
$$0 \times 3 = 0$$

Now we note that as the multipliers (4, 3, 2, 1 and 0) decrease by 1 (move to the left on a number line one unit at a time), the products (12, 9, 6, 3, and 0) decrease by 3 (move to the left on a number line three units at a time). Now let us move one more unit to the left on the number line from 0 to make the multiplier $^-1$. It should seem natural, then (no *proof* implied!), to have the product be obtained by moving another three units to the left on the number line from 0 to $^-3$. Thus it should seem reasonable to children that we should take $^-1 \times 3 =\ ^-3$ and, similarly, $^-2 \times 3 =\ ^-6$, $^-3 \times 3 =\ ^-9$, etc.

   Now let us consider the sequence of products

4 × ⁻3 = ⁻12
3 × ⁻3 = ⁻9
2 × ⁻3 = ⁻6
1 × ⁻3 = ⁻3
0 × ⁻3 = 0

and note that as the multipliers decrease by 1 (move to the left one unit on a number line), the products *increase* by 3 (move to the right three units on a number line). It thus should seem natural (again, no proof implied) that as the multiplier moves another unit to the left (to ⁻1), the product will move another three units to the right (to 3). That is, we arrive at the conclusion that ⁻1 × ⁻3 = 3. Similarly, ⁻2 × ⁻3 = 6, etc.

Another way of developing these rules and a formal proof of their validity is given in the following programed lesson.

Programed Lesson 7.4

1. Complete the following multiplication "rabbits".

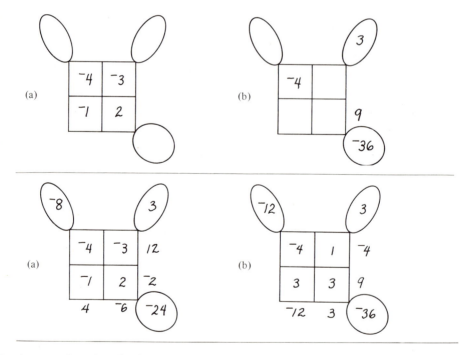

2. Assume that the distributive property holds for multiplication of integers, that n × 0 = 0 for any integer n, and that the additive inverse of an integer is unique; that is, if n + m = 0, then m = ⁻n. Then, if a and b are any integers,

7.5 **Division of Integers** **161**

i) $(a \times b) + (a \times {}^-b) = a \times (\underline{} + \underline{})$, by the _____ property.

b, ${}^-b$, distributive

ii) Hence, since $b + {}^-b = \underline{}$, we have $(a \times b) + (a \times {}^-b) = a \times \underline{} = \underline{}$.

0, 0, 0

iii) Since $(a \times b) + (a \times {}^-b) = 0$, we conclude that $a \times {}^-b$ is the _____ inverse of $a \times b$. By the uniqueness of the additive inverses, we conclude that $a \times {}^-b = {}^-(a \times b)$. For example, if $a = 2$ and $b = 3$, we have $2 \times {}^-3 = {}^-(\underline{} \times \underline{}) = {}^-\underline{}$.

additive, 2, 3, 6

3. Similarly,
   i) $({}^-a \times b) + ({}^-a \times {}^-b) = {}^-a \times (\underline{} + \underline{}) = {}^-a \times \underline{} = \underline{}$.

b, ${}^-b$, 0, 0

ii) Hence ${}^-a \times {}^-b$ is the _____ inverse of _____. But we already know, from the work of Exercise 2, that ${}^-a \times b$ and $a \times b$ are additive inverses. By the uniqueness of additive inverses, it follows that ${}^-a \times {}^-b = \underline{}$.

additive, ${}^-a \times b$, $a \times b$

7.5 DIVISION OF INTEGERS

Division of integers presents no new problems if we accept the same definition of division for integers as we did for whole numbers, i.e., for all integers a, b, and q with $b \neq 0$,

$$a \div b = q \qquad \text{if and only if } a = b \times q.$$

Under this definition it follows, for example, that

$$6 \div {}^-2 = {}^-3 \qquad \text{because } 6 = {}^-2 \times {}^-3$$
$${}^-6 \div 2 = {}^-3 \qquad \text{because } {}^-6 = 2 \times {}^-3$$

and

$${}^-6 \div {}^-2 = 3 \qquad \text{because } {}^-6 = {}^-2 \times 3$$

In general, if a and b are any integers with $b \neq 0$, we have

$${}^-a \div b = a \div {}^-b = {}^-(a \div b) \qquad \text{and} \qquad {}^-a \div {}^-b = a \div b$$

162   The Integers   7.6

Programed Lesson 7.5

1. $12 \div {}^-3 =$ _____ because $12 = {}^-3 \times$ _ _____ .

    ${}^-4, {}^-4$

2. ${}^-15 \div 5 =$ _____ because ${}^-15 = 5 \times$ _____ .

    ${}^-3, {}^-3$

3. ${}^-21 \div {}^-3 =$ _____ because ${}^-21 = {}^-3 \times$_____ .

    $7, 7$

Exercises 7.5

Follow the instructions for the exercises of Section 7.3, except that now "$+$", "$-$", "$\times$", or "$\div$" may be put in the frames.

1. $({}^-3 \,\square\, {}^-5) \triangle 10 = 5$

2. $({}^-6 \,\square\, {}^-3) \triangle 4 = 6$

3. $(3 \,\square\, {}^-1) \triangle 4 = {}^-12$

4. $({}^-6 \,\square\, 2) \triangle 3 = {}^-4$

5. $({}^-6 \,\square\, 3) \triangle {}^-2 = 0$

6. $[({}^-2 \,\square\, 3) \triangle ({}^-5 \,\bigcirc\, 1)] \,\diamondsuit\, 2 = {}^-6$

7. $[(8 \,\square\, {}^-2) \triangle ({}^-6 \,\bigcirc\, {}^-3)] \triangle 10 = 12$

8. $[({}^-5 \,\square\, {}^-3) \triangle 2] \bigcirc ({}^-4 \,\bigcirc\, {}^-2) = 2$

9. $[({}^-5 \,\square\, {}^-2) \triangle 2] \bigcirc (2 \,\diamondsuit\, 4) = 1$

10. $10 \,\square\, [(3 \triangle 4) \,\bigcirc\, {}^-6] \,\diamondsuit\, 8 = 0$

7.6 A FORMAL APPROACH TO INEQUALITIES FOR INTEGERS

Our treatment of inequalities for both whole numbers and integers has been an informal one based upon locations on a number line. That is, we said that $a < b$ (or, equivalently, $b > a$) if the point corresponding to a was to the left of the point corresponding to b.

We now give a more formal approach based on the definition:

If a and b are integers, then $a < b$ if and only if there exists a *positive* integer n such that $a + n = b$.

7.6   A Formal Approach to Inequalities for Integers   163

For example,

$2 < 5$ because $2 + 3 = 5$ and 3 is a positive integer;

$^-2 < 0$ because $^-2 + 2 = 0$ and 2 is a positive integer;

and

$^-3 < 5$ because $^-3 + 8 = 5$ and 8 is a positive integer.

From this definition we can prove the following properties of inequalities:

If a, b, and c are any integers, then:

1. $a = b$, $a < b$, or $b < a$,
and only one of these can hold.   (**Trichotomy** property of inequalities)

2.   If $a < b$ and $b < c$, then $a < c$.   (**Transitive** property of inequalities)

3.   If $a < b$, then $a + c < b + c$.   (**Additive** property of inequalities)

4(i).   If $a < b$ and $c > 0$, then $c \times a < c \times b.$ $\left.\begin{matrix} \\ \\ \end{matrix}\right\}$ (**Multiplicative** properties

4(ii). If $a < b$ and $c < 0$, then $c \times a > c \times b.$ of inequalities)

To prove property (1), we let $b - a = m$. Then if $m = 0$, we have $a = b$. If m is a positive integer, we have $b = a + m$ so that, taking $n = m$ in our definition, we have $a + n = b$. Thus $a < b$ by our definition. Finally, if m is a negative integer, we take $n = {}^-m$ so that n is a positive integer. Then from $b - a = m$, we have $b - m = b + (^-m) = b + n = a$ so that, by our definition, $b < a$. Since $b - a = m$ is either 0, a positive integer, or a negative integer (and only one of these), our conclusion follows.

For (2) we note first that $a < b$ and $b < c$ means that there exist positive integers n and m such that $a + n = b$ and $b + m = c$. Replacing b in $b + m = c$ by $a + n$, we have

$$(a + n) + m = c$$

so that

$$a + (n + m) = c. \quad \text{(Why?)}$$

But since n and m are both positive integers, their sum, $n + m$, is also a positive integer, and hence $a < c$ by our definition. For example, $2 < 6$ and $6 < 11$, and so we can conclude that $2 < 11$.

For (3) we observe that $a < b$ means that there exists a positive integer n such that $a + n = b$. Then

$$(a + n) + c = b + c$$
$$a + (n + c) = b + c \quad \text{(Why?)}$$
$$a + (c + n) = b + c \quad \text{(Why?)}$$
$$(a + c) + n = b + c \quad \text{(Why?)}$$

164   The Integers   **7.6**

Thus $a + c < b + c$ by our definition. For example, $2 < 6$ and so we can conclude that $2 + 5 < 6 + 5$ and $2 + (^-7) < 6 + (^-7)$.

For property 4(i) we begin as before, with $a + n = b$ for some positive integer n. Then

$$c \times (a + n) = c \times b$$

$$(c \times a) + (c \times n) = c \times b \qquad \text{(Why?)}$$

But we are given that $c > 0$, so that both c and n are positive integers. Thus $c \times n$ is a positive integer and so $c \times a < c \times b$ by our definition. For example, $2 < 6$ and $5 > 0$ so we can conclude that $5 \times 2 < 5 \times 6$.

A proof of 4(ii) is considered in the following programed lesson.

Programed Lesson 7.6

1. If a and b are integers, then $a < b$ if and only if there exists a _____ integer n such that $a + $ ____ $ = $ ____.

positive, n, b

2. $5 < 7$ because $5 + $ ____ $ = 7$ and ____ is a _____ integer.

2, 2, positive

3. Since $9 + $ ____ $ = 12$ and ____ is a _____ integer, we can conclude that ____ $<$ ____.

3, 3, positive, 9, 12

4. The fact that if a and b are any integers, then one and only one of $a = b$, $a < b$, and $b < a$ is true, is called the ____ _____ property of _____.

trichotomy, inequalities

5. The fact that if a, b, and c are any integers such that $a < b$ and $b < c$, then $a < c$ is known as the _____ property of _____.

transitive, inequalities

6. The additive property of inequalities states that for all integers a, b, and c, if $a < b$, then ____ $<$ ____.

$a + c$, $b + c$

7. We know that $4 < 9$, and $3 > 0$. We can thus conclude that $3 \times 4 < $ ____ $\times$ ____ by one of the _____ properties of _____.

3, 9, multiplicative, inequalities

7.6   **A Formal Approach to Inequalities for Integers**   **165**

8. If a, b, and c are integers such that $a < b$ and $c = 0$, we can conclude that $c \times a$ ____ $c \times b$.

$=$

9. Rewrite properties (1) through (4) in terms of ">".

1. $a = b$, $a > b$, or $b > a$.
2.   If $a > b$ and $b > c$, then $a > c$.
3.   If $a > b$, then $a + c > b + c$.
4(i). If $a > b$ and $c > 0$, then $c \times a > c \times b$.
4(ii). If $a > b$ and $c < 0$, then $c \times a < c \times b$.

10. If $a < b$ and $c < 0$, we know that there exists a positive integer n such that ____ $= b$.

$a + n$

11. So $c \times ($____$) = c \times b$.

$a + n$

12. Thus $(c \times a) + ($____$) = c \times b$ by the _____ property.

$c \times n$, distributive

13. Since $c < 0$, c is a _____ integer and so $c \times n$ is a _____ integer.

negative, negative

14. Since $c \times n$ is a negative integer, we can write $c \times n = {}^-m$ where m is a _____ integer.

positive

15. Thus $(c \times a) + {}^-m = c \times b$ and so

$[(c \times a) + {}^-m] + m = c \times b +$ ____.

m

16. It follows that $(c \times a) + ({}^-m + m) = c \times b +$ ____ by the _____ property of _____.

m, associative, addition

17. Then, since ${}^-m + m =$ ____ and $(c \times a) + 0 =$ ____, we have $c \times a = (c \times b) + m$, so that $(c \times b) +$ ____ $= c \times a$.

0, $c \times a$, m

166   The Integers 7.7

18. Since m is a positive integer, it follows from our definition that $c \times b < c \times a$, so that $c \times a \underline{\hspace{1cm}} c \times b$.

$>$

7.7 ABSOLUTE VALUE

There are many times when it is very important to distinguish between an integer and its opposite: dressing for a temperature of $^-25°$ is quite a different matter than dressing for a temperature of $25°$! In other situations, however, we may have no real interest in the distinction. Thus if we ask for the distance from 0 of the points corresponding to 2 and $^-2$ on a number line, the answer is 2 in both cases.

We say that the **absolute value** of both 2 and $^-2$ is 2, and write $|2| = |^-2| = 2$. Similarly, $|^-6| = |6| = 6$ and, in general, $|a| = |^-a|$ for all integers a. If $a = 0$, we define $|a| = 0$.

More formally, if a is any integer, we say that

$$|a| = a \quad \text{if } a \geq 0 \quad (\text{i.e., if } a > 0 \text{ or } a = 0),$$

and

$$|a| = {}^-a \quad \text{if } a < 0.$$

Thus, for example,

$$|2| = 2 \quad \text{because } 2 > 0,$$
$$|0| = 0 \quad \text{because } 0 = 0,$$

and

$$|^-2| = {}^-(^-2) = 2 \quad \text{because } ^-2 < 0.$$

Programed Lesson 7.7

1. If $a \geq 0$, then $|a| = \underline{\hspace{1cm}}$.

a

2. If $a < 0$, then $|a| = \underline{\hspace{1cm}}$.

^-a

3. $|5 - 2| = \underline{\hspace{1cm}}; \ |^-3 + 5| = \underline{\hspace{1cm}}; \ |^-3 - 5| = \underline{\hspace{1cm}}; \ |6 - 6| = \underline{\hspace{1cm}}.$

3, 2, 8, 0

4. Is it true that, for all integers a and b, $|a \times b| = |a| \times |b|$?

7.7   **Absolute Value**   **167**

Yes

---

5. Is it true that, for all integers a and b, $|a + b| = |a| + |b|$?

No. For example, $|3 + {}^-5| = |{}^-2| = 2$, whereas $|3| + |{}^-5| = 3 + 5 = 8$. It is true only if a and b are both positive integers, or both negative integers, or one or both are zero. It *is* true that we always have $|a + b| \le |a| + |b|$.

Chapter Test

In preparation for this test you should review the meaning of the following terms:

Integer (150)   Trichotomy property of inequalities (163)
Negative integer (150)   Transitive property of inequalities (163)
Positive integer (150)   Additive property of inequalities (163)
Opposite of an integer (149)   Multiplicative properties of inequalities (163)
Negative of an integer (149)   Absolute value (166)

You should also review how to:

1. Show addition of integers on a number line (153);

2. Subtract (156), multiply (159), and divide (161) integers;

3. Complete addition and multiplication "rabbits" involving integers (154 and 160);

4. Fill in frames with the correct symbols of operation (158 and 162).

The first eight exercises are taken from a chapter review in a sixth-grade text.

1. List the integers from $^-15$ to 15.

2. Give the opposite of each integer.
   [A] 3   [B] $^-3$ [c] 9   [D] $^-7$ [E] 75   [F] $^-75$ [G] $^-183$

3. Give the sign $(<, >, =)$ for each ▦.
   [A] 6 ▦ $^-6$ [B] 0 ▦ $^-6$ [c] $^-30$ ▦ 7   [D] $^-9$ ▦ $^-7$

4. List the integers greater than $^-10$ and less than $^-5$.

5. Find the sums.
   [A] $9 + {}^-5$ [c] $^-9 + {}^-8$ [E] $11 + {}^-11$ [G] $17 + {}^-4$ [ı] $^-4 + 9$
   [B] $^-8 + 3$ [D] $^-8 + 12$ [F] $^-7 + 12$ [H] $^-12 + {}^-5$ [J] $8 + {}^-15$

6. Find the missing addends.
   [A] $4 - {}^-3 = n$ [B] $^-4 - 6 = n$ [c] $^-16 - {}^-7 = n$ [D] $0 - {}^-8 = n$

168 The Integers

7. Use integers to write an addition equation for this problem.
You lose $9 and then find $12. How much do you have?

8. Complete each sentence.
[A] A negative integer is _?_ than a positive integer.
[B] _?_ is the only integer that is neither positive nor negative.
[C] An integer is _?_ than another integer that is to the right of it on the number line.

9. ⁻3 is called the _____ of three, _____ three, or the _____ _____ of three.

10. The set {..., ⁻4, ⁻3, ⁻2, ⁻1, 0, 1, 2, 3, 4, ...} is called the set of _____.

11. The set of natural numbers is also called the set of _____ integers.

12. For all integers a, ⁻(⁻a) = _____.

13. Show the addition ⁻4 + 6 = 2 on a number line.

14. Show the addition 5 + ⁻3 = 2 on a number line.

15. For all integers a, ⁻a + a = _____.

16. If a is a negative integer, then ⁻a is a _____ integer.

17. Complete the following multiplication "rabbit".

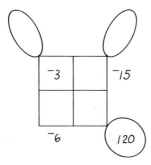

18. Fill in the frames with +, −, ×, or ÷ to make a true statement
[(3 □ ⁻4) △ 6] ○ (12 ○ 6) = ⁻3

19. If a and b are any integers, then $a < b$ if and only if there exists a _____ integer n such that $a + \_ = \_$.

20. The fact that if a, b, and c are any integers such that $a < b$ and $b < c$, then $a < c$ is called the _____ property of _____.

21. If a and b are integers and $a < b$, then $b \_ a$.

22. The trichotomy property of inequalities states that either _____, _____, or _____, and that only one of these statements is true.

Chapter Test 169

23. If a, b, and c are any integers such that $a < b$ and $c > 0$, then $a \times c \underline{} b \times c$.

24. If a, b, and c are any integers such that $a < b$ and $c < 0$, then $a \times c \underline{} b \times c$.

25. $|7 - 2| = \underline{}$, $|2 - 7| = \underline{}$, $|8 + {}^-8| = \underline{}$.

TEST ANSWERS

1. ${}^-15$, ${}^-14$, ${}^-13$, ${}^-12$, ${}^-11$, ${}^-10$, ${}^-9$, ${}^-8$, ${}^-7$, ${}^-6$, ${}^-5$, ${}^-4$, ${}^-3$, ${}^-2$, ${}^-1$, 0, 1, 2, 3, 4, 5, 6, 7, 8, 9, 10, 11, 12, 13, 14, 15.
2. [A] ${}^-3$, [B] 3, [C] ${}^-9$, [D] 7, [E] ${}^-75$, [F] 75, [G] 183
3. [A] >, [B] >, [C] <, [D] <
4. ${}^-9$, ${}^-8$, ${}^-7$, ${}^-6$
5. [A] 4, [B] ${}^-5$, [C] ${}^-17$, [D] 4, [E] 0, [F] 5, [G] 13, [H] ${}^-17$, [I] 5, [J] ${}^-7$
6. [A] 7, [B] ${}^-10$, [C] ${}^-9$, [D] 8
7. $3; ${}^-9 + 12 = 3$
8. [A] less, [B] zero, [C] less
9. Opposite, negative, additive inverse
10. Integers
11. Positive
12. a
13.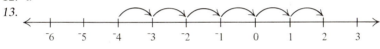
14.
15. 0
16. Positive
17. [diagram]
18. $+, \times, \div, \div$
19. Positive, n, b
20. Transitive, inequalities
21. >
22. $a = b$, $a < b$, $b < a$ (in any order)
23. <
24. >
25. 5, 5, 0

Grisly grids

by William G. Mehl and David W. Mehl

Reprinted from *The Arithmetic Teacher*, Vol. 16 (1969), pp. 357–359.

The multiplication lattice grid is familiar to most teachers and many students throughout the country. A more challenging version of this grid type might include directed numbers. The writers have modified the traditional presentation and introduced a grid incorporating a new set of "rules." The grid can be used to stimulate and to entertain the most ambitious prealgebra or algebra student.

Let's look at the grid's anatomy. Each small square shall be called a cell. Each cell consists of two congruent triangular regions. The upper region is designated *UR* and the lower region *LR* (Fig. 1). The total grid may be in the form of a square or other rectangle, thus enabling the teacher to regulate the number of factors and cells for any given presentation. All marginal entries at the top and right of the grid shall represent factors, while the entries at the left and bottom shall represent sums. Each cell contains a product. All entries consist of integers, and all marginal entries are integers x such that $^-10 < x < {}^+10$.

Cell   **Figure 1**

The following rules regulate the operation of the grids:

1. All tens are recorded in *UR*, while all ones are entered in *LR*.
2. A negative product consisting of two digits is always recorded such that the tens digit is negative and the ones digit is *not* negative.
3. A positive product consisting of two digits is always recorded such that neither digit may be negative.
4. The integers in the margins opposite any cell may have one digit only.
5. Marginal sums are found by adding obliquely from *right* to *left*, starting in the lower right corner of the rectangle.
6. The solution consists of any necessary marginal or tabular entries, and in some cases requires that the marginal sums be totaled and entered in a frame as directed by an arrow (Fig. 4).

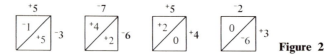

Figure 2

The teacher might begin simply by presenting a grid such as Fig. 3. A variation of the first presentation might appear as shown in Fig. 4. Note the required answer recorded in the frame at the head of the arrow.

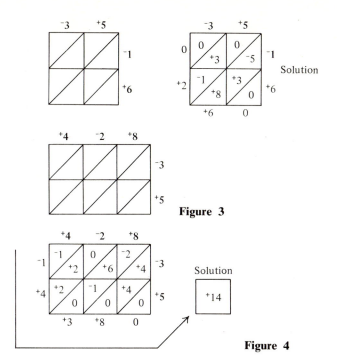

Figure 3

Figure 4

It is suggested that the teacher formulate several grids of the type previously shown to provide sufficient orientation for students and as a prerequisite to attempting the grid types that follow. The student must be thoroughly familiar with the multiplication, division, and addition of directed numbers, some minor rules for divisibility, and have a talent for observation and deductive reasoning. Of course, the teacher may elect to have students attempt some simpler solutions than those shown here, or confine the entire presentation to merely determining products.

Complete the grid (Fig. 5) and place your final answer in the frame at the head of the arrow.

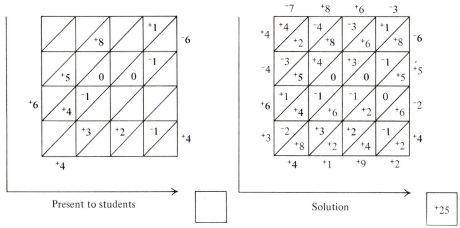

Figure 5

172   The Integers

As the students attempt the solution of this and similar grids, many observations may be noted, and the writers have found that successful students are most eager to share them. These observations include the following:

1. No product can exceed 81 or be less than ⁻81.
2. No factor can consist of two digits.
3. If *UR* is positive or negative, then *LR* is positive.
4. If *UR* is positive, the signs of both factors are alike.
5. If *UR* is negative, the signs of both factors are unlike.
6. If *UR* is 0, then *LR* is 0, negative or positive.
7. If *LR* is negative, then *UR* is 0.
8. If *LR* is even, then either the upper or righthand marginal entry for that cell is even.
9. If *LR* is 0 or 5 and *UR* ≠ 0, then either the upper or righthand marginal entry for that cell is 5 or ⁻5.
10. If "carrying" is necessary while adding obliquely, a positive sum "carries" a positive tens digit, while a negative sum "carries" a negative tens digit.

The reader may enjoy attempting the next grid (Fig. 6) while paying particular attention to the observations previously noted.

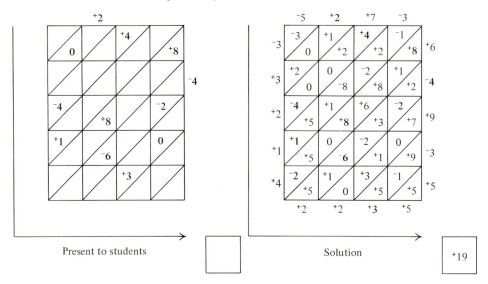

Figure 6

Perhaps some explanation or advice may be useful to the teacher who desires to formulate grids of his own. For best results, the writers recommend that a completed grid be formulated at the beginning. The teacher would then contribute to the difficulty of the solution by removing appropriate factors, products, sums, and partial products. Here it is advisable for the teacher to attempt his own solution in order to decide whether or not he may have abbreviated the original grid beyond the point of effecting a complete solution.

An alternate method of presenting and solving grids of our type follows.

Complete the grid in accordance with the mathematical instructions below:

1. A is not positive.
2. $d = 5$.
3. $^-3 < g < {}^-1$.
4. k is not negative.
5. $(A)(F) > 0$.
6. $B \neq 7$.
7. $7 < p < 9$.
8. $i > 0$.
9. $|A| + |B| = (C)(D)$.
10. $D - A = j$.
11. $q \not< 0$.

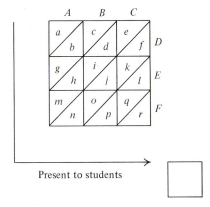

Present to students

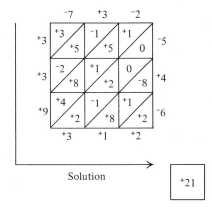

Solution

Figure 7

REFERENCES FOR FURTHER READING

1. Ashlock, R. B., and T. A. West, "Physical representations for signed number operations," *AT*, **14** (1967), pp. 549–553. ("Signed numbers" as used here is old-fashioned terminology for the set of integers! This article also includes an extensive bibliography on the topic.)
2. Cohen, L. S., "A rationale in working with signed numbers," *AT*, **12** (1965), pp. 563–567.
3. Coon, L. H., "Number line multiplication for negative numbers," *AT*, **13** (1966), pp. 213–217.
4. Cotler, S., "Charged particles: A model for teaching operations with directed numbers," *AT*, **16** (1969), pp. 349–353. ("Directed numbers" is still another terminology sometimes used to refer to integers.)
5. Hollis, L. Y., "Multiplication of integers," *AT*, **14** (1967), pp. 555–556.
6. Kelley, J. L., and D. Richert, *Elementary Mathematics for Teachers*. Holden-Day: San Francisco, 1970 (Chapter 4).
7. Mauthe, A., "Climb the ladder," *AT*, **16** (1969), pp. 354–356. Describes a way of presenting negative numbers and operations on them to children.
8. Sherzer, L., "Adding integers using only the concepts of one-to-one correspondence and counting," *AT*, **16** (1969), pp. 360–362.

Chapter 8

More on Geometry

8.1 INTRODUCTION

In Section 1.6 I pointed out that contemporary mathematics programs introduce geometrical concepts in the very early grades. I suggest you review that material at this time.

Further work on geometry continues throughout primary school. The topics introduced, the level at which they are introduced, and the treatment of them varies considerably from one text series to the next. (Much more variation exists than in the treatment of the arithmetic component.) Indeed, we can expect considerable changes in the geometry component as new texts and revisions of present texts appear. In particular, there is likely to be more stress on geometrical *activities* rather than on developing a formal vocabulary concerning geometry. (See the article at the end of this chapter for samples of such activities.)

In any case, there seems little point in devoting a great deal of time and space here to an extensive review of such vocabulary and a listing of formal definitions. Much of this you should already know from your previous work in mathematics and the rest can easily be picked up in teaching the material. Instead, after some brief comments and a few exercises concerning vocabulary in the next section and the introduction of the concepts of congruence and similarity, we will proceed to a discussion of some activities that can be used to enhance children's interest in geometry and, at the same time, increase their understanding.

We are concerned here only with nonmetric geometry (i.e., geometry that does not involve measurement). The important topic of measurement will be discussed in Chapter 11. We will also defer until Chapter 15 a discussion of transformation geometry—a topic rich in activity possibilities that is beginning to find its way into primary-school mathematics programs.

8.2 DEFINITIONS IN GEOMETRY

We have already considered in an informal fashion, in Section 1.6, the following terms: line segment, ray, line, curve, simple curve, closed curve, open curve, triangle, square, rectangle, and circle. To formalize these concepts, however, is not always easy. Consider, for example, the concept of a triangle. Now even young children will have absolutely no difficulty in recognizing that Fig. 8.1 consists of various triangles, whereas no triangles are shown in Fig. 8.2. From this point of view, then, we can say that they understand what a triangle is. But what about a formal definition? When (and if) a formal definition is desired for some geometrical concept, it is often a good idea to ask children to try to frame a definition themselves, rather than just giving them one.

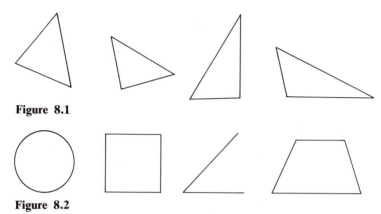

Figure 8.1

Figure 8.2

Such a procedure for the triangle might result in the following attempts at a definition:

1. *Pupil:* "A triangle is a three sided figure."
   So you say, "Like this?" and draw:

2. *Pupil:* "No, the sides must be line segments."
   So you say, "Like this?"

   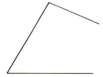

3. *Pupil:* "No, the line segments must intersect."
So you say, "Like this?"

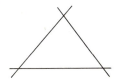

(Here you could find that some children might regard this as a triangle—but the majority will reject the "extra" parts of the line segments extending beyond the points of intersection.)

4. *Pupil:* "No, the line segments must end when they come together."

Now we still don't have a very formal definition of a triangle compared with, for example, "a triangle is a simple closed curve that is the union of three line segments". The above remarks, however, go a long way toward formalizing the concept of a triangle. Furthermore, if we use such procedures in considering definitions, children will learn a lot more than they would if we simply present them with ready-made definitions—no matter how elegant.

As the Teacher's Guide for Book T of the Southampton [England] Mathematics Project states: "In general, a definition sums up an experience and should not precede it."!

Exercises 8.2

1. We present in two columns below some terms that are introduced in various contemporary primary-school mathematics programs.

    a) Try to form your own definitions of these terms.

    b) Consult a dictionary for a definition. (You are likely to find rather inadequate definitions there from a mathematician's point of view.)

    c) Find definitions of these terms in a contemporary:

    i) Primary school series (perhaps at more than one grade level);

    ii) High-school geometry text.

    (The varied results should provide good material for a class discussion!)

Angle	Rectangle
Right angle	Rhombus
Polygon	Trapezoid
Regular polygon	Triangle
Hexagon	Isosceles triangle
Parallelogram	Equilateral triangle
Quadrilateral	Right triangle
Square	

2. From the information you have gained in doing Exercise 1 and your previous work in geometry, classify the following figures in as many ways as possible. (For example, a trapezoid is also a quadrilateral and a hexagon is also a polygon.)

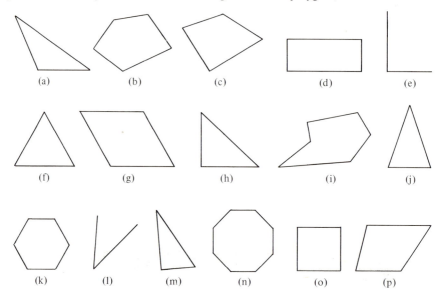

8.3 CONGRUENCE AND SIMILARITY

Congruent figures are sometimes defined as figures that have the "same size and shape". But what do we mean by "size" and by "shape"? It turns out that this a somewhat sticky question. (Indeed, in advanced treatments of geometry, congruence is taken as an undefined term with certain specified properties.) From the point of view of the work on this topic in the primary school, however, we can consider two figures to be congruent if a tracing of one can be made to fit exactly over the other. Figure 8.3 shows this approach to congruence in a fifth-grade text.

Figure 8.3

DISCUSSION EXERCISES

1. Explain how you could use a tracing to show whether or not these two angles are congruent to each other.

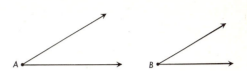

2. For each exercise, trace one of the triangles on a thin sheet of paper. Use this tracing to tell whether or not the two triangles are congruent.

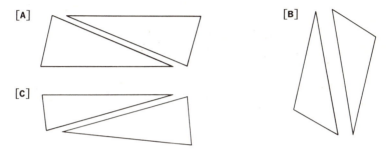

Figure 8.3 (concluded)

We use the symbol "≅" to indicate congruence and write, for example, $\overline{AB} \cong \overline{CD}$, $\angle ABC \cong \angle DEF$, and $\triangle ABC \cong \triangle DEF$ for the segments, angles, and triangles shown in Fig. 8.4.

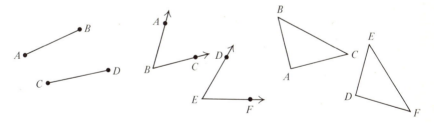

Figure 8.4

Note that we don't use "=" here because "=" means *identity* and certainly $\overline{AB}$ is not the *same* line segment as $\overline{CD}$, $\angle ABC$ is not the *same* angle as $\angle DEF$, and $\triangle ABC$ is not the *same* triangle as $\triangle DEF$. (It is true, however, that $AB = CD$ where AB and CD are the *lengths* of $\overline{AB}$ and $\overline{CD}$, respectively; $m(\angle ABC) = m(\angle DEF)$ where $m(\angle ABC)$ and $m(\angle DEF)$ are the *measures* of $\angle ABC$ and $\angle DEF$, respectively; and *area* of $\triangle ABC$ = *area* of $\triangle DEF$.)

In Chapter 11 we will consider this topic of measurement, and in Chapter 15 we will consider another way of looking at congruence.

Congruence of geometrical figures is a special case of **similarity** of geometrical figures. In similarity we are concerned only with the *shape* and not with the size of the figures: two figures are similar if they have the the same shape.

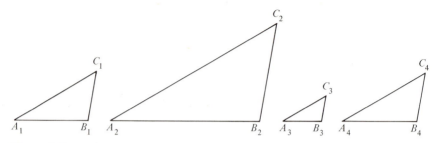

Figure 8.5

Figure 8.5 shows triangle $A_1B_1C_1$ and the triangles $A_2B_2C_2$, $A_3B_3C_3$, and $A_4B_4C_4$ similar to it. In $\triangle A_2B_2C_2$ we have $A_2B_2 = 2 \cdot A_1B_1$, $A_2C_2 = 2 \cdot A_1C_1$ and $B_2C_2 = 2 \cdot B_1C_1$; in $\triangle A_3B_3C_3$ we have $A_3B_3 = \frac{1}{2} \cdot A_1B_1$, $A_3C_3 = \frac{1}{2} \cdot A_1C_1$, and $B_3C_3 = \frac{1}{2} \cdot B_1C_1$; and in triangle $A_4B_4C_4$ we have $A_4B_4 = 1 \cdot A_1B_1$, $A_4C_4 = 1 \cdot A_1C_1$, and $B_4C_4 = 1 \cdot B_1C_1$ so that $\triangle A_4B_4C_4$ is also *congruent* to $\triangle A_1B_1C_1$. To indicate similarity we use the symbol "~". Thus $\triangle A_1B_1C_1 \sim \triangle A_2B_2C_2$ is read "triangle $A_1B_1C_1$ is similar to triangle $A_2B_2C_2$".

Similarity of polygons will be discussed further in Chapter 10 when the concept of ratio is introduced.

Programed Lesson 8.3

1. If two figures are congruent, are they necessarily similar?

Yes

2. If two figures are similar, are they necessarily congruent?

No

3. Are all squares necessarily similar?

Yes

4. Are all circles necessarily similar?

Yes

5. Are all triangles necessarily similar?

No

6. Are all right triangles necessarily similar?

No

7. Are all equilateral triangles necessarily similar?

Yes

8. Are all isosceles triangles necessarily similar?

No

9. Which of the following figures appear to be congruent?

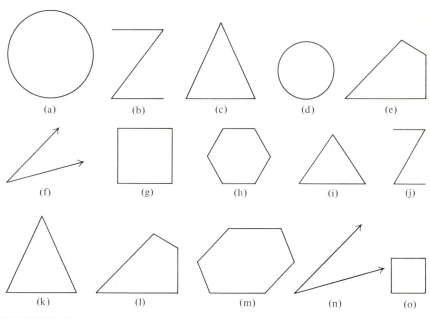

(c) and (k),   (e) and (l),   (f) and (n)

10. Which of the figures of Exercise 9 appear to be similar?

(a) and (d),   (b) and (j),   (c) and (k),   (e) and (l),
(f) and (n),   (g) and (o)

11. Given △ABC, is △ABC ≅ △ABC?

Yes

12. Given △ABC and △DEF, if △ABC ≅ △DEF, is △DEF ≅ △ABC?

Yes

13. Given △ABC, △DEF, and △GHI, if △ABC ≅ △DEF and △DEF ≅ △GHI, is △ABC ≅ △GHI?

Yes

182 More on Geometry   8.4

14. Answer the questions of Exercises 11 through 13 if "≅" is replaced by "∼".

   Yes, in all three cases

8.4 TESSELLATIONS

Rather than doing a set of exercises at the end of this section, you are to answer the questions given throughout the text.

You have all seen floors or walls constructed with ceramic, asphalt, or vinyl tiles. This is nothing new; the Romans made extensive use of ceramic tiles, decorating the floors, walls, and ceilings of buildings, and even the sidewalks, in geometric patterns, or in scenes showing plants, animals, and people. The small tiles that were used were called "tessellae," and we now use the word "tessellation" to describe ways of filling plane regions with various shapes (such as is done in tiling a floor, where no gaps are allowed).

Children will enjoy attempting to fill plane regions with various geometric figures, and in doing so, can learn a great deal about the relationship of various geometric figures.

Tessellations can be studied by drawing figures on paper, but it is easier and more interesting for children to use blocks of the shapes shown in Fig. 8.6. (You, as the teacher, can have a set to show on an overhead projector.) Sets of such blocks can be bought commercially (under various names such as Pattern Blocks), or they can be constructed using wood or heavy cardboard. For each group of four to six pupils, you should provide, if possible, 50 (equilateral) triangles, 50 trapezoids, 50 narrow rhombi (plural of rhombus), 50 wide rhombi, 25 squares, and 25 hexagons. If you make a set of about a fifth as many, you will find it very useful in answering the questions raised.)

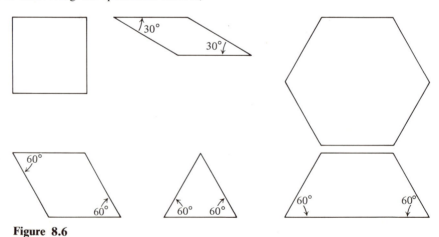

Figure 8.6

The sides of each block are all the same length (usually about one inch or three centimeters) except the long side of the trapezoid, which is twice the length of the other sides. The commercial versions come with each shape a different color, and home-made sets can be so painted or stained.

You have certainly seen a floor covered with square tiles (a tessellation of squares).

1. Can you tile a floor using only equilateral triangles? only hexagons? other figures from our set? (Of course, for all tiling except those of squares, the tiles that fit against the wall must be cut if the floor is of a rectangular shape.)

The point at which the corners of a figure meet in a tessellation is called a **vertex.** In a tessellation of squares, four squares meet at each vertex, as shown in Fig. 8.7.

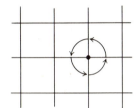

Figure 8.7

2. For the tessellations that you were able to construct in answering Question 1, how many figures meet at a vertex?

Now cut out several regular pentagons like the one shown in Fig. 8.8.

Figure 8.8

3. Can you make a tessellation consisting only of these pentagons?

Study Fig. 8.9, a tessellation of equilateral triangles.

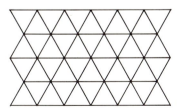

Figure 8.9

4. Can you see, in this figure, any other of the six basic shapes?

5. What other tessellations of the basic shapes can you see in the figure? (Imagine the figure extended in the same pattern.)

6. Can you find larger triangles? Larger other shapes? (Imagine the figure extended in the same pattern.)

More on Geometry

Figure 8.10

Another way to approach tessellations is to make a figure with a compass, such as the one shown in Fig. 8.10. By joining with line segments the pairs of points where the circles intersect, you can make different tessellations.

7. Can you make a tessellation of equilateral triangles in this way?
8. Can you make a tessellation of hexagons in this way?
9. Can you make a tessellation of trapezoids in this way?
10. Can you make a tessellation of right triangles in this way?

You know that a larger square can be made from four smaller squares.

11. What is the minimum number of squares needed to make a still larger square?
12. The next larger square?
13. Do you see a pattern here?
14. What is the minimum number of triangles needed to make a larger triangle?
15. The next larger triangle?
16. A still larger triangle?
17. Do you see a pattern here?

See what else can be done with our basic shapes.

18. Can you make a wide rhombus from other wide rhombi?
19. Can you make a narrow rhombus from other narrow rhombi?
20. Can you make a hexagon from other hexagons?
21. Can you make a trapezoid out of other trapezoids? Will it be the same shape as (similar to) the given trapezoid?
22. Can you make a hexagon out of triangles?

23. Can you make a hexagon out of wide rhombi?
24. Can you make a hexagon out of narrow rhombi?
25. Can you make a hexagon out of trapezoids?
26. In how many ways can you make an equilateral triangle using 3 wide rhombi and 3 triangles?

There are endless ways in which you can make figures out of the basic shapes. Experiment and see what you can do. Figure 8.11 shows one example.

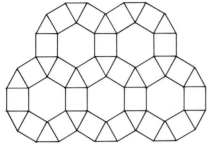

Figure 8.11

8.5 THE GEOBOARD

A geoboard, which can be bought commercially or easily made, is simply a board on which pegs or nails are arranged in square arrays. Figure 8.12 shows a geoboard with rubber bands around pegs to show a triangle and a rectangle.

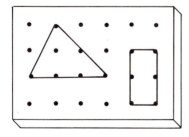

Figure 8.12

The geoboard is a very versatile tool for investigations in geometry. As early as the first grade it can be used to introduce children to, for example, the many different-shaped figures called triangles, which, whatever their shape, share the common property of having three sides.

This is not a methods text, however, and we refer you to the various books and articles listed in the bibliography for suggestions for using the geoboard at early grade levels. What we will do here will be to point out, through the exercises, some uses at a more advanced level that should be of interest to you either as review or as extensions of what you already know. (Using an actual geoboard will be helpful here.)

Exercises 8.5

1. What theorem concerning right triangles does the following geoboard configuration suggest? (For simplicity, the rubber bands are represented here and in the other exercises by simple straight-line segments.) How could we verify this theorem from the geoboard in this special case?

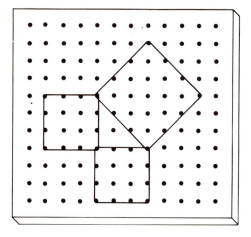

2. What is the minimum number of pegs that either touch a rubber band or are inside the rubber band if we form a triangle with a single rubber band? Answer the same question for a quadrilateral (4-sided polygon), a pentagon (5-sided polygon), a hexagon (6-sided polygon) and a septagon (7-sided polygon). Display your results in the following table. Can you now predict the minimum number of pegs needed to form an n-sided polygon?

	Number of sides				
	3	4	5	6	7
Minimum number of pegs					

3. Suppose that children have learned the formula for the area of a rectangle, $b \times h$ (b = length of base, h = length of side). How can they be led to the formula for the area of a parallelogram, $b \times h$ (b = length of base, h = length of altitude) by placing an additional rubber band on the geoboard pictured below?

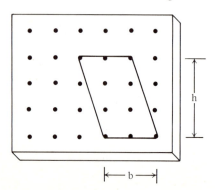

4. After the formula for the area of a parallelogram has been developed, show how children can be led to the formula for the area of a triangle $\frac{1}{2}b \times h$ (b = length of base, h = length of altitude), by placing additional rubber bands on the geoboards pictured below.

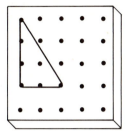
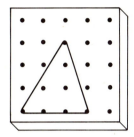
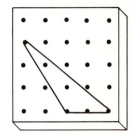

5. The areas of the figures shown on the geoboards pictured below can easily be found by counting squares or half squares. Find these areas in terms of the number of squares, and list them in the table following the figures. Now count the number, b, of pegs touching the rubber band, and complete the table. What relationship is there between b and the area, A, that holds for each figure?

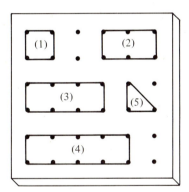

	(1)	(2)	(3)	(4)	(5)
A	1				
b	4				

6. The area of any polygon formed on a geoboard can be found by dividing the area up into rectangles and triangles and applying the formulas for the areas of these figures. This is illustrated in the figure below. The total area is:

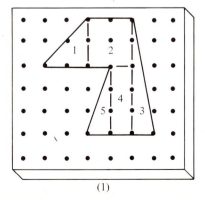

(1)

area (1) + area (2) + area (3) + area (4) + area (5)

$$= (\tfrac{1}{2} \times 2 \times 2) + (2 \times 2) + (\tfrac{1}{2} \times 1 \times 5) + (1 \times 3) + (\tfrac{1}{2} \times 1 \times 3)$$
$$= 2 + 4 + \tfrac{5}{2} + 3 + \tfrac{3}{2} = 13.$$

We can also observe that the number of pegs, b, touching the rubber band is 12, and the number of pegs, i, inside the rubber band but not touching it, is 8. This gives us the first entry in the following table.

	(1)	(2)	(3)	(4)	(5)
A	13				
b	12				
$\tfrac{1}{2}b - 1$	5				
i	8				

Complete this table by filling in the entries for the following figures.

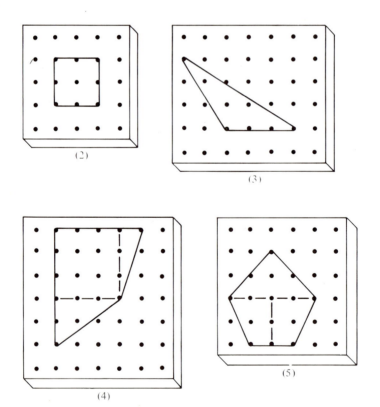

What relationship is there between A, b, and i that holds for each figure? (A relationship discovered and proved by G. Pick in 1899.)

8.6 GEOMETRY OF SPACE

We live in a three-dimensional world and yet, all too frequently, attention in geometry in the early years of primary school has been confined to two dimensions. Almost from the very beginning children should be taught to think about the geometry of three-space. Looking around the classroom we immediately see **planes** (the walls, the floor, the ceiling, etc.) intersecting in pairs in **straight lines** and, in threes, in **points** (the corners of the room). A ball provides an example of a **sphere,** a tin can of a **right circular cylinder,** a matchbox of a **rectangular parallelopiped,** and an ice cream cone of a **right circular cone.**

Models of various other solids can be made from drinking straws fastened together with small lumps of clay, or from various commercial refinements of them. They can also be constructed from thin cardboard as indicated in Fig. 8.13 for a **cube.** The cardboard is to be creased and bent along the dotted lines and the appropriate edges fastened together with Scotch tape. (Alternatively, tabs can be cut out as shown and glue or paste used on these tabs.)

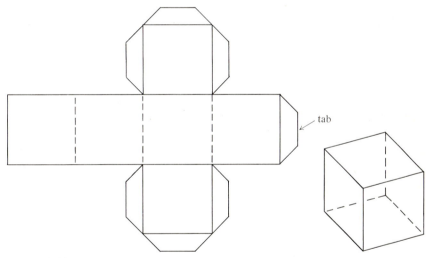

Figure 8.13

Other examples of models that can be made in this way are shown in Fig. 8.14 along with patterns for some of them.

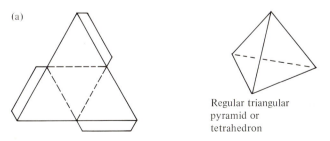

Regular triangular pyramid or tetrahedron

Figure 8.14

190 More on Geometry

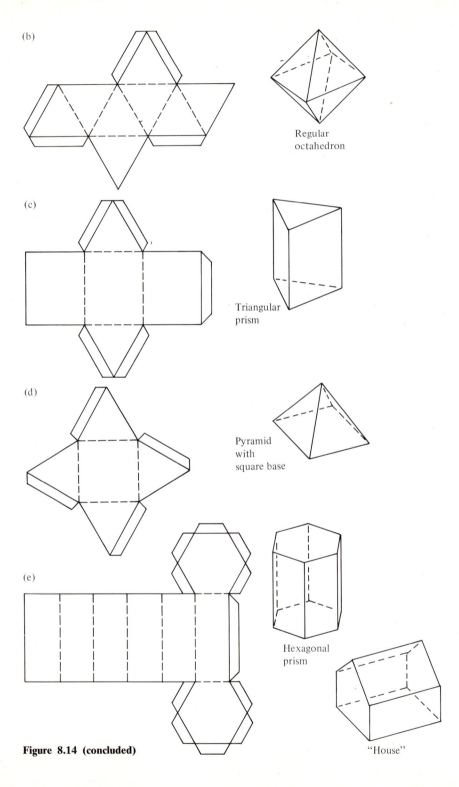

Figure 8.14 (concluded)

Exercises 8.6

1. Consider the number of **faces** (F), **edges** (E), and **vertices** (V), of solids whose faces are polygons (**polyhedra**). For example, a cube has 6 faces, 12 edges, and 8 vertices. (See figure below.) These results can be used to begin the following table. Complete this table by referring to the text figures or to models. Can you find a formula relating F, E, and V that holds for all cases? If you can, you have almost certainly rediscovered a formula due to Leonard Euler (1707–1783).

Solid	Cube	Tetra-hedron	Octa-hedron	Tri-angular prism	Square-base pyramid	Hexagonal prism	"House"
F	6						
V	8						
E	12						

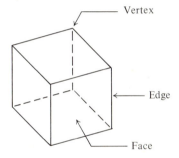

2. Make models (using either straws or cardboard) of some of the solids shown in the text.

8.7 DISCOVERY IN GEOMETRY

As some of the exercises in the previous section have indicated, there are many opportunities for discovery-type activities in geometry. Indeed most, if not all, of the theorems proved in a high-school geometry course can be discovered by children through drawings and measurements. Figure 8.15, from a third-grade text, shows how children can investigate the relationship between the angles formed when two **parallel** lines are intersected by a third line (a **transversal**).

In the following exercises, other examples of the discovery approach to geometry are considered—some concern facts with which you may not be familiar. Some of the exercises, like the first exercises of the Chapter 7 test, are taken from primary-school texts, and the inclusion of such exercises will be continued in several exercise sets in the remaining chapters of this text. Such exercises are certainly not presented with the anticipation of provoking a great deal of mental effort on your part! They can, however, serve as a quick review of important facts and concepts and, furthermore, they emphasize once again the

192   More on Geometry   **8.7**

close relationship of what you are studying here to the mathematics taught in primary school.

Using the Ideas

1. Draw a line that crosses two parallel lines.
Number the eight angles that are formed as shown below.
Color the insides of angles 1 and 2 and cut them out.

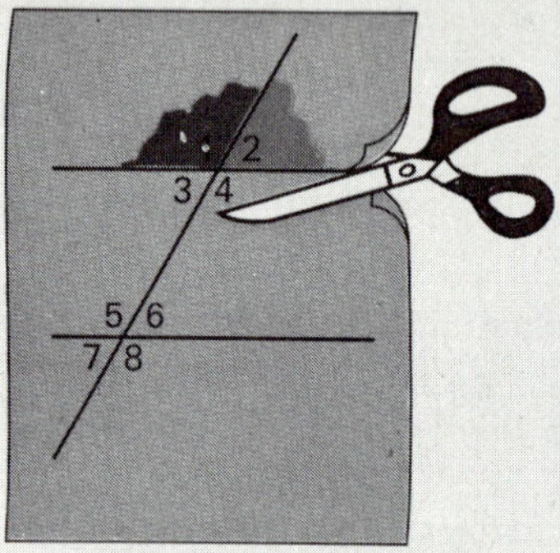

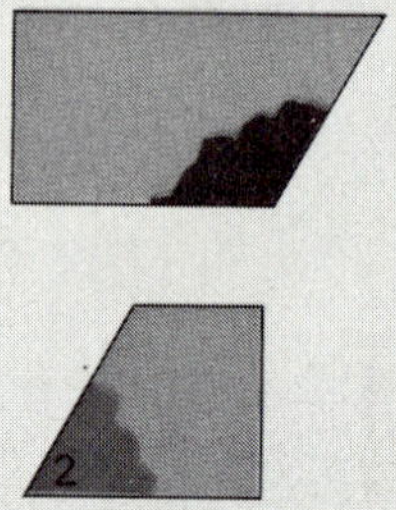

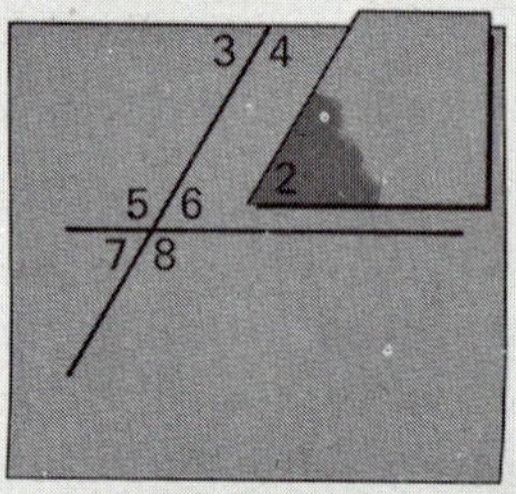

Which of the other angles (3, 4, 5, 6, 7, or 8) are the same size as angle 2 ? Use angle 2 to help you find out.

2. Which of the other angles are the same size as angle 1 ?
Use angle 1 to help you find out.

3. Draw a pair of lines that cross each other. Letter the angles as in the figure.
  A Which angle is the same size as angle *A* ?
  B Which angle is the same size as angle *B* ?

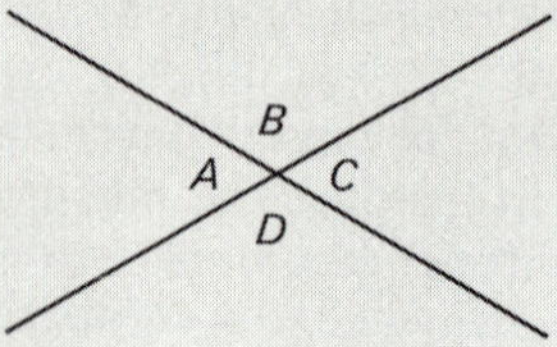

Figure 8.15

Exercises 8.7

1. Do the following exercises from a third-grade text.

Polygons and diagonals

  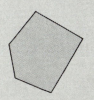 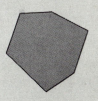

Triangle   Quadrilateral   Pentagon   Hexagon

POLYGONS

The dotted lines in the figure show the **diagonals** of the polygon. How many diagonals does this polygon have? How many diagonals are there from each **vertex**?

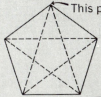

This point is one of the **vertices** of the polygon. It is called a **vertex** of the polygon.

1. Draw a 4-sided polygon (**quadrilateral**) on your paper.

   [A] Draw all the diagonals.
   [B] How many diagonals are there?
   [C] How many are there from each vertex?

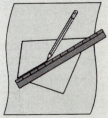

2. Draw a 5-sided polygon (**pentagon**) on your paper. Choose one vertex and label it A.

   [A] How many diagonals can you draw from A?
   [B] Can you draw the same number of diagonals from any vertex?

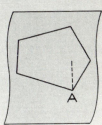

3. Before you try it on your paper, guess how many diagonals you can draw from one vertex of a 6-sided polygon (**hexagon**). Now draw this on your paper and see how well you guessed.

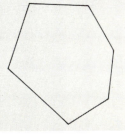

(Continued on page 194)

194 More on Geometry   **8.7**

(Exercise 1 continued)

4. Here is a table showing what you have found so far.

Polygon	Number of diagonals from 1 vertex
3 sides	0
4 sides	1
5 sides	2
6 sides	3
7 sides	

  [A] How many diagonals can you draw from one vertex of a 7-sided polygon?

  [B] Draw a figure to show this.

5. [A] If a polygon has 10 sides, how many diagonals can you draw from each vertex?

  [B] How many diagonals can you draw from one vertex of a 15-sided polygon?

  [C] If you could draw just 10 diagonals from each vertex of a polygon, how many sides would it have?

  [D] If you could draw just 15 diagonals from each vertex of a polygon, how many sides would it have?

6. Give the number of diagonals for each figure.

  [A] pentagon   [B] hexagon   [C] a 10-sided polygon

2. Do the following exercises from a third-grade text.

3. Draw a large 4-sided figure on your paper.

  [A] Cut it out.

  [B] Find the midpoint of each side by folding.

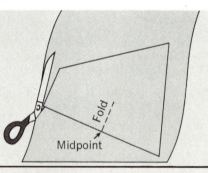

4. [A] Connect the midpoints with segments as in the figure.

  [B] What kind of a figure do you think you have?

  [C] Do you think this would work if you started over with a different 4-sided figure?

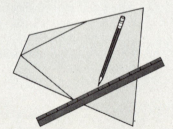

3. Do the following exercises from a fourth-grade text.

Central and inscribed angles

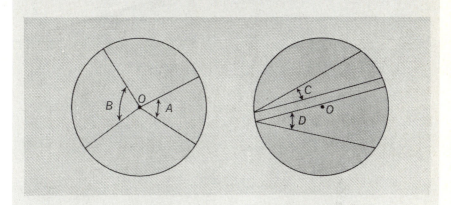

Angles *A* and *B* are **central angles** of the circle.

Angles *C* and *D* are **inscribed angles** of the circle.

1. Draw a circle with a 2½-inch or 3-inch radius.

[A] Mark the center of your circle and label it *O*.

[B] Mark two points on your circle about 2 or 3 inches apart. Label these points *A* and *B*.

[C] Draw the central angle *AOB*.

[D] Mark two other points on your circle.

[E] Draw the inscribed angles to *A* and *B* and label them *L* and *M* as in the picture.

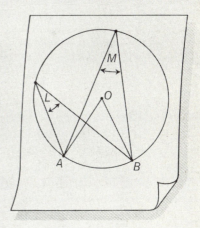

(Continued on page 196)

196 More on Geometry 8.7

(Exercise 3 continued)

2. Color the inside of angle *L* one color and the inside of angle *M* another color.

   [A] Cut your paper about where the dotted line shows in the picture.

   Do not cut any part of angle *AOB*.

   Cut as much of angle *L* and *M* as you can.

   [B] Cut out angles *L* and *M*.

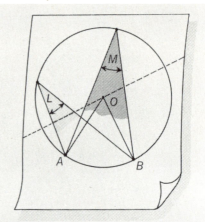

3. Place angles *L* and *M* on each other. How do they compare?

4. Paste angles *L* and *M* on central angle *AOB*. How do they fit?

5. Do you think this would work for other angles *L* and *M* you might choose?

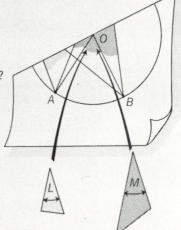

4. Do the exercises on pages 197 and 198 (taken from a fourth-grade text).

Inscribed circles

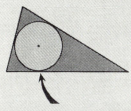

This circle is **inscribed** in the triangle.

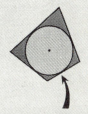

This circle is **inscribed** in the quadrilateral.

A circle is **inscribed** in figures such as those above when each side of the figure is tangent to the circle.

1. You can find the circle inscribed in a triangle in the following way.

    [A] Draw and cut out a large triangle. The sides of your triangle should be at least 12 centimeters long.

    [B] Fold one corner of your triangle as shown in the figure.

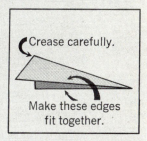

Crease carefully.

Make these edges fit together.

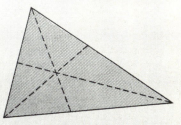

[C] Now fold each of the other 2 corners as you did the first. If you did your work carefully, your triangle should look something like this. (The three creases should pass through one point.)

(Continued on page 198)

(Exercise 4 continued)

[D] Paste your triangle on another sheet of paper. Then use your book and mark the point right below the point where the folds intersect.

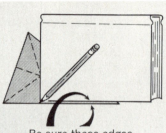

[E] Decide where to place your compass, and draw the circle inscribed in the triangle.

Be sure these edges fit upon each other.

2. Trace this square on your paper. Figure out a way to draw a circle that is inscribed in the square. You should begin by searching for the center of the circle.

5. Do the following exercises from a fifth-grade text.

DISCUSSION EXERCISES

1. [A] What can you say about the opposite sides of a parallelogram?
   [B] Would you say that the opposite sides of a rectangle are parallel?
   [C] Is a rectangle also a parallelogram?

2. [A] Each angle of a rectangle is what kind of angle?
   [B] What special properties does a square have?

3. What special properties does a rhombus have?

EXERCISES

1. Give the missing words.

   [A] A 4-sided polygon with all congruent sides is either a _?_ or a _?_.

   [B] A parallelogram with all right angles is either a _?_ or a _?_.

8.7   **Discovery in Geometry**

6. How many triangles are there in the figure below? How many would there be if the other two diagonals were drawn? (Try to find a systematic way of determining your answer to the second question.)

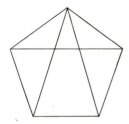

7. The block of wood shown below is a cube with edges 4 centimeters long. It is made by putting together a number of smaller cubes, each with edges one centimeter long. How many of these smaller cubes were needed to make the large cube? How many of them are on the outside of the large cube, and how many are hidden inside? Try to generalize your result to a cube n centimeters long made with the small cubes.

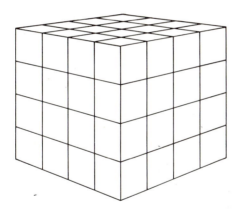

8. How many different shapes can you cut from squared paper using two squares? 3 squares? 4 squares? 5 squares? (The figure below shows some possibilities for 2, 3, 4, and 5.)

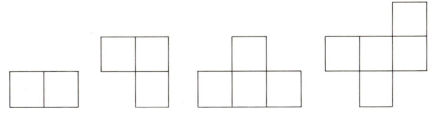

9. Let *ABCD* be any quadrilateral. Construct squares on each side and locate *P*, *Q*, *R*, and *S*, the centers of these squares, as shown in the figure at the top of page 200. What can you conclude about the relationship of $\overline{PR}$ and $\overline{QS}$?

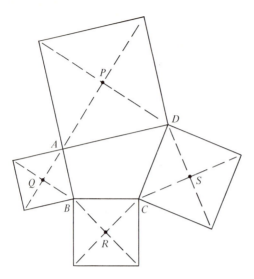

10. The line segment (**chord**) joining any two points on a circle divides the circular region into two parts, as shown in the first figure below. The second figure shows the situation for three points, and the third for four points. This gives us a start on the following table:

Number of points	2	3	4	5	6
Number of regions	2	4	8	?	?

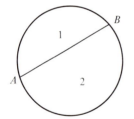 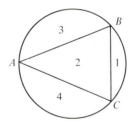 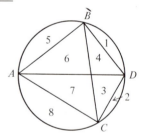

Draw a figure for five points, determine the number of regions, and then make a guess as to the number of regions that would be obtained with 6 points. Now check your guess by drawing a figure for 6 points and counting the number of regions. (*Note.* The problem involves the *maximum* number of regions. Thus, you must choose your points so that no more than two chords intersect at any one point. For example, the figure below would not do for the 6-point case since $\overline{AD}$, $\overline{BE}$, and $\overline{CF}$ all intersect at O.)

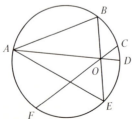

Chapter Test

In preparation for this test, you should review the meaning of the following terms and symbols. (It is not necessary that you be able to give a formal definition in every case, but you should be able to recognize and name the geometrical figures listed.)

Angle (177)	Plane (189)
Right angle (177)	Sphere (189)
Polygon (177)	Right circular cylinder (189)
Regular polygon (177)	Rectangular parallelopiped (189)
Hexagon (177)	Right circular cone (189)
Parallelogram (177)	Cube (189)
Quadrilateral (177)	Parallel lines (191)
Rectangle (177)	Transversal (191)
Rhombus (177)	Congruent figures (178)
Square (177)	Similar figures (179)
Trapezoid (177)	Tessellation (182)
Triangle (177)	Geoboard (185)
Isosceles triangle (177)	≅ (179)
Equilateral triangle (177)	∡ (179)
Right triangle (177)	~ (180)

1. Which of the following statements are true and which are false?
   a) Every hexagon is a polygon.
   b) Every square is a parallelogram.
   c) The sides of any regular polygon are congruent.
   d) The three angles of any isoceles triangle are congruent to each other.
   e) Two similar rhombi are necessarily congruent.
   f) If the sides of a polygon are congruent, the polygon is a regular polygon.
   g) No parallelogram is a trapezoid.

2. Answer the questions in the following material taken from a fifth-grade text.

   Investigating the Ideas

   This tiling pattern could completely cover this page.

   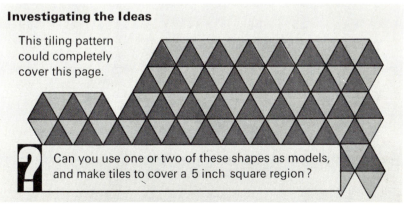

   ? Can you use one or two of these shapes as models, and make tiles to cover a 5 inch square region?

   (See the next page for the shapes mentioned in the above question.)

(Exercise 2 continued)

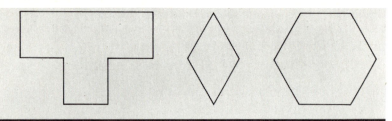

Discussing the Ideas

1. A pattern that covers a plane with tiling is called a **tessellation** of the plane.
   A Can you tessellate a plane with a regular octagon? Explain your answer.
   B Can you use another shape with the octagon to tessellate a plane?

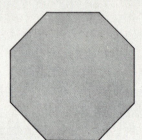

2. Which regular polygons could be used to tessellate a plane?

3. Answer the question on the "activity card" shown below, taken from a third-grade text. ("Dot paper" is simply paper on which dots are placed in a square array, as are the pegs on a geoboard. The Teacher's Guide for this text, however, suggests that the best results are obtained if each child has his own geoboard to work with.)

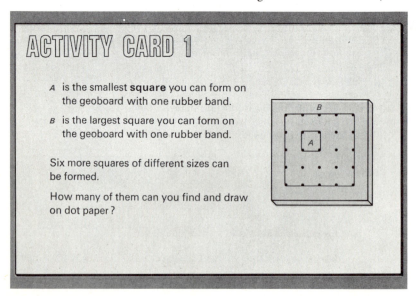

4. The following exercises are taken from a fifth-grade text and introduce some additional vocabulary in regard to solid geometry.

EXERCISES

For each space figure on the left, give the letter of the physical object which best reminds you of that figure.

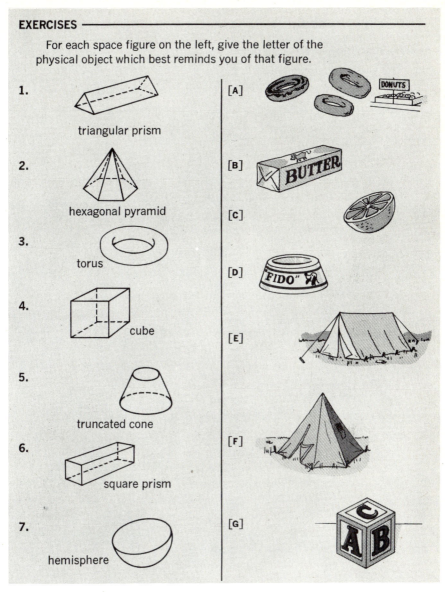

1. triangular prism
2. hexagonal pyramid
3. torus
4. cube
5. truncated cone
6. square prism
7. hemisphere

5. A strip of paper can be cut into four parts with 3 cuts by cutting along the dotted lines shown in the figure below. Can you cut a strip of paper into four parts with only 2 cuts? Can you cut a strip of paper into three parts with only 1 cut?

6. The following exercises are taken from a seventh-grade text and introduce the important geometrical notions of **convex** and **concave.** (These concepts are introduced earlier in some series.)

Convex and Concave Figures

Investigating the Ideas

Suppose each drawing below represents a floor plan of a room and you are standing inside the room.

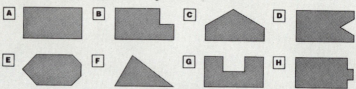

 In which rooms could you see all parts of the room no matter at what point you stood in the room?

Discussing the Ideas

1. The rooms in which you can see all parts of the room from any inside point are **convex** shaped rooms.
   The others are **non-convex** shaped rooms.
   Is your classroom convex or non-convex in shape?

2. Can you draw a non-convex triangle?

3. Is the statement true or false?
   All 4-sided geometric figures are convex.
   Draw some pictures that would illustrate your answer.

4. Another way in which you can think about convex and non-convex figures is illustrated at the right.

   A figure is **convex** if every segment joining any two points on the boundary of the figure contains only points in the interior of the figure or on the boundary.
   Non-convex figures are **concave** figures.

   A Draw a 6-sided figure that is convex.
   B Draw a 6-sided figure and show that it is concave.

TEST ANSWERS

1. a) T   b) T   c) T   d) F   e) F   f) F   g) T
2. Tessellations are possible with the T-shaped figure, and, of course, with the rhombus and the hexagon.
   1(A). No
   1(B). Octagons, together with squares with sides of length equal to the lengths of the sides of the octagons, can be used to tessellate a plane.
   2.   Equilateral triangle, square, hexagon

3. The four obvious ones are:

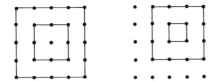

The four that are less obvious are:

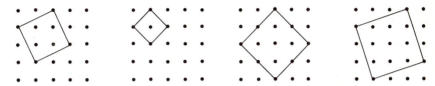

4. (1) E (2) F (3) A (4) G (5) D (6) B (7) C
5. By cutting the paper into two parts and then placing one strip above the other and cutting again, we obtain 4 parts. If we fold the paper before we cut it, one cut will produce 3 pieces.
6. ? A, C, E, F
   1. Depends on the classroom (probably convex!)
   2. No
   3. No. For example,

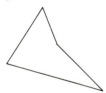

   is not convex.
   4(A). For example, a regular hexagon
   4(B). For example,

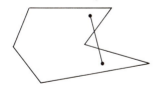

Geometric activities for early childhood education

by George Immerzeel

Reprinted from *The Arithmetic Teacher*, Vol. 20 (1973), pp. 438–443.

A class of first graders was working on a geometry lesson in which they were to arrange a basic set of geometric blocks like the arrangements shown on a screen in front of the classroom. Three students in the group were noticeably successful in completing the task. They were not only highly successful in the interpretive aspects of the activity, but they were also able to verbally justify their selection of particular blocks, even when the decision required considerable deductive reasoning. Their teacher pointed out that these three students were all in the lowest reading group in the class.

Activities like these, which have produced a high level of student success and excitement, have provided the motivation for further experimentation with the learning of geometric concepts in the elementary school. Many compelling reasons for introducing young children to geometric experiences have become evident as a result. Such early childhood experiences—

- provide a conceptual framework that helps the child interpret his environment (a successful experience at age 6 has a six-year advantage over the same experience at age 12);
- build a background for many of the basic models for mathematical concepts such as number lines, arrays, measurement models for addition, models for fractions, and graphs;
- contribute to the child's ability to recognize and identify patterns and to organize materials and information (steps in the important process of generalization);
- provide an opportunity for success for many children who do not succeed with highly symbol-oriented experiences;
- capitalize on the optimum time for developing geometric concepts.

A variety of activities have proved to be highly successful in developing geometric concepts with active, curious, young children. The most successful activities have had the following characteristics in common:

1. The child is a participant, not an observer. If there is hardware, each child uses the hardware.

2. Children can experience various levels of success.

3. The experience is self-reinforcing—the child can tell when he has a correct solution.

4. The experience contributes to the child's knowledge of significant content.

5. The concepts are built first and the more complex generalizations are left for future experiences.

Geometric Activities for Early Childhood Education 207

6. The concept is developed before the vocabulary is introduced—the focus is on the *idea*, not the terminology.

Some activities that have been found to be very successful with young children are described in the following paragraphs. The examples that are given are not intended to be a sequence of lessons; they are simply samples from experiences that are appropriate for young children.

Using Blocks

The following experiences are built around a set of 10 blocks (see Fig. 1). A useful set of blocks can be made by a carpenter (or a teacher who is handy with tools) at a cost of approximately $1.50 a set.

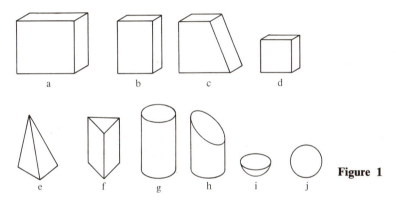

Figure 1

Objective: Given a set of blocks, the student can classify the blocks into at least three subsets.

Materials: Set of blocks for each student.

Students are told to do the following tasks in sequence:

1. Pick out the blocks that have eight corners.
2. Pick out the blocks that have six sides.
3. Pick out the blocks that have twelve edges.
4. How many blocks have flat sides?
5. Which blocks can be stacked up?
6. Which blocks are taller than ▢ ?

   Objective: Given a block, the student can describe it so that the other students can pick out a block shaped like the block that is being described.

   Materials: Set of blocks for each student.

The student sits with his back to the class. He pretends that he is talking on the telephone and describes a block in the set. Each member of the class is to pick out a block like the one that the student decribes. The student's success is measured by the number of other students who can select the right block from his description.

Objective: Given a picture of a subset of the blocks, the student can translate the picture into a three-dimensional model.

Materials: 35 mm slides of subsets of blocks. Set of blocks for each student.

The teacher shows a slide of a subset of blocks like that shown in Fig. 2. Each student chooses the appropriate subset of blocks and arranges them according to the picture. As the sequence progresses, the arrangements of blocks that are pictured become more complex.

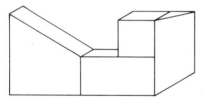

Figure 2

Objective: Given a picture of a geometric figure, the student can identify a block with a face like the picture.

Materials: Collection of pictures of polygons. Set of blocks for each student.

Each student is given a series of pictures of rectangles, squares, triangles, and other polygons that may or may not represent the faces of some of the blocks. The student picks out the blocks that have faces like the pictures.

Making Polygons

As described here, these activities are paper-and-pencil activities. They can be duplicated on the overhead projector, however.

Objective: Given a drawing of a right triangle, the student can trace it to make composite polygons such as rectangles and triangles.

Materials: A picture of a right triangle that can be traced. (See Fig. 3.) Several sheets of tracing paper. Worksheet with drawings of composite polygons.

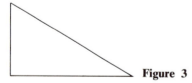

Figure 3

Each student makes a tracing of the right triangle. He then moves the tracing paper so that the original triangle and the tracing form a rectangle and traces the original triangle again. The student uses the same procedure to draw figures like those shown in Fig. 4.

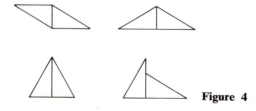

Figure 4

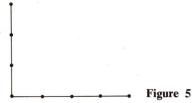

Figure 5

Objective: Given a drawing of a right angle, the student can trace it to make certain geometric figures.

Materials: A picture of a right angle like that shown in Fig. 5. Sheet of tracing paper.

The student makes tracings of the angle to draw a square, a rectangle, and a triangle whose sides are 3, 4, and 5 units.

Recognizing Perimeter

The sequence of activities described here would be appropriate for seven- and eight-year-olds. For older students, the same type of activity could be extended to such figures as trapezoids or irregular quadrilaterals.

Objective: Given a rectangle, the student can recognize the perimeter.

Materials: Twelve straws or small sticks that are all the same length.

The student is told to make a rectangle with 8 sticks. He is asked to describe his rectangle with a pair of numbers as shown in Fig. 6. The student is then given an activity sheet that includes questions like the following:

1. How many sticks does it take to make a (3, 2) rectangle?
2. How many different rectangles can you make from 12 sticks?
3. How many sticks does it take to make each of the following rectangles?

   (1, 1)   (2, 1)   (3, 1)
   (2, 2)   (3, 2)   (3, 3)
   (4, 1)   (4, 2)   (5, 1)

   (*Answers:* 4, 6, 8; 8, 10, 12; 10, 12, 12.)

4. Which of the rectangles are squares?
   (*Answers:* (1, 1), (2, 2), (3, 3).)

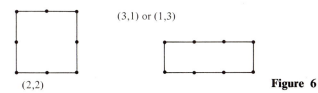

Figure 6

The student can also be given worksheets with similar questions for triangles, 5-sided polygons, and even three-dimensional figures.

Using Cubes

There are many ways that cubes can be used. Pictures of geometric figures could be drawn on each of the six faces and the figures could be classified by certain characteristics. Cubes could be glued together like "Soma" blocks and the various shapes could be pictured for the students to construct. Cubes also make excellent units of measure for volumes of small boxes.

> *Objective: Given a picture of a solid, the student can identify the number of cubes it would take to make it. (Primitive experience with volume.)*
>
> *Materials:* Twenty-seven one-inch cubes for each student. Worksheets.

The student is given a worksheet with pictures of solids like those shown in Fig. 7. He is asked how many cubes it would take to make each solid. The student then makes each solid with his set of cubes.

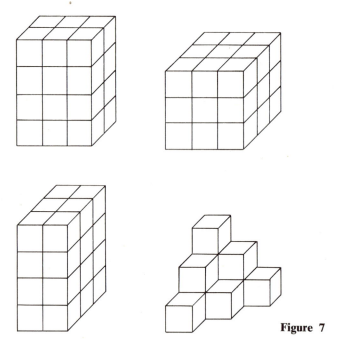

Figure 7

> *Objective: Given a picture of a polygon, the student can identify the small polygons that make it up.*
>
> *Materials:* Worksheet.

The student completes a worksheet like the one shown in Fig. 8.

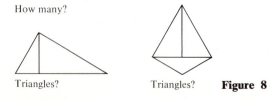

Figure 8

Geometric Activities for Early Childhood Education   211

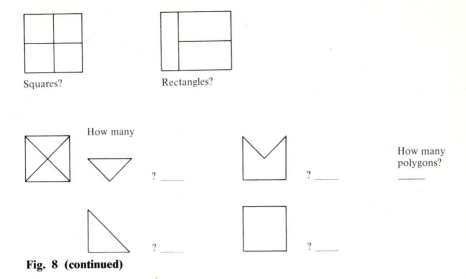

Fig. 8 (continued)

Using Geoboards

Geoboards are good sources for geometric experiences and there are many descriptions of their use in the classroom. If a set of geoboards is not available, a worksheet can be used to develop many of the same experiences.

> *Objective:* Given a rectangular set of points (a lattice), the student can produce specific geometric shapes.
>
> *Materials:* Geoboard (or worksheet) for each child.

The student completes a worksheet with activities like the ones shown in Fig. 9.

Figure 9

Finding Measures

Measuring can be a geometric activity and there are many opportunities for measuring experiences in the classroom. For example, the geoboard or worksheets can be used as shown in Fig. 10.

More on Geometry

```
•     •     •      1. Which points are
A     B     C         3 units from A?
                   2. Which points are
•     •     •         closer to A than I?
D     E     F      3. Which points are
                      the same distance
•     •     •         from A as from I?
G     H     I      4. Which points are
                      one unit from E?
```

Figure 10

An activity that helps to establish the importance of a standard measure uses two sticks of the same length but of different colors. The sticks might be about 30 inches long. A student is given one of the sticks and told to find something in the room that is the same length as the stick. After several students have successfully completed the task, the other stick, of a different color, is substituted. Again, students are told to find something in the room that is the same measure. At first students do not compare the sticks, so they duplicate the previous trials. Finally a student compares the sticks and makes use of the information found in the early experience.

Two sticks of different lengths can also be used. For example, two students are selected to measure the living rooms in their homes, but they are given sticks of different lengths. The following day the students report their results and the class is asked to compare the living rooms. The class recognizes the difficulty in the units of measure and concludes by laying out the lengths on the classroom floor.

If measurement experiences are worthwhile, the student should find out something worthwhile about what he is measuring. For example, in the next activity the student learns something about rectangles. Students are given small sticks with masking tape handles, for convenience in handling the sticks, and a worksheet with a picture of a rectangle, as shown in Fig. 11. The students are told to measure various line segments in the figure and they are asked questions that lead them to various conclusions: If AB is 2 sticks long, how long is FE? Or, after finding the length of AB and the length of BC, they are asked what other measurements they know. Children can measure many things around a school. Of course it is necessary to build the primitive concepts first, but nine-year-olds really get involved when they are given steel tapes and are asked to measure various things on the school playground.

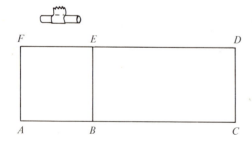

Figure 11

The experiences that have been shared in this article are but a few of the activities that are possible in teaching geometry in early childhood. Children live in a world filled with geometry. These experiences bring some of that world into the elementary classroom.

REFERENCES FOR FURTHER READING

1. Aman, G., "Discovery on a geoboard," *AT*, **21** (1974), pp. 267–272. Shows details as to how children can be led to discover Pick's theorem by using a geoboard. (See Exercise 6 of Exercises 8.5.)

2. Egsgard, J. C., "Geometry all around us, K–12," *AT*, **16** (1969), pp. 437–445. Describes how teachers can use shapes occurring in nature to motivate the study of geometry at all levels.

3. Gurau, P. K., "Discovering precision," *AT*, **13** (1966), pp. 453–456. Describes how definitions can be drawn out of students. (Also see Section 8.2.)

4. Liedtke, W., "Geoboard mathematics," *AT*, **21** (1974), pp. 273–277. A general survey of various activities using the geoboard.

5. Lulli, H., "Polyhedra construction," *AT*, **19** (1972), pp. 127–130.

6. Niman, J., and R. Postman, *Mathematics on the Geoboard*, Cuisenaire Company of America: New York, 1974.

7. Nuffield Mathematics Project. *Shape and Size*, John Wiley and Sons: New York, 1971. The Nuffield mathematics project of the United Kingdom features an activity approach to mathematics. *Shape and Size* is just one of many of their publications that feature the actual work of children.

8. Skidell, A., "Polynominoes and symmetry," *AT*, **14** (1967), pp. 353 and 358. Polynominoes are generalizations of dominoes, and you worked with them in Exercise 8 of Exercises 8.7.

9. Sullivan, J. J., "Pick's theorem on a geoboard," *AT*, **20** (1973), pp. 673–675. More sophisticated than the article by Aman listed above.

Chapter 9

Rational and Irrational Numbers

9.1 WHAT ARE RATIONAL AND IRRATIONAL NUMBERS?

Figure 9.1, taken from a fifth-grade text, shows how some points on a number line that do not correspond to whole numbers can be labeled by **fractions**. Fractions such as these are also called **positive rational numbers.** (In Section 9.9, however, we will note a distinction that is sometimes made between fractions and rational numbers.) We can also consider **negative** rational numbers (**negative** fractions) on a number line, as shown in Fig. 9.2. Here, as you can see, the use of the raised dash becomes quite awkward, and so, from now on, we shall write $-\frac{5}{2}, -\frac{11}{8}, -\frac{3}{4}$, as well as $-1, -2, -3$, etc.

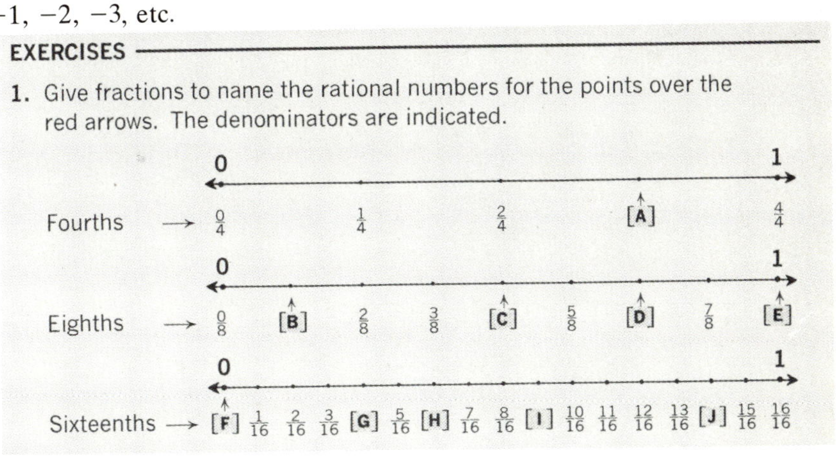

EXERCISES

1. Give fractions to name the rational numbers for the points over the red arrows. The denominators are indicated.

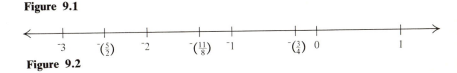

Fourths: $\frac{0}{4}$, $\frac{1}{4}$, $\frac{2}{4}$, [A], $\frac{4}{4}$

Eighths: $\frac{0}{8}$, [B], $\frac{2}{8}$, $\frac{3}{8}$, [C], $\frac{5}{8}$, [D], $\frac{7}{8}$, [E]

Sixteenths: [F], $\frac{1}{16}$, $\frac{2}{16}$, $\frac{3}{16}$, [G], $\frac{5}{16}$, [H], $\frac{7}{16}$, $\frac{8}{16}$, [I], $\frac{10}{16}$, $\frac{11}{16}$, $\frac{12}{16}$, [J], $\frac{15}{16}$, $\frac{16}{16}$

Figure 9.1

Figure 9.2

Rational numbers can also be described as numbers that can be written in the form $\frac{a}{b}$ or $-\frac{a}{b}$, where a and b are nonnegative integers and $b \neq 0$. Note that $0 = \frac{0}{1}$, $1 = \frac{1}{1}$, $2 = \frac{2}{1}$, $3 = \frac{3}{1}$, ..., and, in general, for every whole number a, $a = \frac{a}{1}$. (For convenience in printing, we often find $\frac{a}{b}$ written as a/b.)

Figure 9.3

Every rational number corresponds to a point on a number line. For example, to locate the point corresponding to the rational number $\frac{12}{7}$, we divide the segment from 0 to 1 into 7 congruent parts, and then lay off 12 of these parts beginning from 0, and going to the right, as shown in Fig. 9.3. Similarly, to locate $-\frac{17}{12}$ on a number line, we divide the interval from 0 to -1 into 12 congruent parts, and then lay off 17 of these parts beginning from 0 and going to the left, as shown in Fig. 9.4.

Figure 9.4

Although every rational number corresponds to a point on a number line, it is *not* true that every point on a number line corresponds to a rational number. To show this, consider Fig. 9.5, where we have drawn $\overline{AP}$ perpendicular to our number line and of length 1, and then, with a compass with point at 0 and opening of length OP, drawn an arc of a circle intersecting the number line at B. By the Pythagorean theorem, we have

$$(OP)^2 = 1^2 + 1^2 = 2.$$

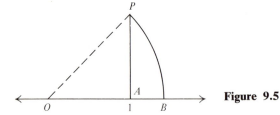

Figure 9.5

Suppose now that OP were a rational number, $\frac{a}{b}$. Then we would have

$$\left(\frac{a}{b}\right)^2 = \frac{a^2}{b^2} = 2,$$

so that

$$a^2 = 2b^2.$$

Thus, if OP is a rational number, we must have

1. $\qquad a \times a = 2 \times b \times b$

for some natural numbers a and b. We will show that this is impossible and hence show that OP is not a rational number.

To do this, we let m be the number of times that 2 occurs as a factor on the lefthand side, $a \times a$, of (1), and n the number of times that 2 occurs as a factor on the righthand side, $2 \times b \times b$, of (1). Then if (1) is to hold, we must certainly have $m = n$.

Now what are the possible values of m? We might have $m = 0$, as would be the case if $a = 17$; we might have $m = 2$, as would be the case if $a = 18$ ($18 \times 18 = 2 \times 2 \times 9 \times 9$); we might have $m = 4$, as would be the case if $a = 36$ ($36 \times 36 = 2 \times 2 \times 2 \times 2 \times 9 \times 9$). In any case, m must be an *even* number. (Remember that 0 is an even number.) But now, because the righthand side of (1) is $2 \times b \times b$, a similar analysis leads to the conclusion that n is an *odd* number. Since no even number is equal to an odd number, we conclude that $m \neq n$ and hence that OP is not a rational number. Numbers that are not rational numbers are called **irrational** numbers.

We write $\sqrt{2}$ for OP and, in general, $\sqrt{a}$ for the positive number whose square is a and read $\sqrt{a}$ as "the square root of a". Thus $\sqrt{2}$ is an irrational number. Other irrational numbers are $\sqrt{3}$, $\sqrt{5}$, $\sqrt[3]{2}$ (the number whose cube is 2), and π (the ratio of the circumference of any circle to its diameter).

The entire set of numbers corresponding to points on a number line is called the set of **real numbers**. That is,

$$\{x: x \text{ is a real number}\}$$
$$= \{x: x \text{ is a rational number}\} \cup \{x: x \text{ is an irrational number}\}.$$

Notice that

$$\{x: x \text{ is a rational number}\} \cap \{x: x \text{ is an irrational number}\} = \emptyset.$$

That is, no rational number is also an irrational number.

Programed Lesson 9.1

1. Locate and label the points corresponding to $\frac{3}{5}$, $-\frac{1}{5}$, $-\frac{7}{10}$, $-\frac{9}{5}$, $\frac{11}{5}$, and $\frac{5}{2}$ on a number line.

2. To prove that $\sqrt{3}$ is an irrational number, we suppose that $\sqrt{3} = \frac{a}{b}$, where a and b are _____ _____ .

   positive integers (or natural numbers)

218   Rational and Irrational Numbers   9.2

3. Then $3 = \underline{\hspace{1cm}}$, so that $a^2 = \underline{\hspace{1cm}}$.

$$\frac{a^2}{b^2}, 3b^2$$

---

4. Thus $a \times a = 3 \times \underline{\hspace{1cm}}$.

$$b \times b$$

---

5. The number of factors of 3 on the lefthand side of the equation $a \times a = 3 \times b \times b$ must be _____, whereas the number of factors of 3 on the righthand side must be _____. This is _____, and so we conclude that $\sqrt{3}$ is not a _____ number and hence must be a _____ number.

---

even, odd, impossible, rational, irrational

---

6. The set of real numbers is the union of the set of _____ numbers and the set of _____ numbers.

---

rational, irrational (in either order)

9.2 EQUIVALENT FRACTIONS

You know that any whole number can be written in many ways. For example, $2 = 1 + 1 = 1 \times 2 = 2 \div 1 = \sqrt{4}$, etc. Likewise, any rational number can be written in many ways. Of particular usefulness in computation with rational numbers is the fact that, for example,

$$\frac{1}{2} = \frac{3 \times 1}{3 \times 2} = \frac{3}{6}, \qquad \frac{5}{8} = \frac{7 \times 5}{7 \times 8} = \frac{35}{56}, \qquad 7 = \frac{7}{1} = \frac{5 \times 7}{5 \times 1} = \frac{35}{5},$$

$$\frac{7}{4} = \frac{6 \times 7}{6 \times 4} = \frac{42}{24},$$

and, in general,

$$\frac{a}{b} = \frac{m \times a}{m \times b}$$

for any natural number m. We will call this fact the **fundamental principle** for fractions.

The set of fractions

$$\left\{ \frac{a}{b}, \frac{2 \times a}{2 \times b}, \frac{3 \times a}{3 \times b}, \frac{4 \times a}{4 \times b}, \cdots \right\}$$

is called the set of (positive) fractions **equivalent** to $\frac{a}{b}$. As you will see in Chapter 10, the concept of equivalent fractions is of considerable use in doing computations with fractions.

9.2 Equivalent Fractions 219

The fact that $\frac{a}{b} = \frac{m \times a}{m \times b}$ for any natural number m can be made plausible to children by such pictures as those shown in Fig. 9.6, taken from a third-grade book.

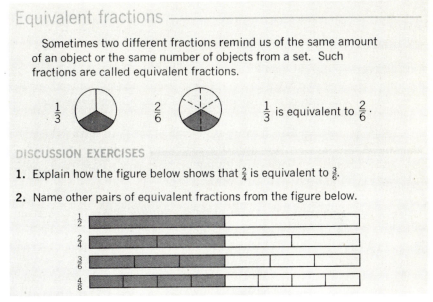

Figure 9.6

Exercises 9.2

1. Answer the questions in the following exercise taken from a third-grade text.

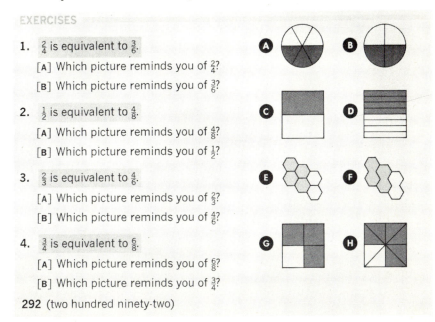

292 (two hundred ninety-two)

2. Which of the following fractions are equivalent to $\frac{5}{8}$?

$$\frac{10}{16}, \quad \frac{12}{18}, \quad \frac{6}{9}, \quad \frac{25}{40}, \quad \frac{500}{800}, \quad \frac{750}{1000}, \quad \frac{625}{1000}$$

9.3 COMMENTS ON OUR DEFINITIONS

Our description of real numbers—both rational and irrational—has been basically a geometrical one and is certainly not a precise definition. Indeed, giving a precise definition of real numbers turns out to be a quite difficult task that was not fully accomplished until the late nineteenth century.

From a nongeometrical point of view, all that we have said about rational numbers is that they are numbers that can be written in the form $\frac{a}{b}$ or $-\frac{a}{b}$, where a and b are nonnegative integers with $b \neq 0$. That is, we have really only described *numerals* (symbols) for these numbers.

To see the difference between describing what something looks like (its form) and what it is, suppose that you were teaching English to a man in a remote village in Ethiopia where no one has seen or heard of electrical apparatus of any kind. You show this man a picture such as that shown in Fig. 9.7 and tell him that it is a picture of a toaster. Now saying just this about a toaster would no doubt enable him to recognize an actual toaster if he ever saw one. But, from the picture alone, he would have, for example, no idea of what it was made of and would identify a plastic model of a toaster as the genuine article.

Figure 9.7

Likewise our "definition" of a rational number as a number that "can be written in the form $\frac{a}{b}$ or $-\frac{a}{b}$, where a and b are nonnegative integers with $b \neq 0$" certainly would enable anyone to recognize that $\frac{1}{2}$, $-\frac{3}{4}$, $\frac{10}{3}$, and $\frac{0}{4}$ are rational numbers, whereas $\frac{\sqrt{2}}{3}$, $-\frac{\sqrt{3}}{\sqrt{2}}$, and $\frac{\pi}{4}$ are not rational numbers.

But what about, for example,

$$\frac{1}{2} + \frac{1}{3}, \quad \frac{1}{2} \times \frac{3}{4}, \quad \frac{1}{2} \div 2, \quad \text{and} \quad \frac{\sqrt{4}}{\sqrt{9}}?$$

They don't "look like" rational numbers since they are certainly not of the prescribed form. But remember that we have said "can be written in the form . . ."

Thus, since

$$\frac{1}{2} + \frac{1}{3} = \frac{5}{6}, \qquad \frac{1}{2} \times \frac{3}{4} = \frac{3}{8}, \qquad \frac{1}{2} \div 2 = \frac{1}{4}, \qquad \text{and} \qquad \frac{\sqrt{4}}{\sqrt{9}} = \frac{2}{3},$$

we can conclude that these are indeed rational numbers.

Now it is possible to get away from the unsatisfactory word "form," but doing so is no easy task. (See Section 9.9.) Fortunately, however, from the point of view of teaching primary-school mathematics there isn't any particular problem in using this rather vague approach to rational numbers via the number line and "form." What is needed, of course, for successful teaching about rational numbers in primary school, is a realization of how children can develop an intuitive feeling for the concept of a rational number. A number line is certainly a valuable tool here and is widely used in primary-school mathematics texts. In addition, a good teacher will use sticks broken into parts, string cut into pieces, folded strips of paper, etc.

As logically unsatisfactory as our definition of rational number is, the situation in regard to irrational numbers is even less logically satisfactory. We have said essentially that irrational numbers are real numbers that are not rational numbers, without, however, defining real numbers. So, when we turn around and say that the set of real numbers is the union of the set of rational numbers and the set of irrational (not rational) numbers, we are really using a circular definition. (Imagine defining a woman as a human being that is not a man, and then defining the set of human beings as the union of the set of men and the set of women ("not men")!)

9.4 TERMINOLOGY

We have seen that it is not easy to give a precise definition of a rational number although, fortunately, it is not too difficult to convey the intuitive idea of a fraction to children. There are similar difficulties in regard to the terminology traditionally used in connection with rational numbers, as well as with terminology used in contemporary texts.

For example, some contemporary primary-school texts distinguish between "fraction" and "fractional number", where "fractional number" refers to the abstract idea of a fraction and "fraction" refers to ways of writing fractional numbers (i.e., to numerals). (One such text using this terminology comments in the teacher's guide that "...correct usage of the terms *fraction* and *fractional number* may sometimes be awkward... In many cases, it will be more meaningful to children if you say fraction when you really mean fractional number"!)

Furthermore, the concept of a fraction undergoes a certain amount of transformation as we progress in a mathematics sequence. Initially, of course, the word "fraction" may be used synonymously with "positive rational number": $\frac{1}{2}$, $\frac{3}{4}$, and $\frac{10}{3}$ are certainly considered as fractions. Later, however, it is customary to label as "fractions" such expressions as $\frac{\sqrt{2}}{2}$, $\frac{\sqrt{5}}{\sqrt{6}}$, $\frac{\pi}{3}$, etc. Indeed, we often apply the term "fraction" to such expressions as

222   **Rational and Irrational Numbers**   **9.4**

$$\frac{x}{y}, \quad \frac{x+y}{x-y}, \quad \text{and} \quad \frac{5x^2 - 3x + 1}{2x},$$

rather than using the more precise but cumbersome term of "rational algebraic expressions."

The words **numerator** and **denominator** are words that will cause no difficulty to children: "The numerator is the number on the top, and the denominator is the number on the bottom." But numbers, being abstractions, can't be put on "top" or "bottom"! Perhaps, then, they are numerals, which, being, symbols, *can* be put on "top" or "bottom"? Alas, we want to be able to say, for example, "multiply the numerators," in describing multiplication of fractions and it is numbers, not numerals, that can be multiplied. Again, it is fortunate that this point is not one that causes difficulty for children, and we leave further comment on it for Section 9.9.

Much more could be said about difficulties and conflicts of opinion among authors in establishing a vocabulary in regard to fractions and rational numbers. If you are interested in a further discussion of these problems, you can consult the first four articles listed in the References for further reading. The brief discussion given here should at least serve to alert you to the problem and cause you to pay careful and critical attention to the vocabulary on this topic in whatever primary-school text you will be using.

Some additional terminology:

1. A fraction $\frac{a}{b}$ is said to be written in **lowest terms** if a and b are relatively prime, i.e., if a and b have no common divisors except 1. Thus $\frac{1}{2}, \frac{3}{7}$, and $\frac{15}{8}$ are all fractions written in lowest terms but $\frac{2}{4}, \frac{6}{14}$, and $\frac{30}{16}$ are not.

2. A fraction $\frac{a}{b}$ is said to be a **proper** fraction if $a < b$, and an **improper** fraction if $a \geq b$. Thus $\frac{1}{2}, \frac{3}{4}$, and $\frac{5}{6}$ are all proper fractions, whereas $\frac{7}{2}, \frac{9}{4}$, and $\frac{6}{6}$ are all improper fractions. (This is ancient terminology which many mathematicians would prefer to eliminate!)

3. Expressions such as $1\frac{3}{4}$ and $2\frac{1}{2}$ are said to be **mixed** numbers. We have $1\frac{3}{4} = 1 + \frac{3}{4} = \frac{4}{4} + \frac{3}{4} = \frac{7}{4}$ and $2\frac{1}{2} = 2 + \frac{1}{2} = \frac{4}{2} + \frac{1}{2} = \frac{5}{2}$. Mixed fractions are certainly unnecessary (from an abstract point of view) but are useful in practical problems where comparison of size is involved. Thus it is easier to see immediately that $1\frac{3}{4} < 2\frac{1}{2}$ than it is to see that $\frac{7}{4} < \frac{5}{2}$.

Programed Lesson 9.4

1. In the fraction $\frac{5}{6}$, the numerator is __ and the denominator is __.

    5, 6

---

2. Which of the fractions $\frac{15}{24}$ and $\frac{15}{26}$ is written in lowest terms?

    $\frac{15}{26}$

---

3. A fraction $\frac{a}{b}$ is said to be written in lowest terms if a and b are _____ _____, i.e., if a and b have no _____ _____ except ___.

   relatively prime, common divisors, 1

4. Which of the fractions $\frac{10}{3}$ and $\frac{5}{7}$ is a proper fraction?

   $\frac{5}{7}$

5. A fraction $\frac{a}{b}$ is said to be a proper fraction if _____ and an improper fraction if _____.

   $a < b, a \geq b$

6. $4\frac{7}{8}$ is said to be a _____ fraction.

   mixed

7. Every rational number can be written in the form $\frac{a}{b}$ or $-\frac{a}{b}$, where a and b are whole numbers and b _____.

   $\neq 0$

9.5 DECIMAL FRACTIONS

What we have called fractions are sometimes referred to as **common** fractions (and, at one time, **vulgar** fractions!). Thus $\frac{1}{3}, \frac{2}{10}, \frac{5}{6}, \frac{23}{10}$, and $\frac{1304}{100}$ are all examples of common fractions. We can also write some common fractions as **decimal** fractions. Thus, for example,

$$\frac{2}{10} = 0.2, \qquad \frac{23}{10} = 2.3, \qquad \text{and} \qquad \frac{1304}{100} = 13.04.$$

Decimal fractions can easily be represented on an abacus, as illustrated in Fig. 9.8. (And, in Chapter 10, we will find the abacus a useful tool for visualizing addition and subtraction of decimal fractions.)

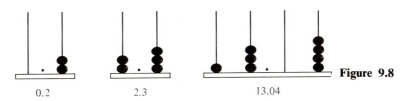

0.2   2.3   13.04   Figure 9.8

Just as we can write whole numbers in expanded form (Section 3.7), we can write decimal fractions in expanded form. We have, for example,

$$1.23 = 1 + \left(2 \times \frac{1}{10}\right) + \left(3 \times \frac{1}{10^2}\right)$$

224 Rational and Irrational Numbers 9.5

and

$$41.103 = (4 \times 10) + 1 + \left(1 \times \frac{1}{10}\right) + \left(0 \times \frac{1}{10^2}\right) + \left(3 \times \frac{1}{10^3}\right).$$

Using this expanded form and the fundamental principle of Section 9.2, we can show how decimal fractions can be written as common fractions. For example,

$$1.23 = 1 + \left(2 \times \frac{1}{10}\right) + \left(3 \times \frac{1}{10^2}\right) = 1 + \frac{2}{10} + \frac{3}{10^2}$$

$$= \frac{1 \times 10^2}{10^2} + \frac{2 \times 10}{10^2} + \frac{3}{10^2} = \frac{100 + 20 + 3}{10^2} = \frac{123}{100}.$$

In practice, of course, we don't need to go through such a complicated procedure. We simply "remove" the decimal point and use the whole number obtained as the numerator with the appropriate power of 10 ($10^1 = 10$, $10^2 = 100$, etc.) as the denominator. If there are n digits to the right of the decimal point, the proper power of 10 is 10^n. For example, in 41.103 there are 3 digits to the right of the decimal point so that we get

$$41.103 = \frac{41,103}{10^3} = \frac{41,103}{1000}.$$

What about the converse problem? That is, is it always possible to find a decimal fraction equal to a given common fraction? Clearly, in order for a common fraction to be equal to a decimal fraction, it must be possible to express it as a common fraction with the denominator a power of 10. To do this, when possible, we again use the fact that

$$\frac{a}{b} = \frac{m \times a}{m \times b},$$

where a, b, and m are whole numbers with $b \neq 0$ and $m \neq 0$. Thus, for example,

$$\frac{1}{5} = \frac{2 \times 1}{2 \times 5} = \frac{2}{10} = 0.2, \qquad \frac{3}{25} = \frac{4 \times 3}{4 \times 25} = \frac{12}{100} = 0.12.$$

Since the only factors of 10 are 1, 2, 5, and 10, it should be clear that the only common fractions that can be written as decimal fractions are those that can be written with denominators of the form $2^m \times 5^n$, where m and n are nonnegative integers. (Recall that $2^0 = 5^0 = 1$, so that we can consider as denominators $2^0 \times 5^1 = 1 \times 5 = 5$ and $2^1 \times 5^0 = 2 \times 1 = 2$.) For example,

$$\frac{1}{50} = \frac{1}{2 \times 5^2} = \frac{2 \times 1}{2 \times (2 \times 5^2)} = \frac{2}{2^2 \times 5^2} = \frac{2}{100} = 0.02$$

and

$$\frac{21}{16} = \frac{21}{2^4} = \frac{5^4 \times 21}{5^4 \times 2^4} = \frac{625 \times 21}{625 \times 16} = \frac{13,125}{10,000} = 1.3125$$

What about, for example, $\frac{3}{24}$? $24 = 2^3 \times 3$ and so the denominator of $\frac{3}{24}$ is not of the form $2^n \times 5^m$. However,

9.5 **Decimal Fractions** **225**

$$\frac{3}{24} = \frac{3 \times 1}{3 \times 8} = \frac{1}{8} = \frac{1}{2^3} = \frac{5^3}{5^3 \times 2^3} = \frac{125}{1000} = 0.125*$$

Thus a common fraction should be written in lowest terms before testing to see whether it can be written as a decimal fraction.

We have previously said (Section 7.5) that if a, b, and q are integers, then $a \div b = q$ if and only if $a = b \times q$. For example, $6 \div 3 = 2$ because $6 = 3 \times 2$. Notice now that $2 = \frac{2}{1} = \frac{3 \times 2}{3 \times 1} = \frac{6}{3}$, so that $6 \div 3 = \frac{6}{3}$. Similarly, $28 \div 7 = 4$ and $4 = \frac{4}{1} = \frac{7 \times 4}{7 \times 1} = \frac{28}{7}$, so that $28 \div 7 = \frac{28}{7}$. In general, whether or not b is a multiple of a, we have

$$\frac{a}{b} = a \div b$$

(We will return to this point in Chapter 10.) Thus another way of showing that $\frac{1}{5} = 0.2$, $\frac{3}{25} = 0.12$, $\frac{1}{50} = 0.02$, $\frac{21}{16} = 1.3125$, etc. is to perform a division. (The "why" of such divisions will be discussed in Chapter 10; here I am assuming that you can recall the mechanics of the process!) For example,

$$
\begin{array}{r}
0.12 \\
25 \overline{\smash)3.00} \\
\underline{2\,5} \\
50 \\
\underline{50} \\
0
\end{array}
\qquad \text{and} \qquad
\begin{array}{r}
1.3125 \\
16 \overline{\smash)21.0000} \\
\underline{16} \\
50 \\
\underline{48} \\
20 \\
\underline{16} \\
40 \\
\underline{32} \\
80 \\
\underline{80} \\
0
\end{array}
$$

What happens if we attempt to find a decimal fraction equal to $\frac{1}{3}$ by this division process? We have

$$
\begin{array}{r}
0.333.\,.\,. \\
3 \overline{\smash)1.000.\,.\,.} \\
\underline{9} \\
10 \\
\underline{9} \\
10 \\
\underline{9} \\
1 \\
\cdot \\
\cdot \\
\cdot
\end{array}
$$

---

* It is customary to write .02 as 0.02, .125 as 0.125, etc., in order to make it clear that no digits to the left of the decimal point have just been forgotten.

226   Rational and Irrational Numbers   **9.5**

and continue to get "3's" without end. We get a decimal, but it is a **nonterminating** decimal in contrast to 0.12 and 1.3125, which are **terminating** decimals.

Let us consider another example of a nonterminating decimal. Dividing 1 by 7 we have

$$
\begin{array}{r}
0.142857142857 \\
7\,\overline{)1.000000000000} \\
\underline{7} \\
30 \\
\underline{28} \\
20 \\
\underline{14} \\
60 \\
\underline{56} \\
40 \\
\underline{35} \\
50 \\
\underline{49} \\
\longrightarrow 10 \\
\underline{7} \\
30 \\
\underline{28} \\
20 \\
\underline{14} \\
60 \\
\underline{56} \\
40 \\
\underline{35} \\
50 \\
\underline{49} \\
10 \\
\vdots
\end{array}
$$

So we get $\frac{1}{7} = 0.142857142857\ldots$ Notice that, although the decimal is nonterminating, we have a repetition of the sequence of digits 1, 4, 2, 8, 5, and 7. This is because, at the point indicated by the arrow, we are, so to speak, beginning our division over again, and so the cycle is bound to repeat. We will call decimals such as $0.333\ldots$ and $0.142857142857\ldots$, **repeating** decimals.

Will this repetition of digits always occur in any such division that does not produce, at some stage, a remainder of 0? Certainly. When we are dividing 1 by 7 the possible remainders at each stage of the division are 0, 1, 2, 3, 4, 5, and 6 (and 0 cannot occur since $7 \neq 2^m \times 5^n$). Thus, at some stage, after at most six "partial" divisions, we must repeat a remainder and, from that point on, will get a repetition. (Notice that for 1/7 we encountered all possible remainders before repeating any of them.) On the other hand, in writing 1709/4950 as a repeating decimal, there conceivably might be as many as 4949 different remainders! However, as you should check, only the remainders 2240, 2600, and 1250 appear, and we have

$$\frac{1709}{4950} = 0.34525252 \cdots$$

Notice that here the first two digits (3 and 4) are not part of the repeating cycle; repetition begins with the cycle 52.

We write

$$\frac{1}{3} = 0.\bar{3} \qquad \frac{1}{7} = 0.\overline{142857} \qquad \text{and} \qquad \frac{1709}{4950} = 0.34\overline{52},$$

where the "bar" indicates the repeating cycle.

We have seen, then, that every common fraction can be expressed as a terminating decimal, or as a nonterminating but repeating decimal. Since we can also consider any terminating decimal as a repeating decimal simply by "adding on" zeros as in

$$0.2 = 0.2000 \cdots = 0.2\bar{0},$$

we can say that every common fraction is equal to a repeating decimal.

What about the converse of this last statement? That is, is it true that every repeating decimal is equal to a common fraction? We have already seen that repeating decimals that involve only a repetition of 0 are equal to common fractions. Thus, for example,

$$1.23\bar{0} = 1.23 = \frac{123}{100} \qquad \text{and} \qquad 41.103\bar{0} = 41.103 = \frac{41{,}103}{1000}.$$

The following four examples should serve to make plausible the fact that *every* repeating decimal is indeed equal to a common fraction.

Example 1. To find a common fraction equal to $0.\bar{4}$, we write $x = 0.\bar{4} = 0.4444 \cdots$ Then we have

$$
\begin{array}{rl}
10x = 4.4444 \cdots & = 4.\bar{4} \\
(-) \quad x = 0.444 \cdots & = 0.\bar{4} \\
\hline
9x & = 4 \\
\end{array}
$$

$$x = \frac{4}{9}$$

Here we multiply by $10^1 \,(= 10)$ because there is one digit in the repeating cycle.

228   Rational and Irrational Numbers   **9.5**

Example 2.   To find a common fraction equal to $0.\overline{371}$, we write $x = 0.\overline{371} = 0.371371371\cdots$ Then we have

$$1000x = 371.371371\cdots = 371.\overline{371}$$
$$(-)\quad x = \quad 0.371371\cdots = \quad 0.\overline{371}$$
$$999x \qquad\qquad\qquad = 371$$
$$x = \frac{371}{999}$$

Here we multiply by 10^3 ($= 1000$) because there are three digits in the repeating cycle.

Example 3.   Our first two examples considered decimals where all of the digits involved were part of the repeating cycle. Now to find a common fraction equal to $0.4\overline{12}$, where the first digit is not part of the repeating cycle, we proceed as follows:

$$1000x = 412.\overline{12}$$
$$(-)\quad 10x = \quad 4.\overline{12}$$
$$990x = 408$$
$$x = \frac{408}{990} = \frac{6 \times 68}{6 \times 165} = \frac{68}{165}$$

Here we multiply by 10^1 ($= 10$) to get $4.\overline{12}$, because there is one nonrepeating digit (4); then multiply by $10^1 \times 10^2$ ($= 1000$) to get $412.\overline{12}$, because there are two digits in the repeating cycle; and, finally, subtract $4.\overline{12}$ from $412.\overline{12}$.

Example 4.   To find a common fraction equal to $4.32\overline{7}$ we proceed as follows:

$$1000x = 4327.\overline{7}$$
$$(-)\ 100x = \quad 432.\overline{7}$$
$$900x = 3895$$
$$x = \frac{3895}{900} = \frac{779}{180}$$

Here we multiply by 10^2 ($= 100$) to get $432.\overline{7}$, because there are two nonrepeating digits to the right of the decimal point, and then by $10^2 \times 10^1$ ($= 1000$) to get $4327.\overline{7}$, because there is one digit in the repeating cycle.

Note that it is always possible to check the answers to problems like these by performing a division. For example,

$$\begin{array}{r}
4.3277\cdots \\
180\overline{\smash{)}779.0000\cdots} \\
\underline{720} \\
590 \\
\underline{540} \\
500 \\
\underline{360} \\
1400 \\
\underline{1260} \\
1400 \\
\underline{1260} \\
140
\end{array}$$

·
·
·

We have shown, in an informal fashion, that the set of repeating decimals is equal to the set of common fractions. What about nonrepeating decimals? Consider, for example, the nonterminating and nonrepeating decimal

$$0.101001000100001 \cdots$$

where, at each stage of writing, we insert an additional "0" before writing another "1". Since this decimal is not a repeating decimal, it cannot, by the result we have just announced, be equal to a common fraction. It is therefore not a rational number, and so is an irrational number.

On the other hand, we already know (Section 9.1) that $\sqrt{2}$ is an irrational number. Hence $\sqrt{2}$ is not a rational number, and hence the decimal representation of $\sqrt{2}$, $\sqrt{2} = 1.4142135 \cdots$ is a nonrepeating and nonterminating decimal. Likewise, π is represented by the nonrepeating and nonterminating decimal $3.14159 \cdots$ (Note that 3.14, 3.142, and $3\frac{1}{7}$, etc., are all only *approximations* to π.) We can summarize the results of this section as follows:

1. $\{x: x \text{ is a rational number}\} = \{x: x \text{ is a repeating decimal}\}$

2. $\{x: x \text{ is an irrational number}\}$
$$= \{x: x \text{ is a nonrepeating and nonterminating decimal}\}$$

3. $\{x: x \text{ is a real number}\}$
$$= \{x: x \text{ is a rational number}\} \cup \{x: x \text{ is an irrational number}\}$$
$$= \{x: x \text{ is a repeating decimal}\} \cup \left\{ \begin{array}{l} x: x \text{ is a nonrepeating and} \\ \text{nonterminating decimal} \end{array} \right\}$$
$$= \{x: x \text{ is a decimal}\}$$

(Remember here that we are considering all decimals as being repeating or nonrepeating with, for example, $0.2 = 0.2\overline{0}$.)

Programed Lesson 9.5

1. Fractions such as $\frac{2}{3}$, $\frac{5}{7}$, and $\frac{10}{9}$ are called _____ fractions in contrast to 0.2, 1.31, and 0.03, which are called _____ fractions.

   common, decimal

2. Represent 24.013 by a picture of an abacus.

   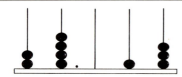

3. What number is represented by the abacus pictured below?

   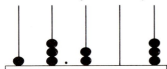

   13.203

4. The fundamental principle for fractions asserts that $\frac{a}{b} = \frac{m \times a}{\_\_\_}$, where a, b, and m are whole numbers with b ___ and m ___.

   $m \times b$, $\neq 0$, $\neq 0$

5. A common fraction, $\frac{a}{b}$, in lowest terms, can be written as a terminating decimal if and only if $b = 2^m \times$ ___, where ___ and ___ are _____ integers.

   5^n, m, n, nonnegative

6. $2.34\overline{267}$ is called a _____ decimal.

   repeating

7. We can express the terminating decimal 0.34 as a repeating decimal by writing 0.34 = ___.

   $0.34\overline{0}$

8. Write 31.248 in expanded form.

   $(3 \times 10) + 1 + \left(2 \times \frac{1}{10}\right) + \left(4 \times \frac{1}{10^2}\right) + \left(8 \times \frac{1}{10^3}\right)$

9. $\{x : x \text{ is a rational number}\} = \{x : x \text{ is a } _____ \text{ decimal}\}$.

9.6 Fractions in Bases other than Ten **231**

repeating

---

10. $\{x: x \text{ is an irrational number}\} = \{x: x \text{ is a } \underline{\hspace{2cm}} \text{ and } \underline{\hspace{2cm}} \text{ decimal}\}.$

nonrepeating, nonterminating (in either order)

---

11. $\sqrt{5}$ and π are examples of real numbers that are also $\underline{\hspace{2cm}}$ numbers.

irrational

Exercises 9.5

1. Convert the following common fractions to terminating decimals.

a) $\frac{3}{25}$ b) $\frac{7}{32}$ c)   d) $\frac{23}{8}$ e) $\frac{12}{15}$

2. Convert the following common fractions to repeating decimals.

a) $\frac{4}{9}$ b) $\frac{17}{15}$ c) $\frac{3}{11}$ d) $\frac{3}{13}$

3. Convert the following repeating decimals to common fractions.

a) $0.\overline{43}$ b) $0.3\overline{12}$ c) $0.12\overline{34}$ d) $3.1\overline{405}$ e) $14.24\overline{762}$

4. Multiply 142,857 first by 2, then by 3, then by 4, then by 5, and then by 6. What interesting fact did you observe about these products? Now multiply 142,857 by 7. Try to explain the results you obtained by considering the repeating decimal for $\frac{1}{7}$.

5. Try to develop a similar result by considering the repeating decimal for $\frac{1}{13}$.

9.6 FRACTIONS IN BASES OTHER THAN TEN

There is no real difficulty in considering common fractions in any base. We have, for example,

$$\left(\frac{3}{12}\right)_{\text{five}} = \left(\frac{3}{7}\right)_{\text{ten}},$$

since $12_{\text{five}} = 7_{\text{ten}}$ and $3_{\text{five}} = 3_{\text{ten}}$;

$$\left(\frac{15}{13}\right)_{\text{six}} = \left(\frac{11}{9}\right)_{\text{ten}},$$

since $15_{\text{six}} = 11_{\text{ten}}$ and $13_{\text{six}} = 9_{\text{ten}}$;

$$\left(\frac{15}{12}\right)_{\text{seven}} = \left(\frac{3 \times 4}{3 \times 3}\right)_{\text{seven}} = \left(\frac{4}{3}\right)_{\text{seven}} = \left(\frac{4}{3}\right)_{\text{ten}},$$

232   Rational and Irrational Numbers   9.6

since, in base seven, $15 = 3 \times 4$ and $12 = 3 \times 3$; and also $4_{\text{seven}} = 4_{\text{ten}}$ and $3_{\text{seven}} = 3_{\text{ten}}$.

The situation in regard to decimal fractions in bases other than ten is somewhat more complicated. Indeed, we can't really use the term "decimal" since "deci" refers to ten! Let us call them **basimal** fractions. We have, for example, using the expanded form for basimals analogous to the expanded form for decimals (Section 9.5),

$$243.12_{\text{five}} = \left(2 \times 10^2 + 4 \times 10 + 3 + 1 \times \frac{1}{10} + 2 \times \frac{1}{10^2}\right)_{\text{five}}$$

$$= \left(2 \times 5^2 + 4 \times 5 + 3 + 1 \times \frac{1}{5} + 2 \times \frac{1}{5^2}\right)_{\text{ten}}$$

$$= (50 + 20 + 3 + 0.2 + 0.08)_{\text{ten}}$$

$$= 73.28_{\text{ten}}$$

Note that, for example, $(\frac{1}{2})_{\text{ten}} = 0.5_{\text{ten}}$ is a terminating decimal in base ten, but in base five we have

$$
\begin{array}{r}
0.222 \cdots \\
2\overline{\big)\,1.000 \cdots} \\
4 \\
\overline{10} \\
4 \\
\overline{10} \\
4 \\
\overline{10} \\
\cdot \\
\cdot \\
\cdot
\end{array}
$$

Thus $(\frac{1}{2})_{\text{ten}} = (\frac{1}{2})_{\text{five}} = (0.222 \cdots)_{\text{five}} = 0.\overline{2}_{\text{five}}$ is a repeating basimal in base five. On the other hand, $\frac{1}{3}$, for example, is not a terminating decimal in base ten: $(\frac{1}{3})_{\text{ten}} = 0.\overline{3}_{\text{ten}}$. But, since $(\frac{1}{3})_{\text{ten}} = (\frac{2}{6})_{\text{ten}} = (\frac{2}{10})_{\text{six}} = 0.2_{\text{six}}$, it follows that $\frac{1}{3}$ is equal to a terminating decimal in base six.

Exercises 9.6

1. Write each of the following basimals in the expanded form in the given base.

    a) 162.5_{seven} **b)** 325.321_{six} **c)** 101.032_{five}

2. Write each of the following basimals as a common fraction (base ten) in lowest terms.

    a) 0.3_{eight} **b)** 0.4_{eight} **c)** 0.02_{six}
    d) 1.22_{three} **e)** 2.4_{nine} **f)** 3.2_{five}

3. Write $(\frac{1}{2})_{\text{ten}}$ as a basimal fraction in:

  a) base four **b)** base six **c)** base three **d)** base seven

4. Write 0.3_{ten} as a basimal fraction in base six.

9.7 NEGATIVE RATIONAL NUMBERS

We have restricted our attention so far to rational numbers that can be written as $\frac{a}{b}$ or $-\frac{a}{b}$, where a is a nonnegative integer (i.e., a whole number) and b is a positive integer. It is also convenient to allow a and b to be negative integers. The question is: What meaning shall we give to such expressions as

$$\frac{-3}{4}, \quad \frac{3}{-4}, \quad \text{and} \quad \frac{-3}{-4}?$$

Now if a is a negative integer and b is a positive integer, there is a natural interpretation for $\frac{a}{b}$. Thus, for example, $\frac{-3}{4}$ can be considered as the point obtained by dividing the interval from 0 to -1 into 4 equal parts and taking the third division to the left of 0 as the point corresponding to $\frac{-3}{4}$, as shown in Fig. 9.9. But we have previously labeled this point as $-\frac{3}{4}$ and so we will agree that $\frac{-3}{4} = -\frac{3}{4}$. In general, we make the definition

$$\frac{-a}{b} = -\frac{a}{b},$$

where a and b are nonnegative integers and $b \neq 0$.

Figure 9.9

There isn't, however, any direct geometrical interpretation of, for example, $\frac{3}{-4}$. It just doesn't make sense to talk about dividing the interval from 0 to 1 into "negative four parts"! If, however, we extend our fundamental principle (Section 9.1) to say that

$$\frac{a}{b} = \frac{m \times a}{m \times b}$$

for all *integers* a, b, and m with $b \neq 0$ and $m \neq 0$, then it is easy to show that

$$\frac{a}{-b} = -\frac{a}{b}$$

for, using this principle, we have

234 **Rational and Irrational Numbers** 9.8

$$\frac{a}{-b} = \frac{(-1) \times a}{(-1) \times (-b)} = \frac{-a}{b} = -\frac{a}{b}$$

Thus, for example,

$$\frac{3}{-4} = \frac{-3}{4} = -\frac{3}{4}$$

Similarly, we have

$$\frac{-a}{-b} = \frac{(-1) \times (-a)}{(-1) \times (-b)} = \frac{a}{b}$$

Thus, for example,

$$\frac{-3}{-4} = \frac{3}{4}$$

9.8 INEQUALITIES AND ABSOLUTE VALUE

As we did for integers, we can consider inequalities for rational numbers as being based on a number line: $\frac{a}{b} > \frac{c}{d}$ if and only if the point corresponding to $\frac{a}{b}$ is to the right of the point corresponding to $\frac{c}{d}$ and $\frac{c}{d} < \frac{a}{b}$ if and only if $\frac{a}{b} > \frac{c}{d}$. $\frac{a}{b}$ is a positive rational number if $\frac{a}{b} > 0$ and $\frac{a}{b}$ is a negative rational number if $\frac{a}{b} < 0$. We can also say that a rational number is positive if it can be written in the form $\frac{a}{b}$ where a and b are both positive integers. Then, using the approach to inequality for integers given in Section 7.6, we can say that

$$\frac{a}{b} > \frac{c}{d} \quad \text{if and only if there exists a positive rational number } \frac{m}{n}$$

$$\text{such that } \frac{c}{d} + \frac{m}{n} = \frac{a}{b}$$

For example, $\frac{1}{2} > \frac{1}{3}$ because $\frac{1}{3} + \frac{1}{6} = \frac{2}{6} + \frac{1}{6} = \frac{3}{6} = \frac{1}{2}$ and $\frac{1}{6}$ is a positive rational number.

The trichotomy and transitive properties of inequalities given for integers in Section 7.6 also hold for rational numbers. That is, if r, s, and t are any rational numbers, then:

1. $r = s$, $r > s$, or $s > r$ (one only)   **(Trichotomy** property)

2. If $r > s$ and $s > t$, then $r > t$ **(Transitive** property)

It is easy to see, for example, that $\frac{1}{2}$ is greater than $\frac{1}{3}$. But how about $\frac{13}{16}$ and $\frac{14}{17}$? Is $\frac{13}{16} > \frac{14}{17}$ or is $\frac{14}{17} > \frac{13}{16}$, or is $\frac{14}{17} = \frac{13}{16}$? A quick way to check equality or inequality for positive rational numbers written as common fractions is to use the following theorem.

9.8   Inequalities and Absolute Value   **235**

If $\frac{a}{b}$ and $\frac{c}{d}$ are positive rational numbers with $b > 0$ and $d > 0$, then:

1. $\dfrac{a}{b} = \dfrac{c}{d}$ if and only if $a \times d = b \times c$;

2. $\dfrac{a}{b} > \dfrac{c}{d}$ if and only if $a \times d > b \times c$;

and

3. $\dfrac{a}{b} < \dfrac{c}{d}$ if and only if $a \times d < b \times c$

Proof. By the fundamental principle,

$$\frac{a}{b} = \frac{d \times a}{d \times b} \qquad \text{and} \qquad \frac{c}{d} = \frac{b \times c}{b \times d}$$

Since $d \times a = a \times d$ and $b \times d = d \times b$, we have

$$\frac{a}{b} = \frac{a \times d}{d \times b} \qquad \text{and} \qquad \frac{c}{d} = \frac{b \times c}{d \times b}$$

Now

$$\frac{a \times d}{d \times b} \qquad \text{and} \qquad \frac{b \times c}{d \times b}$$

are fractions with the same denominator, $d \times b$. Hence,

$$\frac{a \times d}{d \times b} = \frac{b \times c}{d \times b}$$

if and only if the numerator, $a \times d$, of

$$\frac{a \times d}{d \times b}$$

is equal to the numerator, $b \times c$, of

$$\frac{b \times c}{d \times b},$$

i.e., if and only if $a \times d = b \times c$. Similarly,

$$\frac{a \times d}{d \times b} > \frac{b \times c}{d \times b}$$

if and only if $a \times d > b \times c$ and

$$\frac{a \times d}{d \times b} < \frac{b \times c}{d \times b}$$

if and only if $a \times d < b \times c$.

236   Rational and Irrational Numbers   **9.8**

For example,

$$\frac{21}{48} = \frac{42}{96}$$

because $21 \times 96 = 2016 = 48 \times 42$ and

$$\frac{14}{17} > \frac{13}{16}$$

because $14 \times 16 = 224 > 17 \times 13 = 221$.

Note that the comparison of rational numbers written in decimal form is much simpler. Thus, for example, to compare 7.81435 and 7.81429, all we need to do is to look at the first digit where the two numerals differ (3 and 2 in this case) and compare these two digits. Since $3 > 2$ we conclude that $7.81435 > 7.81429$.

The absolute value of a rational number is defined in the same way as for integers (Section 7.7). That is,

$$\left|\frac{a}{b}\right| = \frac{a}{b} \quad \text{if} \quad \frac{a}{b} \geq 0$$

and

$$\left|\frac{a}{b}\right| = -\frac{a}{b} \quad \text{if} \quad \frac{a}{b} < 0$$

For example, $\left|\frac{3}{4}\right| = \frac{3}{4}$, $\left|-\frac{8}{7}\right| = -\left(-\frac{8}{7}\right) = \frac{8}{7}$, $\left|2.\overline{314}\right| = 2.\overline{314}$, and $\left|-16.15\right| = -(-16.15) = 16.15$.

Programed Lesson 9.8

1. If r and s are any two rational numbers, we say that $r > s$ if the point corresponding to r is to the _____ of the point corresponding to s.

     right

---

2. If r and s are any rational numbers, then $r < s$ if and only if s __ r.

     $>$

---

3. A rational number r is said to be positive if $r >$ __ and negative if $r <$ __.

     $0, 0$

---

4. We can also say that a rational number is positive if it can be written in the form $\frac{m}{n}$ where m and n are _____ integers.

     positive

9.8 Inequalities and Absolute Value 237

5. If r and s are rational numbers, then $r > s$ if and only if there exists a _____ rational number t such that ___ + t = ___.

   positive, s, r

6. The _____ property for rational numbers states that, for any rational numbers r and s, one and only one of the statements $r = s$, $r > s$, $s > r$ is true.

   trichotomy

7. The transitive property of inequality for rational numbers states that if r, s, and t are any rational numbers such that $r > s$ and $s > t$, then _____.

   $r > t$

8. $\frac{a}{b} = \frac{c}{d}$ if and only if _____ = _____.

   $a \times d$, $b \times c$ (either order)

9. Use the result of Exercise 8 to determine whether or not $\frac{12}{11} = \frac{13}{12}$.

   $\frac{12}{11} \neq \frac{13}{12}$ since $12 \times 12 = 144 \neq 11 \times 13 = 143$

10. If $\frac{a}{b}$ and $\frac{c}{d}$ are rational numbers with $b > 0$ and $d > 0$, then $\frac{a}{b} > \frac{c}{d}$ if and only if _____ > _____.

    $a \times d$, $b \times c$

11. Use the result of Exercise 10 to show that $\frac{17}{15} > \frac{15}{14}$.

    Since $17 \times 14 = 238 > 15 \times 15 = 225$, $\frac{17}{15} > \frac{15}{14}$.

12. Is $5.31928 > 5.319859$? Give a reason for your answer.

    No. The digits 5, 3, 1, and 9 are the same in both numerals. Since $2 < 8$, we have $5.31928 < 5.319859$.

13. Show $\frac{1}{2}$, $\frac{3}{2}$, $-\frac{1}{2}$, and $-\frac{3}{2}$ on a number line.

    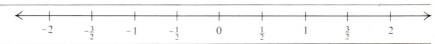

14. From the number line of Exercise 13, we see that $\frac{3}{2}$ ___ $\frac{1}{2}$ and $-\frac{3}{2}$ ___ $-\frac{1}{2}$.

    $>$, $<$

15. If $\frac{a}{b} > \frac{c}{d}$, then $-\frac{a}{b}$ ___ $-\frac{c}{d}$ and if $\frac{a}{b} < \frac{c}{d}$, then $-\frac{a}{b}$ ___ $-\frac{c}{d}$.

238   **Rational and Irrational Numbers**   9.9

$<, >$

---

16. $\left|-\frac{16}{15}\right| = \underline{\quad}, \left|\frac{55}{24}\right| = \underline{\quad}, |-2.\overline{34}| = \underline{\quad}, |15.\overline{3}| = \underline{\quad}.$

$\dfrac{16}{15}, \quad \dfrac{55}{24}, \quad 2.\overline{34}, \quad 15.\overline{3}$

9.9 A FORMAL APPROACH TO RATIONAL NUMBERS

We begin by considering ordered pairs (see Section 5.2) of integers such as $(3, 4)$, $(-5, 2)$, $(5, -2)$, $(-3, -4)$, and $(0, 2)$, where the second element of the ordered pair is not 0. These will correspond to the fractions $\frac{3}{4}, \frac{-5}{2}, \frac{5}{-2}, \frac{-3}{-4}$, and $\frac{0}{2}$, respectively. (Note that the first elements correspond to numerators and now are definitely numbers and not numerals! Similarly, the second elements correspond to denominators and are also definitely numbers.)

Now we make the definition

$$(a, b) \sim (c, d)$$

(read "the ordered pair a, b is equivalent to the ordered pair c, d") if and only if

$$a \times d = b \times c$$

Thus, for example, $(3, 4) \sim (6, 8)$ because $3 \times 8 = 4 \times 6$, whereas $(3, 4)$ is not equivalent to $(6, 9)$ (written $(3, 4) \not\sim (6, 9)$) because $3 \times 9 \neq 4 \times 6$. Note that this definition of $(a, b) \sim (c, d)$ corresponds to $\frac{a}{b} = \frac{c}{d}$ if and only if $a \times d = b \times c$, as discussed in Section 9.8. Here, however, we have $(a, b) = (c, d)$ if and only if $a = c$ and $b = d$. Thus, whereas $(3, 4) \sim (6, 8)$, $(3, 4) \neq (6, 8)$.

Now we define a rational number $\frac{a}{b}$ by

$$\frac{a}{b} = \{(x, y): (x, y) \sim (a, b)\}$$

For example,

$$\frac{3}{4} = \{(x, y): (x, y) \sim (3, 4)\} = \{\cdots(-6, -8), (-3, -4), (3, 4), (6, 8), \ldots\}$$

Thus a rational number is seen as a *set* of ordered pairs of integers, called an **equivalence class** of ordered pairs.

If we use such an approach, we can make a clear distinction between fractions (ordered pairs) and rational numbers (sets of ordered pairs)—a distinction that we have not made previously in this chapter. Whether such a distinction should be made in the primary school is a moot question. Figure 9.10, from a fifth-grade text, illustrates how this text does introduce such a distinction. (But the teacher's guide for this text acknowledges that this distinction is a rather abstract

Figure 9.10

one and says "There is no great harm if the children feel that they are still working with fractions and not with rational numbers.")

Exercises 9.9

Using the definition that $(a, b) \sim (c, d)$ if and only if $a \times d = b \times c$, show that, for all integers a, b, c, d, e, and f ($b \neq 0$, $d \neq 0$, and $f \neq 0$):

1. $(a, b) \sim (a, b)$.

2. If $(a, b) \sim (c, d)$, then $(c, d) \sim (a, b)$.

3. If $(a, b) \sim (c, d)$ and $(c, d) \sim (e, f)$, then $(a, b) \sim (e, f)$.

4. What would be appropriate names for the properties of equivalence of ordered pairs stated in Exercises 1 through 3?

9.10 PERCENT

We have seen that any rational number has many names: $\frac{1}{2} = \frac{2}{4} = \frac{3}{6} = \cdots = 0.5$, for example. Still another name for $\frac{1}{2}$ is 50% (fifty percent). One percent is equal to $0.01 = \frac{1}{100}$. When we use the language of percent, as in "4 is 25% of 16", we can translate "percent" as "hundredth" and "of" as "times" so that "4 is 25% of 16" becomes "$4 = 0.25 \times 16$" or "$4 = \frac{25}{100} \times 16$".

In a problem involving percent, we may have any one of three numbers missing. Thus our example could lead to three problems:

1. What is 25% of 16?

2. 4 is what percent of 16?

3. 4 is 25% of what number?

These questions can then be rephrased as shown on page 240.

240 **Rational and Irrational Numbers** 9.10

1'. $\square = 0.25 \times 16$

2'. $4 = \triangle \times 16$

3'. $4 = 0.25 \times \bigcirc$

In standard algebraic symbols we have

$$P = r \times A$$

and so (1'), (2'), and (3') become, respectively,

1". $P = 0.25 \times 16$

2". $4 = r \times 16$

3". $4 = 0.25 \times A,$

so that $P = 4$, $r = \frac{4}{16} = \frac{1}{4} = 0.25 = 25\%$, and $A = \frac{4}{0.25} = 16$.

The language of percent is, of course, in common use, especially in relation to interest. Thus, if $1000 is invested for one year at $5\frac{1}{2}\%$ ($= 5.5\%$), the interest in dollars is $5\frac{1}{2}\%$ of $1000 = 0.055 \times 1000 = 55$. Three other examples of common use of the word percent are:

1. Joe Doakes' salary was increased from $12,000 to $14,000. By what percent was his salary increased?

Solution. The increase was $2000, and so we want to know what percent 2000 is of 12,000. We have

$$2000 = r \times 12,000$$

so that

$$r = \frac{2000}{12000} = \frac{1}{6} = \left(\frac{100}{6}\right)\% = 16\frac{2}{3}\%$$

2. Mary Worth purchased some stock on January 1, 1975. A year later she sold the stock for 6% less than she paid. If her loss was $3000, what did she pay for the stock?

Solution. We have

$$3000 = 0.06 \times A,$$

so that the purchase price of the stock, in dollars, was

$$A = \frac{3000}{0.06} = 50,000$$

3. After receiving a 6% increase in salary, Janet Doakes' salary was $14,840. What was her salary before the increase?

9.10 Percent **241**

Solution. If the salary before the increase was x dollars, we have

$$x + 0.06x = 14{,}840$$
$$1.06x = 14{,}840$$
$$x = \frac{14{,}840}{1.06} = 14{,}000$$

(See the article at the end of this chapter for further remarks on solving problems involving percent.)

Exercises 9.10

1. Find
  a) 2.5% of 140   **b)** 30% of 2000

  c) 0.3% of 1500.

2. **a)** 15 is what percent of 250?   **b)** 18 is what percent of 96?

  c) 120 is what percent of 80?

3. **a)** 16 is 25% of what?   **b)** 30 is 4% of what?

  c) 120 is 250% of what?

4. A store increased the price of milk from $1.25 a gallon to $1.30 a gallon. By what percent was the price of milk increased?

5. The ABC car company increased all of its prices by 12%. If the price after the increase was $5600 for the VIP model, what was the price of the VIP model before the increase?

6. Sally Doakes took a 6% cut in her $10,000 salary on January 1, 1975. On January 1, 1976, however, she received a 6% raise. What was her salary on January 1, 1976?

7. A saleswoman receives a commission of 10% on her sales. She pays an assistant 10% of her commission. What rate does she clear on her sales?

8. A store advertised "Save $2 on all shirts. 25% off original price." What was the original price of a shirt and what is the sale price?

9. The population of a town increased 12% in the period from 1960 to 1970. In 1970 the population was 23,520. What was the population in 1960?

10. Mrs. Garcia paid $120 in gasoline taxes in 1975. If the tax on gasoline was 24%, how much did she spend on gasoline?

Chapter Test

In preparation for this test, you should review the meaning of the following terms:

Fraction (215)   Improper fraction (222)

242 Rational and Irrational Numbers

Rational number (215)
Irrational number (217)
Real number (217)
Equivalent fractions (218)
Fundamental principle
    for fractions (217)
Numerator (222)
Denominator (222)
Lowest terms (222)
Proper fraction (222)

Mixed fraction (222)
Common fraction (223)
Decimal fraction (223)
Expanded form of a decimal fraction
    (223)
Terminating decimal (226)
Nonterminating decimal (226)
Repeating decimal (226)
Absolute value (236)
Percent (239)

You should also review how to:

1. Form a set of equivalent fractions (218);

2. Reduce a fraction to lowest terms (222);

3. Write a decimal fraction in expanded form (223);

4. Represent a decimal fraction on an abacus (223);

5. Convert a terminating decimal fraction to a common fraction (224);

6. Convert a common fraction to a (terminating or nonterminating but repeating) decimal fraction (224–226);

7. Convert a nonterminating but repeating decimal fraction to a common fraction (227–229);

8. Convert a basimal fraction in a base other than ten to a fraction in base ten (231–232);

9. Test two common fractions or decimals for equality or inequality (234–236);

10. Solve problems involving percent (239–241).

1. Locate and label on a number line the points corresponding to $-\frac{7}{4}$ and $\frac{5}{2}$.

2. The essential reason why $\sqrt{2}$ is not a rational number is that if $\sqrt{2} = \frac{a}{b}$, where a and b are positive integers, then $2b^2 = $ ___. Then the number of factors of 2 on the lefthand side of this equation must be ___, whereas the number of factors on the righthand side must be ___. Since this is impossible, we conclude that $\sqrt{2}$ is an _____ number.

3. In the fraction $\frac{10}{7}$, 10 is called the _____ and 7 is called the _____.

4. Which of the fractions $\frac{10}{4}, \frac{17}{19}, \frac{6}{8}$, and $\frac{8}{5}$ are proper fractions? Which are written in lowest terms?

5. $\frac{16}{9}$ can be written as the mixed fraction _____.

6. $\frac{5}{6}, \frac{9}{10}, \frac{0}{3}$, and $\frac{19}{17}$ are examples of _____ fractions; 1.2, 0.41, and 1.32103 are examples of _____ fractions.

7. 1.2 and 0.41 are examples of _____ decimals; $1.242424\cdots$ and $4.141141114\cdots$ are examples of _____ decimals.

Chapter Test   243

8. 1.242424 ··· is a nonterminating but _____ decimal; 1.141141114 ··· is a nonterminating and _____ decimal.

9. So 1.242424 ··· represents a _____ number whereas 1.141141114 ··· represents an _____ number.

10. Indeed, 1.242424 ··· is equal to the common fraction _____ in lowest terms.

11. We know that $\frac{17}{250}$ is equal to a terminating decimal because 250 = ___ × ___.

12. If a, b, and c are any integers with $b \neq 0$ and $c \neq 0$, then $\frac{c \times a}{c \times b}$ = _____.

13. Express $\frac{3}{14}$ as a repeating decimal.

14. We can write 2.42 as the repeating decimal _____.

15. $\{x: x$ is a repeating decimal$\} = \{x: x$ is a _____ number$\}$.

16. $\{x: x$ is a nonterminating and nonrepeating decimal$\}$
    $= \{x: x$ is an _____ number$\}$.

17. Represent 53.2013 on a picture of an abacus.

18. What number is represented on the abacus pictured below?

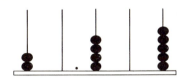

19. If r and s are any rational numbers such that the point corresponding to r on a number line is to the right of the point corresponding to s, we say that r ___ s.

20. If r and s are any rational numbers, then $r > s$ if and only if s ___ r.

21. If r is a rational number such that $r > 0$, we say that r is a _____ rational number.

22. If r and s are rational numbers, t is a positive rational number, and $r + t = s$, then r ___ s.

23. The _____ property of inequality for rational numbers states that if r, s, and t are any rational numbers such that $r > s$ and $s > t$, then $r > t$.

24. If $a \times d = b \times c$ and $b \neq 0$ and $d \neq 0$, then $\frac{a}{b}$ = ___.

25. Is $\frac{16}{17} > \frac{15}{18}$?

26. Is 6.84521 > 6.84498?

27. Is $0.\overline{34} > 0.3434$?

28. What percent of 16 is 5?

244 Rational and Irrational Numbers

29. Jane's salary of $260 a week was increased by 6%. What was her new salary?

30. If r is a rational number and $r < 0$, then $|r| = $ ___.

31. A formal approach to rational numbers can be based on the concept of _____ classes of _____ pairs of _____.

32. Write 231.452_{six} in expanded form in base six.

33. Write 13.23_{five} as a common fraction (base ten) in lowest terms.

34. Write $(\frac{1}{4})_{seven}$ as a repeating basimal numeral in base seven.

TEST ANSWERS

1.

2. a^2, odd, even, irrational
3. Numerator, denominator
4. $\frac{17}{19}$ and $\frac{6}{8}$; $\frac{17}{19}$ and $\frac{8}{5}$
5. $1\frac{7}{9}$
6. Common, decimal
7. Terminating, nonterminating
8. Repeating, nonrepeating
9. Rational, irrational
10. $\frac{41}{33}$
11. 2×5^3
12. $\frac{a}{b}$
13. $0.2\overline{142857}$
14. $2.42\overline{0}$
15. Rational
16. Irrational

17.

18. 20.405
19. $>$
20. $<$
21. Positive
22. $<$
23. Transitive
24. $\frac{c}{d}$
25. Yes
26. Yes
27. Yes
28. 31.25
29. $\$275.60$
30. $-r$
31. Equivalence, ordered, integers
32. $\left(2 \times 10^2 + 3 \times 10 + 1 + 4 \times \dfrac{1}{10} + 5 \times \dfrac{1}{10^2} + 2 \times \dfrac{1}{10^3}\right)_{six}$
33. $\left(\dfrac{213}{25}\right)_{ten}$
34. $0.\overline{15}_{seven}$

The equation method
of teaching percentage

by Rolla V. Kessler

Reprinted from *The Arithmetic Teacher*, Vol. 7 (1960), pp. 90–92.

The equation is the basis of all problems concerning arithmetic, mathematics, chemistry, physics, and engineering. A problem is an incomplete equation. When you get the answer, the equation is complete. Perhaps we should rewrite our elementary arithmetic books and teach children from the beginning to solve equations. In our present method, we let our children run into equations like "hitting a stone wall" in algebra. Why not set up a system of problem solving in the elementary school which is basic to all problems? Ratios will be easier in chemistry and physics when children are taught to solve equations from the very beginning.

Probably most elementary teachers are now ready to quit reading this article, as they feel the equation is too difficult for their children. The easiest combination in the second grade is an equation. The teachers do not think of it as an equation, so why should the children? The equation sign is the most important thing we have in math and science.

Those of us who have tried the equational method in the teaching of percentages know that it is easier than the three-case method. Children understand percentage problems. They can solve all three cases without deciding which case they are solving by some certain process which they might confuse. They would be solving incomplete equations the same way they did in third and fourth grade.

In the equational method, there is only one "case" to solve all percentage problems and that is to solve the incomplete multiplication equation. Let's look at some simple fourth-grade incomplete multiplication equations:

$$3 \times 4 = \_ \_$$
$$3 \times \_\_\_ = 12$$
$$12 = 3 \times \_\_\_$$

Rule: Multiply if you can. If you cannot multiply, you divide one factor into the product. Children must know:

Factor × factor = Product

or

Product = Factor × factor

When you teach percentage, you start out with an English lesson. "Is" is the most important word in the sentence. It is the verb. You cannot have a complete sentence or question without a verb. "Is" in percentage always means "is equal to" and can be read that way. Mathematically "is" or "is equal to" is your equation sign. After all, that is what the English language is telling us. It is quite easy for children to understand "of" when used with fractions, decimals, or percentage. Then "of" means "times" or is the multiplica-

246   Rational and Irrational Numbers

tion sign. "What" or "what number" is your unknown number, which is written as a blank space in the incomplete equation. "Find" when used in percentage begins a command. This type of problem can easily be changed by the children to a question, which is your problem or incomplete equation.

Find 25% of 32.

What is 25% of 32?

What number is equal to 25% of 32?

Now we translate from English to mathematics, putting the math terms in the exact order they are in the question.

What number is equal to 25% of 32?

$$\text{\_\_\_\_} = \quad 25\% \times 32$$
$$\text{\_\_\_\_} = 25\% \times 32$$

If the student does not know how to solve this incomplete equation, we write a simpler one he can solve:

$$\text{\_\_\_\_} = 25\% \times 32$$
$$\text{\_\_\_\_} = 3 \times 4$$

We still use the same rule we used in the third and fourth grade: Multiply if you can. If you cannot multiply, divide the factor into the product. This problem they can multiply and complete the equation.

8 is what % of 32?

8 *is equal to what number* % of 32?

$$8 \quad = \quad \text{\_\_\_\_} \quad \% \times 32$$
$$8 = \text{\_\_\_}\% \times 32$$

Here we divide the factor into the product. If the student has trouble, ask him to write a simpler incomplete equation in the same order:

$$8 = \text{\_\_\_}\% \times 32$$
$$8 = \text{\_\_\_} \times 2$$

He solves the percentage equation the same way he solves his simpler equation.

This problem tells the student to multiply a blank number by another number. The factor is always nearer the multiplication sign. The product stands "alone" on one side of the equation sign. There are two factors and only one product. In this problem, only one factor is known so he cannot multiply. You never multiply the product, but you divide into the product.

8 is 25% of what number?

8 *is equal to* 25% of *what number?*

$$8 \quad = \quad 25\% \times \quad \text{\_\_\_\_}$$

$$8 = 25\% \times \underline{}$$
$$12 = 3 \times \underline{}$$

We only use the simpler equation at the beginning of percentage. Children quickly follow their rule and make no mistakes in trying to decide which case they are solving. Percentage becomes another form of decimal work based on the incomplete equation.

We still have the complex question in percentage. If children haven't studied the complex sentence, then the math teacher is going to have to teach English again. In the complex sentence or question, we have one main sentence and a dependent clause. The clause depends on the main sentence for its meaning and is nothing when written by itself. In the complex percentage questions, we have two questions, one depending on the other for its meaning. Really, both questions have the same unknown number. Your dependent question really has two unknown numbers and means nothing by itself. You cannot solve a multiplication equation when you know only one number.

If 50% of a number is 20, what is 10% of the number?

"If" is your conjunction and tells that you have two incomplete equations of different values.

If: 50% of a number is 20

$$50\% \times \underline{} = 20$$

What is 10% of the number?

$$\underline{} = 10\% \times \underline{}$$

Since one number is common to both equations, let us use the capital N.

$$50\% \times N = 20$$
$$\underline{} = 10\% \times N$$

Your question says that your N's are the same number. You can solve the first equation to find "N." Substitute that number in the second equation in place of "N" and solve.

This method makes the formula method of solving interest problems easier. Your formula is an equation. Your children have already worked with equations. When we have three factors, no matter what kind of problem it is, the rule is: Multiply first; if problem is not complete, you divide into the product. In all interest problems, the amount of interest in dollars and cents is your product.

I have never taught second or third grade arithmetic. Tests going through the principal's office show that children have more trouble with written problems than straight fundamentals. Is it because they cannot restate a problem into a simpler question? Primary teachers can make a similar rule for addition and subtraction. Perhaps written problems in primary grades can be asked or reworded so they become a simple question, which is an incomplete equation requiring addition or subtraction. I believe the addition equation to be [a] basic from which subtraction should develop. Perhaps the primary children need help in transferring from English to math terms into the forms of very simple incomplete addition equations. From my work with children using the multiplication equation, I believe the addition equation can be used with better results in second and third grade.

248   Rational and Irrational Numbers

This is what I mean when I say addition should be the basic form:

$$3 + 4 = \underline{\hspace{1cm}}$$
$$4 + \underline{\hspace{1cm}} = 7$$
$$7 = 3 + \underline{\hspace{1cm}}$$

Primary children can solve incomplete equations with very little help which will start their foundation for all math and science equations.

REFERENCES FOR FURTHER READING

The first four articles deal with the problem of nomenclature for rational numbers (see Section 9.3). You will note that not all of the writers are in agreement!

1. Ballew, H., "Of fractions, fractional numerals, and fractional numbers," *AT*, **21** (1974), pp. 442–444.

2. Botts, T., "Fractions in the new elementary curricula," *AT*, **8** (1961), pp. 234–238.

3. Mueller, F. J., "On the fraction as a numeral," *AT*, **8** (1961), pp. 234–238.

4. Van Engen, H., "Rate pairs, fractions, and rational numbers," *AT*, **8** (1961), pp. 389–399.

5. Alexander, L. D., "One small jump—into repeating decimals and prime numbers," *AT*, **67** (1974), pp. 520–525.

6. Anderson, J. T., "Periodic decimals," *AT*, **67** (1974), pp. 504–509.

7. Beckenbach, E. F., "Geometric proofs of irrationality of $\sqrt{2}$," *AT*, **15** (1968), pp. 244–250.

8. Bohan, H., "Paper folding and equivalent fractions," *AT*, **18** (1971), pp. 245–249. $\frac{a}{b} = \frac{m \times a}{m \times b}$ via paper-folding exercises.

9. Braunfield, P. and M. Wolfe, "Fractions for low achievers," *AT*, **13** (1966), pp. 647–655. Describes a very unusual approach to fractions, involving "stretchers" and "shrinkers", that was first developed by the University of Illinois Committee on School Mathematics (UICSM).

10. Jacobs, J. E. and E. B. Herbert, "Making $\sqrt{2}$ seem 'real'," *AT*, **21** (1974), pp. 133–136. Describes use of the geoboard to help children visualize $\sqrt{2}$ and other square roots.

11. Jones, E., "Historical conflict—Decimal versus vulgar fractions," *AT*, **7** (1960), pp. 184–188.

12. Murray, J. T., "A more elementary view of the irrationality of $\sqrt{2}$," *AT*, **14** (1967), pp. 110–114. A geometric approach.

13. Sowder, L., "Models for fractional numbers—A quiz for teachers," *AT*, **18** (1971), pp. 44–45.

14. Wilson, P., D. Mundt, and F. Porter, "A different look at decimal fractions," *AT*, **16** (1969), pp. 95–98. Discusses converting common fractions to decimal fractions and finding patterns.

Chapter 10

Computations with Rational Numbers

10.1 INTRODUCTION

In discussing computations with rational numbers it is important to note that, with the increased use of the metric system, there will be less and less need for all but the very simplest kinds of common fractions, such as $\frac{1}{2}$, $\frac{2}{3}$, etc. Thus, whereas right now it is natural to talk about a measurement of $\frac{9}{16}$ inches, such a measurement will, in the future, be reported as something like 1.4 centimeters. The *basic principles* of operations with common fractions, however, will still remain of great importance, since an understanding of these is vital to success in dealing with such algebraic computations as

$$\frac{x-y}{x+y} + \frac{x}{x-y} \quad \text{and} \quad \frac{2x}{x^2-y^2} \cdot \frac{x+y}{3x}$$

10.2 ADDITION OF RATIONAL NUMBERS

It is, of course, entirely reasonable to define addition of common fractions with the same denominator by $\frac{a}{b} + \frac{c}{b} = \frac{a+c}{b}$. Thus, for example, $\frac{2}{7} + \frac{3}{7} = \frac{2+3}{7} = \frac{5}{7}$, as can easily be shown on a number line such as the one pictured in Fig. 10.1.

Figure 10.1

To add fractions with different denominators, then, we need only invoke the fundamental principle. Thus,

$$\frac{1}{2} + \frac{1}{3} = \frac{3 \times 1}{3 \times 2} + \frac{2 \times 1}{2 \times 3} = \frac{3}{6} + \frac{2}{6} = \frac{3+2}{6} = \frac{5}{6};$$

$$2 + \frac{1}{4} + \frac{1}{6} = \frac{2}{1} + \frac{1}{4} + \frac{1}{6} = \frac{12 \times 2}{12 \times 1} + \frac{3 \times 1}{3 \times 4} + \frac{2 \times 1}{2 \times 6}$$

$$= \frac{24}{12} + \frac{3}{12} + \frac{2}{12} = \frac{24+3+2}{12} = \frac{29}{12};$$

$$3\tfrac{3}{4} + 5\tfrac{1}{3} = (3 + \tfrac{3}{4}) + (5 + \tfrac{1}{3}) = (3 + 5) + (\tfrac{3}{4} + \tfrac{1}{3})$$

$$= 8 + \left(\frac{3 \times 3}{3 \times 4} + \frac{4 \times 1}{4 \times 3}\right) = 8 + (\tfrac{9}{12} + \tfrac{4}{12})$$

$$= 8 + \tfrac{13}{12} = 8 + (1 + \tfrac{1}{12}) = 9\tfrac{1}{12}$$

Note, in the second example, that we have used the L.C.M., 12, of 1, 4, and 6 (see Section 6.5). When the L.C.M. is so used, it is frequently referred to as the **least common denominator (L.C.D.)**, since it is the L.C.M. of the denominators. Any common multiple of the denominators can be used as a denominator in adding fractions, but using the L.C.D. reduces the amount of computation. Thus if 24 were used as a common denominator in this example, we would have

$$2 + \frac{1}{4} + \frac{1}{6} = \frac{24 \times 2}{24 \times 1} + \frac{6 \times 1}{6 \times 4} + \frac{4 \times 1}{4 \times 6} = \frac{48}{24} + \frac{6}{24} + \frac{4}{24}$$

$$= \frac{58}{24} = \frac{2 \times 29}{2 \times 12} = \frac{29}{12}$$

In general, if $\frac{a}{b}$ and $\frac{c}{d}$ are any rational numbers, we have

$$\frac{a}{b} + \frac{c}{d} = \frac{a \times d}{b \times d} + \frac{b \times c}{b \times d} = \frac{(a \times d) + (b \times c)}{b \times d}$$

This, then, gives us a formula for adding any two fractions. For example,

$$\frac{5}{7} + \frac{3}{8} = \frac{(5 \times 8) + (3 \times 7)}{7 \times 8} = \frac{40 + 21}{56} = \frac{61}{56}$$

It is much better, however, in teaching children how to add fractions, to stress the use of the fundamental principle rather than the formula, and to have them write

$$\frac{5}{7} + \frac{3}{8} = \frac{8 \times 5}{8 \times 7} + \frac{7 \times 3}{7 \times 8} = \frac{40}{56} + \frac{21}{56} = \frac{61}{56}$$

(See the article at the end of this chapter for an approach to addition of fractions via paper folding.)

Addition of decimal fractions presents no new problems to a child who understands the meaning of decimal fractions and the rationale underlying the procedures for the addition of whole numbers. Again, the abacus is a useful tool. In Figs. 10.2 and 10.3, we show two additions of decimal fractions on an abacus. We also show below the addition performed by the method of partial sums (see Section 4.3), and then in the standard format.

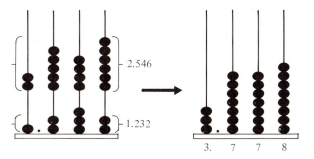

2.546 + 1.232

2.546	2.546
1.232	1.232
0.008	3.778
0.070	
0.700	
3.000	
3.778	

Figure 10.2

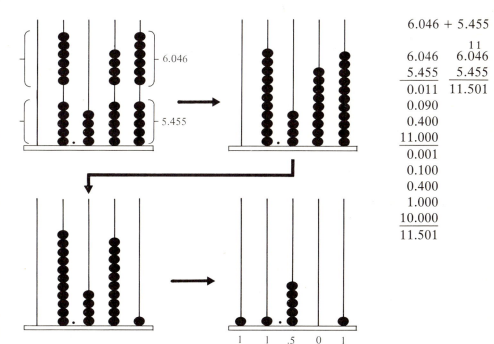

6.046 + 5.455

	11
6.046	6.046
5.455	5.455
0.011	11.501
0.090	
0.400	
11.000	
0.001	
0.100	
0.400	
1.000	
10.000	
11.501	

Figure 10.3

252   Computations with Rational Numbers   10.3

Finally, we note that no new principles are involved in the addition of negative rational numbers whether written as common fractions or as decimal fractions. For example:

$$-\tfrac{1}{2} + (-\tfrac{1}{3}) = -(\tfrac{1}{2} + \tfrac{1}{3}) = -\tfrac{5}{6};$$
$$-0.25 + (-0.75) = -(0.25 + 0.75) = -1;$$
$$-\frac{1}{2} + \frac{1}{3} = -\frac{3}{6} + \frac{2}{6} = \frac{-3}{6} + \frac{2}{6} = \frac{-3+2}{6} = \frac{-1}{6} = -\frac{1}{6};$$
$$-0.25 + 0.75 = 0.5.$$

Exercises 10.2

1. Perform the following additions by using the fundamental principle.
   a) $\tfrac{1}{2} + \tfrac{1}{3} + \tfrac{2}{5}$ b) $\tfrac{2}{5} + \tfrac{7}{15}$ c) $\tfrac{3}{4} + \tfrac{4}{9} + \tfrac{1}{3}$
   d) $-\tfrac{7}{12} + \tfrac{8}{3}$ e) $-\tfrac{12}{7} + (-\tfrac{3}{4})$ f) $1\tfrac{3}{4} + 8\tfrac{2}{3}$

2. Show the following additions on drawings of abaci and then check your results by using the method of partial sums.
   a) 2.314 + 15.702   b) .13.718 + 19.437   c) 43.679 + 2.82

3. Complete the following addition "rabbits".

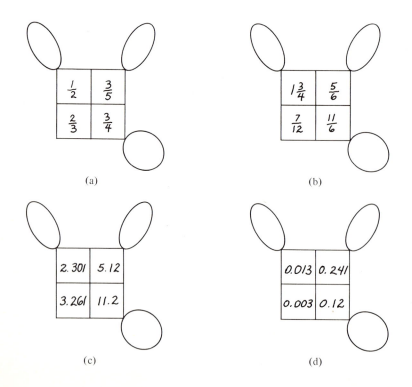

(a)   (b)

(c)   (d)

10.3 SUBTRACTION OF RATIONAL NUMBERS

Subtraction of common fractions is, of course, based upon the fact that

$$\frac{a}{b} - \frac{c}{b} = \frac{a}{b} + \left(-\frac{c}{b}\right) = \frac{a}{b} + \frac{-c}{b} = \frac{a + (-c)}{b} = \frac{a - c}{b}$$

Thus

$$\frac{3}{7} - \frac{1}{7} = \frac{3 - 1}{7} = \frac{2}{7}$$

and

$$\frac{1}{7} - \frac{3}{7} = \frac{1 - 3}{7} = \frac{-2}{7} = -\frac{2}{7}$$

When the denominators are not all the same we again use the fundamental principle. Thus

$$\frac{5}{9} - \frac{3}{6} = \frac{2 \times 5}{2 \times 9} - \frac{3 \times 3}{3 \times 6} = \frac{10}{18} - \frac{9}{18}$$
$$= \frac{10 - 9}{18} = \frac{1}{18}$$

Applying the fundamental principle can also lead, as you should check, to the formula

$$\frac{a}{b} - \frac{c}{d} = \frac{(a \times d) - (b \times c)}{b \times d}$$

Again, however, it is much better to have children use the fundamental principle rather than the formula in doing subtractions of fractions.

Subtraction of decimal fractions can, like addition, be illustrated on an abacus, as shown in Figs. 10.4 and 10.5.

3.574 − 1.532

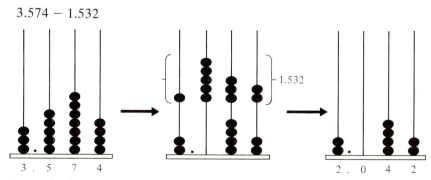

Figure 10.4

254 Computations with Rational Numbers 10.3

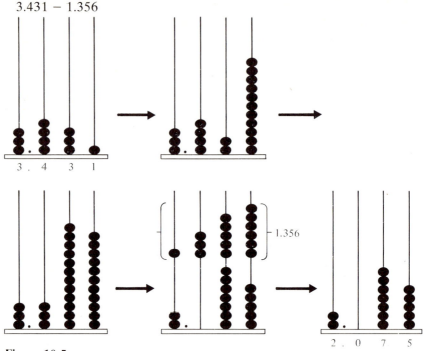

Figure 10.5

These problems can be done without an abacus, as follows:

$$\begin{array}{c} 3.574 \\ -1.532 \\ \hline 2.042 \end{array} \quad \text{and} \quad \begin{array}{c} \overset{\overset{1}{2}1}{3.\cancel{4}\cancel{3}1} \\ -1.356 \\ \hline 2.075 \end{array} \quad \begin{array}{c} \overset{1\;1}{3.431} \\ 4\;6 \\ -1.\cancel{3}\cancel{5}6 \\ \hline 2.075 \end{array}$$

("Borrowing" method)   (Equal additions method)

Finally, note that we can relate subtraction to addition, as we did for integers in Section 7.3, by $r - s = r + (-s)$ for all rational numbers r and s. Thus

$$\begin{array}{l} \tfrac{1}{2} - (-\tfrac{1}{3}) = \tfrac{1}{2} + [-(-\tfrac{1}{3})] \\ \phantom{\tfrac{1}{2} - (-\tfrac{1}{3})} = \tfrac{1}{2} + \tfrac{1}{3} = \tfrac{5}{6} \end{array} \quad \text{and} \quad \begin{array}{l} -0.75 - (-0.25) = -0.75 + [-(-0.25)] \\ = -0.75 + 0.25 = -0.5. \end{array}$$

Exercises 10.3

1. Perform the following subtractions by using the fundamental principle.
   a) $\tfrac{5}{12} - \tfrac{7}{18}$ b) $\tfrac{2}{5} - \tfrac{3}{7}$ c) $-\tfrac{2}{3} - \tfrac{5}{6}$ d) $-\tfrac{5}{4} - (-\tfrac{1}{6})$

2. Show the following subtractions on drawings of abaci and then check your result by direct computation using both the "borrowing" method and the equal additions method.
   a) $15.349 - 7.216$ b) $8.432 - 1.527$ c) $0.7301 - 0.425$

3. Complete the following addition "rabbits".

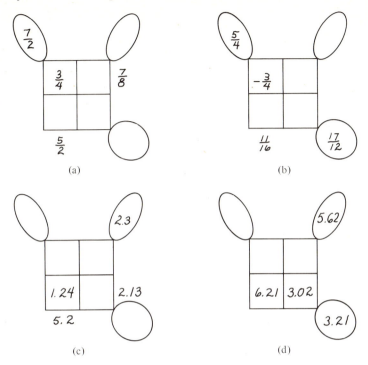

10.4 MULTIPLICATION OF RATIONAL NUMBERS

The rule for multiplying two common fractions is exceedingly simple and natural: Multiply numerators and multiply denominators. That is, for any two rational numbers, $\frac{a}{b}$ and $\frac{c}{d}$,

$$\frac{a}{b} \times \frac{c}{d} = \frac{a \times c}{b \times d}$$

It is interesting to contrast multiplication of fractions with addition of fractions in this connection. In addition of fractions we do not have the "simple and natural" rule.

$$\frac{a}{b} + \frac{c}{d} = \frac{a+c}{b+d}$$

(which many children will think is correct!). But, on the other hand, the correct procedure for adding fractions is easily illustrated using a number line, or by such simple examples as one-half a pie plus one-third of a pie, whereas, although the rule for multiplication of fractions is "simple and natural," it is not as easy to illustrate and justify.

256   Computations with Rational Numbers

What is usually done in the primary school is to use an approach to multiplication of common fractions via a series of diagrams in which we think, for example, of $\frac{1}{2} \times \frac{1}{3}$ as $\frac{1}{2}$ *of* $\frac{1}{3}$. Thus, to show $\frac{1}{2} \times \frac{1}{3} = \frac{1}{6}$, we first represent $\frac{1}{3}$ as the shaded part of the square shown in Fig. 10.6. Then $\frac{1}{2} \times \frac{1}{3}$ is represented by taking $\frac{1}{2}$ of this shaded region, as shown in Fig. 10.7 with heavier shading. Since the square has now been divided into 6 congruent rectangles, the heavily shaded region represents $\frac{1}{6}$ of the total area. So we conclude that $\frac{1}{2} \times \frac{1}{3} = \frac{1}{6}$, and then note that $\frac{1 \times 1}{2 \times 3} = \frac{1}{6}$.

Figure 10.6

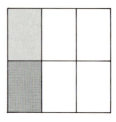
Figure 10.7

Figures 10.8 and 10.9 show the steps in illustrating $\frac{3}{5} \times \frac{3}{4}$. The shaded region in Fig. 10.8 shows $\frac{3}{4}$ of the square and the heavier shading in Fig. 10.9 shows the result of taking $\frac{3}{5}$ of this shaded region. Since the square has now been divided into 20 congruent rectangles and 9 of these are heavily shaded, we conclude that $\frac{3}{5} \times \frac{3}{4} = \frac{9}{20}$, and then note that $\frac{3 \times 3}{5 \times 4} = \frac{9}{20}$.

Figure 10.8

Figure 10.9

As a final example of this procedure, consider Figs. 10.10 and 10.11, which illustrate $\frac{3}{2} \times \frac{5}{4}$. Here our "unit" square has been extended to show $\frac{5}{4}$ in Fig. 10.10 and to show $\frac{3}{2}$ in Fig. 10.11. Now each congruent rectangle represents $\frac{1}{8}$ of the unit square and there are 15 such rectangles. So we conclude that $\frac{3}{2} \times \frac{5}{4} = \frac{15}{8}$, and then note that $\frac{3 \times 5}{2 \times 4} = \frac{15}{8}$.

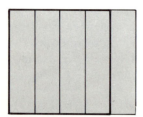
Figure 10.10

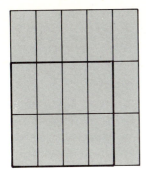

Figure 10.11

Through experiences like these, children can be led to accept as reasonable the rule for multiplying common fractions. Note, however, that we do not present the arguments above as a *proof* of the rule; they are presented only to make the rule *plausible*.

Let us turn now to procedures for multiplying decimal fractions. As you know, we can multiply two decimal fractions by multiplying the two numbers as if the decimal point were not there, and then placing the decimal point in the right place in the answer. Thus to multiply 2.34 by 1.72 we have

$$\begin{array}{r} 234 \\ \times 172 \\ \hline 468 \\ 16380 \\ 23400 \\ \hline 40248 \end{array}$$

and, upon placing the decimal point correctly, we have 4.0248

To see why 2.34 × 1.72 = 4.0248, recall (Section 9.5) that

$$2.34 = \frac{234}{100} = \frac{234}{10^2} \quad \text{and} \quad 1.72 = \frac{172}{100} = \frac{172}{10^2}$$

Thus, by our rule for multiplying common fractions and the fact that $10^m \times 10^n = 10^{m+n}$ for m and n any natural numbers, we have

$$2.34 \times 1.72 = \frac{234}{10^2} \times \frac{172}{10^2} = \frac{234 \times 172}{10^2 \times 10^2} = \frac{40{,}248}{10^4}$$

$$= \frac{40{,}248}{10{,}000} = 4.0248$$

and we see that we "point off" 2 + 2 = 4 decimal places in 40,248 to get our answer of 4.0248.

Similarly,

$$40.3 \times 5.14 = \frac{403}{10} \times \frac{514}{100} = \frac{403}{10^1} \times \frac{514}{10^2} = \frac{403 \times 514}{10^1 \times 10^2} = \frac{207{,}142}{10^3}$$

$$= \frac{207{,}142}{1000} = 207.142$$

258   Computations with Rational Numbers   10.4

and we see that we "point off" $1 + 2 = 3$ decimal places in 207,142 to get our answer of 207.142.

Likewise,

$$0.013 \times 0.004 = \frac{13}{1000} \times \frac{4}{1000} = \frac{13}{10^3} \times \frac{4}{10^3} = \frac{13 \times 4}{10^3 \times 10^3} = \frac{52}{10^6}$$

$$= \frac{52}{1,000,000} = 0.000052$$

and we see that we "point off" $3 + 3 = 6$ decimal places in 52 to get 0.000052.

Note that multiplication with negative rational numbers poses no new problems. We have, for example,

$$-\frac{1}{2} \times \frac{1}{3} = -\left(\frac{1}{2} \times \frac{1}{3}\right) = -\frac{1}{6},$$

$$\left(-\frac{3}{4}\right) \times \left(-\frac{5}{6}\right) = \frac{3}{4} \times \frac{5}{6} = \frac{15}{24} = \frac{3 \times 5}{3 \times 8} = \frac{5}{8},$$

$$(-40.3) \times 5.14 = -(40.3 \times 5.14) = -207.142,$$

and

$$(-0.013) \times (-0.004) = 0.013 \times 0.004 = 0.000052.$$

In particular, note that if a and b are any two nonzero integers, then $\frac{a}{b} \times \frac{b}{a} = 1$. That is, every nonzero rational number $\frac{a}{b}$ has a **multiplicative inverse** $\frac{b}{a}$, also called the **reciprocal** of $\frac{a}{b}$. Note also that, since any integer a can be considered as $\frac{a}{1}$, the reciprocal of a is $\frac{1}{a}$ if $a \neq 0$. For example,

$$\frac{2}{3} \times \frac{3}{2} = \frac{6}{6} = 1, \qquad \frac{-3}{4} \times \frac{4}{-3} = \frac{-12}{-12} = 1, \qquad \frac{5}{-12} \times \frac{-12}{5} = \frac{-60}{-60} = 1,$$

and

$$5 \times \tfrac{1}{5} = 1.$$

In Section 9.5 we gave an argument to show that $\frac{a}{b} = a \div b$ because, for example, $6 \div 3 = 2$ and $\frac{6}{3} = 2$, $28 \div 7 = 4$ and $\frac{28}{7} = 4$. Such an argument is valid when b is a divisor of a, but we stated (and used) the fact that $\frac{a}{b} = a \div b$ even when b is not a divisor of a (as, for example, when we calculated the repeating decimal for 1/7 by dividing 1 by 7). We are now in a position to prove that it is indeed true that

$$\frac{a}{b} = a \div b$$

for all integers a and b with $b \neq 0$. To do this, we simply extend our definition of division to say that:

1. $a \div b = q$ if and only if $a = b \times q$

for some rational number q (previously q was restricted to be an integer). Then, since

$$b \times \frac{a}{b} = \frac{b}{1} \times \frac{a}{b} = \frac{b \times a}{b \times 1} = \frac{a}{1} = a$$

by the fundamental principle, it follows that we can take $q = a/b$ in (1).

Exercises 10.4

1. Complete the following multiplication "rabbits".

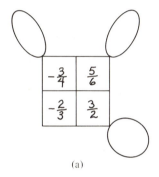
(a)

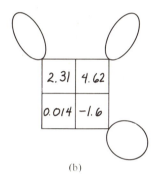
(b)

2. Explain how the figure below shows the product $\frac{2}{3} \times \frac{1}{5}$.

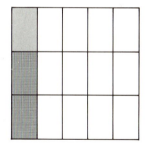

3. Explain how the figure below shows the product $\frac{3}{4} \times \frac{3}{2}$.

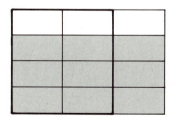

4. Draw figures as in the text to illustrate the following products.
   a) $\frac{1}{2} \times \frac{3}{5}$
   b) $\frac{4}{5} \times \frac{2}{3}$
   c) $\frac{1}{4} \times \frac{5}{2}$
   d) $\frac{5}{3} \times \frac{2}{3}$

260   Computations with Rational Numbers   10.5

5. Find the multiplicative inverse of:
   a) $\frac{2}{3}$ b) $\frac{5}{4}$ c) 3   d) 1

6. Perform the following multiplications by the procedure used in the text to show the proper placement of the decimal point (i.e., by first writing each decimal as a common fraction, etc.).
   a) 1.23×1.4 b) 4.02×0.04 c) 0.012×0.0031 d) 5.001×0.023

7. Formulate a rule for proper placement of the decimal point in the multiplication of two decimal fractions.

10.5 DIVISION OF RATIONAL NUMBERS

We have seen that to multiply two fractions we can multiply their numerators to get the numerator of the product, and multiply their denominators to get the denominator of the product. Would it not seem natural, then, to divide two fractions by performing divisions of numerator and denominator? Is this a correct procedure? Consider

$$\frac{4}{15} \div \frac{2}{3} = \frac{4 \div 2}{15 \div 3} = \frac{2}{5}$$

By the usual "invert and multiply" method for division of fractions we also obtain an answer of $\frac{2}{5}$ since

$$\frac{4}{15} \div \frac{2}{3} = \frac{4}{15} \times \frac{3}{2} = \frac{12}{30} = \frac{6 \times 2}{6 \times 5} = \frac{2}{5}$$

But perhaps it just happens that the first method works for this particular example. Indeed, can we even apply this method at all to the division of $\frac{4}{15}$ by $\frac{3}{7}$? Attempting to apply it leads us to

$$\frac{4}{15} \div \frac{3}{7} = \frac{4 \div 3}{15 \div 7},$$

which certainly is not of the form $\frac{a}{b}$ where a and b are whole numbers. Note, however, that, by the fundamental principle,

$$\frac{4}{15} = \frac{4 \times 3 \times 7}{15 \times 3 \times 7},$$

so that, by this "divide" method, we have

$$\frac{4}{15} \div \frac{3}{7} = \frac{4 \times 3 \times 7}{15 \times 3 \times 7} \div \frac{3}{7} = \frac{(4 \times 3 \times 7) \div 3}{(15 \times 3 \times 7) \div 7} = \frac{4 \times 7}{15 \times 3} = \frac{28}{45}$$

and also, by the usual method,

$$\frac{4}{15} \div \frac{3}{7} = \frac{4}{15} \times \frac{7}{3} = \frac{28}{45}$$

Now I have given this alternative procedure (which will be justified later) for the same reasons as I gave the method of complements for subtraction in Section 4.1. That is, I am pointing out again that there is no such thing as "the" method for doing a computation, and, also, emphasizing the importance of showing why a given computational procedure provides us with correct answers. You are accustomed to the "invert and multiply" procedure, and, whether or not you understand the reasons behind it, tend to regard it as "natural." This alternative procedure, however, is undoubtedly new to you and you certainly should feel that it should have some justification other than that it seems to work. Likewise, your pupils are entitled to more than statements of rules for computation if they are to regard mathematics as the logical system that it is, rather than as a collection of procedures for computation.

Consider now the usual "invert and multiply" rule:

$$\frac{a}{b} \div \frac{c}{d} = \frac{a}{b} \times \frac{d}{c}$$

To justify this rule, we use the same definition of division as we did before for integers. That is, if r and s are any two rational numbers, then

$$r \div s = t \quad \text{if and only if } r = s \times t$$

Thus

$$\frac{a}{b} \div \frac{c}{d} = \frac{u}{v} \quad \text{if and only if } \frac{a}{b} = \frac{c}{d} \times \frac{u}{v}$$

Now, if we take

$$\frac{u}{v} = \frac{a \times d}{b \times c},$$

we have

$$\frac{c}{d} \times \frac{u}{v} = \frac{c}{d} \times \frac{a \times d}{b \times c} = \frac{(c \times d) \times a}{(c \times d) \times b} = \frac{a}{b},$$

and so

$$\frac{a}{b} \div \frac{c}{d} = \frac{u}{v} = \frac{a \times d}{b \times c} = \frac{a}{b} \times \frac{d}{c}.$$

A less abstract way of justifying this rule is as follows: We first write

$$\frac{a}{b} \div \frac{c}{d} = \frac{\dfrac{a}{b}}{\dfrac{c}{d}},$$

262   Computations with Rational Numbers   10.5

by analogy with $x \div y = \frac{x}{y}$ for integers x and y. Then we assume that an extension of the fundamental principle for fractions holds, namely that

$$\frac{r}{s} = \frac{m \times r}{m \times s},$$

not only for all integers m, r, and s with $m \neq 0$ and $s \neq 0$, but also for all rational numbers m, r, and s with $m \neq 0$ and $s \neq 0$. Using this extension of the fundamental principle, we can write

$$\frac{a}{b} \div \frac{c}{d} = \frac{\frac{a}{b}}{\frac{c}{d}} = \frac{\frac{a}{b} \times \frac{d}{c}}{\frac{c}{d} \times \frac{d}{c}} = \frac{\frac{a}{b} \times \frac{d}{c}}{1} = \frac{a}{b} \times \frac{d}{c}$$

This second justification is probably the one most suitable for presentation to children through, however, specific examples rather than the general case. Thus, for example,

$$\frac{2}{3} \div \frac{5}{7} = \frac{\frac{2}{3}}{\frac{5}{7}} = \frac{\frac{2}{3} \times \frac{7}{5}}{\frac{5}{7} \times \frac{7}{5}} = \frac{\frac{2}{3} \times \frac{7}{5}}{1} = \frac{2}{3} \times \frac{7}{5}$$

In a similar fashion, we can justify the "divide the numerator and denominator" rule,

$$\frac{a}{b} \div \frac{c}{d} = \frac{a \div c}{b \div d}$$

by writing

$$\frac{a \div c}{b \div d} = \frac{\frac{a}{c}}{\frac{b}{d}} = \frac{\frac{a}{c} \times (c \times d)}{\frac{b}{d} \times (c \times d)} = \frac{a \times d}{b \times c} = \frac{a}{b} \times \frac{d}{c} = \frac{a}{b} \div \frac{c}{d}$$

Now let us turn to division of decimal fractions. (A review of Section 6.3 on division of whole numbers may be useful here.) Consider

$$1{_\times}23.\overline{\smash{)}6{_\times}46.98}$$

This shows the first stage in the usual procedure in the division of 6.4698 by 1.23, where the decimal points have been "moved" to convert $6.4698 \div 1.23$ to $646.98 \div 123$. It is easy to justify this step. We have

$$6.4698 \div 1.23 = \frac{6.4698}{1.23} = \frac{6.4698 \times 100}{1.23 \times 100} = \frac{646.98}{123}$$

The complete division process then would be as follows.

$$
\begin{array}{r}
5.26 \\
1{,}23.\,\overline{\smash{)}\,6{,}46.98} \\
\underline{6\ 15} \\
31\ 9 \\
\underline{24\ 6} \\
7\ 38 \\
\underline{7\ 38} \\
0
\end{array}
$$

What is really going on? Recall from Section 6.3 that we can consider division as successive subtractions and that the "bring down" idea is not very meaningful. So, to make sense of the above division, we can write

$$
\begin{array}{rl}
1{,}23.\,\overline{\smash{)}\,6{,}46.98} & \\
\underline{6\ 15.00} & 5.00 \\
31.98 & \\
\underline{24.60} & 0.20 \\
7.38 & \\
\underline{7.38} & 0.06 \\
0 & 5.26
\end{array}
$$

If we use this format, "wrong guesses" as discussed in Section 6.3 are allowed. For example, the division of 6.4698 by 1.23 might look like this:

$$
\begin{array}{rl}
1{,}23.\,\overline{\smash{)}\,6{,}46.98} & \\
\underline{3\ 69.00} & 3.00 \\
2\ 77.98 & \\
\underline{2\ 46.00} & 2.00 \\
31.98 & \\
\underline{24.60} & 0.20 \\
7.38 & \\
\underline{6.15} & 0.05 \\
1.23 & \\
\underline{1.23} & 0.01 \\
0 & 5.26
\end{array}
$$

Although it is customary and convenient to move the decimal points so as to obtain a problem where the divisor is a whole number, it is not really necessary to do so. We could, in the problem that we have been discussing, proceed as follows:

$$
\begin{array}{rl}
1.23\,\overline{\smash{)}\,6.4698} & \\
\underline{6.1500} & 5.00 \\
0.3198 & \\
\underline{0.2460} & 0.20 \\
0.0738 & \\
\underline{0.0738} & 0.06 \\
0 & 5.26
\end{array}
$$

264   Computations with Rational Numbers

Below is shown the division of 31.231 by 2.65 in the three different formats we have considered, with the quotient being carried out to three decimal places. Study each computation carefully and note the similarities and the differences.

$$
\begin{array}{r}
11.785 \\
2.65.\overline{)31.23.100} \\
26\ 5 \\
\hline
4\ 73 \\
2\ 65 \\
\hline
2\ 081 \\
1\ 855 \\
\hline
2260 \\
2120 \\
\hline
1400 \\
1325 \\
\hline
75
\end{array}
$$

$$
\begin{array}{rl}
2.65.\overline{)31.23.100} & \\
26\ 50.000 & 10.000 \\
\hline
4\ 73.100 & \\
2\ 65.000 & 1.000 \\
\hline
2\ 08.100 & \\
1\ 85.500 & 0.700 \\
\hline
22.600 & \\
21.200 & 0.080 \\
\hline
1.400 & \\
1.325 & 0.005 \\
\hline
0.075 & 11.785
\end{array}
$$

$$
\begin{array}{rl}
2.65\overline{)31.23100} & \\
26.50000 & 10.000 \\
\hline
4.73100 & \\
2.65000 & 1.000 \\
\hline
2.08100 & \\
1.85500 & 0.700 \\
\hline
0.22600 & \\
0.21200 & 0.080 \\
\hline
0.01400 & \\
0.01325 & 0.005 \\
\hline
0.00075 & 11.785
\end{array}
$$

Note that the remainder appears to be 75 in the lefthand computation, 0.075 in the righthand computation, and 0.00075 in the lower computation! The actual remainder is 0.00075; it appears as 0.075 in the middle computation because we began the computation by moving decimal points to the right two places, and this is equivalent to multiplying 2.65 and 31.231 by 100. Hence, we must divide 0.075 by 100 to get $0.075 \div 100 = 0.00075$ as the true remainder. In the first computation, on the other hand, the actual remainder is almost hopelessly lost because of the total lack of decimal points in the body of the computation.

Fortunately, we rarely need to consider the remainder except to note whether or not it is less than, or greater than, half the divisor when the decimal points are ignored. That is, in this problem, we observe that $75 < \frac{1}{2} \cdot 265$. Hence we know that, if we continued the division, the next digit in the quotient would be less than 5. (It is actually 2, as you should check.) This means that $31.231 \div 2.65 = 11.785$ correct to three decimal places. On the other hand, if the remainder were, say, 200, we would note that $200 > \frac{1}{2} \cdot 265$. Hence, if we were to continue the division in this case, the next digit in the quotient would be greater than 5 (actually 7, as you should check) and so we would have an answer of 11.7857 ... From this we would conclude that the answer to three decimal places should be given as 11.786 rather than 11.785.

As you can plainly see, division of decimal fractions is a rather complicated process to carry out and to explain. In recognition of this fact, most primary-school mathematics series give only a brief introduction to this topic in the sixth grade, and leave further work to the seventh and eighth grades.

As a final item in regard to division of rational numbers, note that division involving negative rational numbers involves no new ideas. Thus, for example,

$$-\tfrac{3}{4} \div \tfrac{5}{6} = -(\tfrac{3}{4} \div \tfrac{5}{6}) \quad \text{and} \quad (-2.32) \div (-0.14) = 2.32 \div 0.14.$$

Exercises 10.5

1. Perform the following divisions by the "division of numerator and denominator" method, and then check your answers by doing the divisions by the "invert and multiply" method. Express your answers in lowest terms.
   a) $\tfrac{8}{21} \div \tfrac{4}{7}$ b) $\tfrac{7}{5} \div \tfrac{3}{8}$ c) $\tfrac{19}{3} \div \tfrac{8}{5}$
   d) $\tfrac{14}{3} \div \tfrac{21}{4}$ e) $\tfrac{13}{4} \div \tfrac{3}{8}$

2. Sometimes people remember the rule "invert and multiply" but can't remember whether it is the first or second fraction that should be "inverted". Show that we can invert the first fraction and then multiply by the second *providing* we then take the answer obtained and "invert" it.

3. Complete the following multiplication "rabbits".

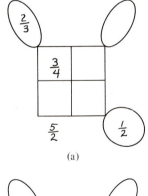

(a)

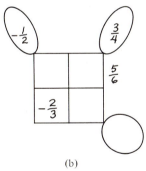

(b)

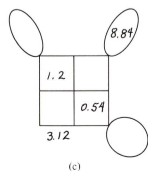

(c)

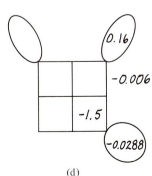

(d)

266 **Computations with Rational Numbers** 10.6

4. Perform the following divisions by using each of the three formats given on page 264. (All of these divisions have a remainder of 0 at some point.)

a) $461.02 \div 7.4$ b) $16.8448 \div 3.29$

c) $0.28704 \div 0.046$ d) $0.0014136 \div 0.38$

10.6 COMPUTATIONS WITH RATIONAL NUMBERS WRITTEN IN BASES OTHER THAN TEN

1. If you understand the basic principles that underlie the rules for computations with common fractions and decimal fractions written in base ten;

2. If you can recall how to do computations with whole numbers when they are written in bases other than ten (Sections 4.6 and 6.4); and

3. If you understand the meaning of common fractions and basimals in bases other than ten (Section 9.6);

then you should be able to do computations with common fractions and basimals in any base.

Here are some examples of such computations, where all numbers are written in base five and all computations are done in base five.

Example 1. $\dfrac{3}{4} + \dfrac{2}{3} = \dfrac{3 \times 3}{3 \times 4} + \dfrac{4 \times 2}{4 \times 3} = \dfrac{14}{22} + \dfrac{13}{22} = \dfrac{32}{22}$

Is this answer in lowest terms? There is an almost irresistible temptation to say "no" and to conclude that numerator and denominator are both divisible by 2! But $32_{\text{five}} = 17_{\text{ten}}$ and $22_{\text{five}} = 12_{\text{ten}}$. So 22_{five} is divisible by 2 but 32_{five} is not divisible by 2. (Indeed, 32_{five} is a prime number!)

Example 2. $\begin{array}{r} 41.32 \\ +30.42 \\ \hline 122.24 \end{array}$

Example 3. $\dfrac{3}{4} - \dfrac{1}{3} = \dfrac{3 \times 3}{3 \times 4} - \dfrac{4 \times 1}{4 \times 3} = \dfrac{14}{22} - \dfrac{4}{22} = \dfrac{14 - 4}{22} = \dfrac{10}{22}.$

(Check that this answer is in lowest terms.)

Example 4. $\begin{array}{r} {}^{2}1 \; {}^{3}1 \\ \cancel{3}.2\,\cancel{4}\,1 \\ -1.\,3\,2\,3 \\ \hline 1.\,4\,1\,3 \end{array}$ or $\begin{array}{r} {}^{1}\quad{}^{1} \\ 3.2\,4\,1 \\ {}^{2}\quad{}^{3} \\ -\cancel{1}.3\,\cancel{2}\,3 \\ \hline 1.\,4\,1\,3 \end{array}$

Example 5. $\dfrac{4}{3} \times \dfrac{12}{10} = \dfrac{4 \times 12}{3 \times 10} = \dfrac{103}{30}$

10.7 **Ratio and Proportion** **267**

Example 6.
$$
\begin{array}{r}
2.32 \\
\times 1.43 \\
\hline
1301 \\
20330 \\
23200 \\
\hline
10.0331
\end{array}
$$

Example 7.
$$\frac{4}{3} \div \frac{12}{10} = \frac{4}{3} \times \frac{10}{12}$$
$$= \frac{4 \times 10}{3 \times 12} = \frac{40}{41}$$

Example 8. $4.1243 \div 1.23$

$$
\begin{array}{r}
2.41 \\
1{.}23. \overline{)\ 4{.}12.43} \\
3\ 01 \\
\hline
1\ 114 \\
1\ 102 \\
\hline
123 \\
123 \\
\hline
0
\end{array}
$$

Exercises 10.6

1. Perform the following additions in base five. Express your answers to (a), (b), and (c) in lowest terms.
   a) $\frac{1}{2} + \frac{1}{3}$ **b)** $\frac{1}{3} + \frac{1}{2}$ **c)** $\frac{3}{4} + \frac{1}{11}$
   d) $2.31 + 3.24$ **e)** $40.142 + 3.02$

2. Perform the following subtractions in base five. Express your answers to (a), (b), and (c) in lowest terms.
   a) $\frac{3}{2} - \frac{1}{4}$ **b)** $\frac{3}{4} - \frac{2}{3}$ **c)** $\frac{4}{3} - \frac{1}{2}$
   d) $2.311 - 1.402$ **e)** $12.234 - 3.14$

3. Perform the following multiplications in base five. Express your answers to (a), (b), and (c) in lowest terms.
   a) $\frac{3}{2} \times \frac{3}{4}$ **b)** $\frac{4}{3} \times \frac{11}{10}$ **c)** $\frac{12}{13} \times \frac{4}{14}$
   d) 2.3×4.1 **e)** 14.2×0.013

4. Perform the following divisions in base five. Express your answers to (a), (b), and (c) in lowest terms.
   a) $\frac{3}{2} \div \frac{10}{12}$ **b)** $\frac{3}{4} \div \frac{11}{12}$ **c)** $\frac{13}{12} \div \frac{4}{3}$
   d) $20.02 \div 2.4$ **e)** $0.1331 \div 1.02$

10.7 RATIO AND PROPORTION

If a bookshelf contains 12 whodunits and 9 westerns, we can say that the **ratio** of whodunits to westerns is 12 to 9 or, since $\frac{12}{9} = \frac{4}{3}$, 4 to 3. We often write 12 to 9 as $12:9$ and, in general, for a and b any positive rational numbers, represent the ratio of a to b as $a:b$. We say that a ratio $c:d$ is in **lowest terms** if c and d are

268   Computations with Rational Numbers   **10.7**

natural numbers whose only common divisor is 1. Any ratio is equal to a ratio in lowest terms. Thus, for example, $2\frac{1}{2}:\frac{1}{2} = 5:1$, since

$$\frac{2\frac{1}{2}}{\frac{1}{2}} = \frac{\frac{5}{2}}{\frac{1}{2}} = \frac{5}{2} \div \frac{1}{2} = \frac{5}{2} \times \frac{2}{1} = \frac{5}{1}$$

An equality between two ratios is called a **proportion.** Thus

$$12:9 = 4:3$$

is a proportion. We read $12:9 = 4:3$ as "twelve is to nine as four is to three." Although there is a technical difference between a ratio and a fraction, in practice we can and do translate the statement of a proportion into a statement about equality of fractions. Thus

$$12:9 = 4:3 \qquad \text{can be written as} \qquad \frac{12}{9} = \frac{4}{3}.$$

Here are two examples of problems involving proportions.

1. A 3-inch-by-4-inch photograph is to be enlarged so that the width will be 1 foot. What will be the length of the enlargement?

Solution. We have $3:4 = 1:x$ or

$$\frac{3}{4} = \frac{1}{x},$$

where x is the length of the enlargement in feet. Hence,

$$4x \cdot \frac{3}{4} = 4x \cdot \frac{1}{x},$$
$$3x = 4,$$
$$x = \tfrac{4}{3}.$$

Thus the length of the enlargement will be $\frac{4}{3}$ feet $= 1\frac{1}{3}$ feet $= 1$ foot, 4 inches.

2. Suppose that the owner of the 12 whodunits and the 9 westerns adds 8 whodunits to his collection. How many westerns must he add to keep the ratio 4 to 3?

Solution. If we let x be the number of westerns to be added, we see that we must have

$$(12 + 8):(9 + x) = 4:3,$$

so that

$$\frac{12 + 8}{9 + x} = \frac{4}{3}$$

10.7   **Ratio and Proportion**   **269**

Thus,

$$3(9 + x) \cdot \frac{20}{9 + x} = 3(9 + x) \cdot \frac{4}{3}$$

$$3 \cdot 20 = 4(9 + x),$$

$$60 = 36 + 4x,$$

$$4x = 24,$$

$$x = 6$$

Check. He now has 20 whodunits and 15 westerns for a ratio of 20 to 15, or, since $\frac{20}{15} = \frac{4}{3}$, of 4 to 3.

The concept of ratio provides us with another way of looking at similarity of polygons (Section 8.3). If two polygons are similar, it is possible to match their sides in such a way that the ratio of corresponding sides is a constant (called the **ratio of similitude**). For example, the triangles shown in Fig. 8.5 are similar. We have

$$A_1B_1 : A_2B_2 = A_1C_1 : A_2C_2 = B_1C_1 : B_2C_2 = \tfrac{1}{2};$$

$$A_1B_1 : A_3B_3 = A_1C_1 : A_3C_3 = B_1C_1 : B_3C_3 = 2;$$

and

$$A_1B_1 : A_4B_4 = A_1C_1 : A_4C_4 = B_1C_1 : B_4C_4 = 1.$$

Exercises 10.7

1. Find x if:
   a) $x : 4 = 6 : 5$
   b) $4 : x = 7 : 12$
   c) $5 : 3 = 2 : (1 + x)$
   d) $7 : 6 = (x + 4) : (x + 3)$

2. At Pickwick College there are 1,020 male students and 940 female students. What, in lowest terms, is the ratio of male to female students?

3. A rectangular map, measuring 1 m by 2 m is to be reduced in size so that its length is 40 cm. What will be the width of this smaller map? (1 m = 100 cm)

4. The ratio of gold to copper in a certain alloy weighing 100 grams is $4 : 1$. If 30 grams of gold are added to the alloy, how much copper must be added to keep the same ratio of gold to copper?

5. The lengths of the sides of two squares are in the ratio of $2 : 3$. What is the ratio of their areas?

6. The ratio of similitude of two similar triangles is $5 : 2$. What is the ratio of their areas?

270   Computations with Rational Numbers   **10.9**

10.8 PROPERTIES OF INEQUALITIES

The addition and multiplication properties of inequalities for rational numbers are similar to those given in Section 7.6 for integers. That is, for all rational numbers r, s, and t,

1. If $r > s$, then $r + t > s + t$ (**Addition property** of inequalities for rational numbers)

2. **a)** If $r > s$ and $t > 0$, then $r \times t > s \times t$ $\left.\right\}$ (**Multiplication properties** of
   b) If $r > s$ and $t < 0$, then $r \times t < s \times t$ inequalities for rational numbers)

   For example,

   $$\tfrac{3}{4} > \tfrac{2}{3} \quad \text{and so} \quad \tfrac{3}{4} + \tfrac{7}{5} > \tfrac{2}{3} + \tfrac{7}{5}$$

   and also

   $$\tfrac{3}{4} + \left(-\tfrac{7}{5}\right) > \tfrac{2}{3} + \left(-\tfrac{7}{5}\right);$$

   $$\tfrac{3}{4} > \tfrac{2}{3} \quad \text{and so} \quad \tfrac{3}{4} \times \tfrac{4}{7} > \tfrac{2}{3} \times \tfrac{4}{7}$$

   whereas

   $$\tfrac{3}{4} \times \left(-\tfrac{4}{7}\right) < \tfrac{2}{3} \times \left(-\tfrac{4}{7}\right)$$

10.9 THE FIELD PROPERTIES

Let us collect together here the various properties of the system of rational numbers under addition and multiplication. They are: For all rational numbers r, s, and t,

Addition	**Multiplication**

1. Closure properties:

   $r + s$ is a rational number $r \times s$ is a rational number

2. Commutative properties:

   $r + s = s + r$ $s \times r = r \times s$

3. Associative properties:

   $r + (s + t) = (r + s) + t$ $r \times (s \times t) = (r \times s) \times t$

4. Existence of identities:

   $r + 0 = 0 + r = r$ $r \times 1 = 1 \times r = r$

10.9 Chapter Test **271**

5. Existence of inverses:

$$r + (-r) = (-r) + r = 0 \qquad \text{If } r = \tfrac{a}{b} \neq 0, \text{ then } \tfrac{a}{b} \times \tfrac{b}{a} = 1$$

6. Distributive property

$$r \times (s + t) = (r \times s) + (r \times t)$$

A set on which are defined two operations having the above properties is called a **field.** The set of real numbers also forms a field, as does the set of complex numbers, which you may have studied about previously. In Section 13.3 we will consider another example of a field.

Chapter Test

In preparation for this test, you should review the meaning of the following terms:

Least common denominator (L.C.D.) (250)

Multiplicative inverse (258)

Reciprocal (258)

Ratio (267)

Proportion (268)

Lowest terms for a ratio (268)

Ratio of similitude (269)

Properties of inequalities (270)

Field properties (270 and 271)

Field (271)

You should also review how to:

1. In base ten, add (249–250), subtract (252–253), multiply (255), and divide (260–262) common fractions;

2. Represent the product of two common fractions by shaded rectangles (256);

3. In base ten, add (251), subtract (253–254), multiply (257–258), and divide (262–265) decimal fractions;

4. Perform computations with rational numbers written in bases other than ten (266–267);

5. Complete addition and multiplication "rabbits" involving rational numbers (259 and 265);

6. Solve problems involving ratio and proportion (268–269).

1. Do the following addition by using the fundamental principle, and write your answer as a mixed fraction.

$$\frac{3}{5} + \frac{2}{3} + \frac{7}{12}$$

2. Show how the subtraction $4.054 - 1.206$ can be carried out on an abacus.

3. Draw a figure, as in the text, to illustrate $\dfrac{2}{3} \times \dfrac{5}{6}$.

4. Multiply 0.032 by 2.6 by first converting each decimal to a common fraction.

272 Computations with Rational Numbers

5. Do the division, $\frac{2}{3} \div \frac{5}{7}$, by the "divide numerator and denominator" method, and check your answer by using the "invert and multiply" rule.

6. Do the division, $0.79756 \div 0.314$, by using each of the three formats given on page 264.

7. Find x if $x : 13 = 4 : 7$.

8. With 7 games played, the ratio of games won to games lost by the Podunk Ponies is $4 : 3$. If they now lose 3 games and win 5, what will be their win–lose ratio?

9. If a, b, and c are any rational numbers such that $a < b$, what can be said about $a + c$ and $b + c$? About $a \times c$ and $b \times c$?

10. Perform the following computations in base five. Express your answer to (a) in lowest terms.
    a) $(\frac{2}{3} \times \frac{4}{11})_{\text{five}}$
    b) $2.31_{\text{five}} \times 0.14_{\text{five}}$

TEST ANSWERS

1. $1\frac{51}{60}$

2.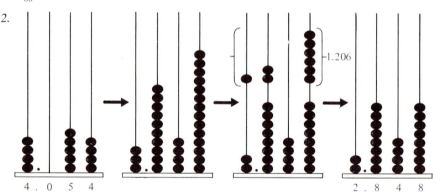

3.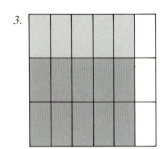

4. $0.032 \times 2.6 = \dfrac{32}{1000} \times \dfrac{26}{10} = \dfrac{832}{10{,}000} = 0.0832$

5. $\dfrac{2}{3} \div \dfrac{5}{7} = \dfrac{(2 \times 5 \times 7) \div 5}{(3 \times 5 \times 7) \div 7} = \dfrac{2 \times 7}{3 \times 5} = \dfrac{14}{15}$; $\dfrac{2}{3} \div \dfrac{5}{7} = \dfrac{2}{3} \times \dfrac{7}{5} = \dfrac{14}{15}$

6.

$$
\begin{array}{r}
2.54 \\
{}_{\times}314.\overline{\smash{\big)}\,{}_{\times}797.56} \\
628 \\
\hline
1695 \\
1570 \\
\hline
1256 \\
1256 \\
\hline
0
\end{array}
$$

$$
\begin{array}{r r}
{}_{\times}314.\overline{\smash{\big)}\,{}_{\times}797.56} & \\
628.00 & 2.00 \\
\hline
169.56 & \\
157.00 & 0.50 \\
\hline
12.56 & \\
12.56 & 0.04 \\
\hline
0 & 2.54
\end{array}
$$

$$
\begin{array}{r r}
.314\overline{\smash{\big)}\,0.79756} & \\
0.62800 & 2.00 \\
\hline
0.16956 & \\
0.15700 & 0.50 \\
\hline
0.01256 & \\
0.01256 & 0.04 \\
\hline
0 & 2.54
\end{array}
$$

7. $x = \frac{52}{7}$

8. $3:2$

9. $a + c < b + c$; $a \times c < b \times c$ if $c > 0$; $a \times c > b \times c$ if $c < 0$; $a \times c = b \times c$ if $c = 0$.

10. a) $\left(\frac{4}{14}\right)_{\text{five}}$

   b) 0.4334_{five}

Addition of unlike fractions

by Mitsuo Adachi

Reprinted from *The Arithmetic Teacher*, Vol. 15 (1968), pp. 221–223.

The teaching of addition of unlike fractions can be a very uninteresting process. Is there a more creative approach to the teaching, one that can challenge students to discovery of generalizations? Can we structure instances so that students themselves will develop the algorithm, "Change the fraction to equivalent fractions with common denominators and proceed as in the addition of like fractions"?

The teaching of unlike fractions (rational numbers of arithmetic) is usually introduced to the students in the following sequence:

1. Addition of like fractions (common denominators)
2. Equivalent fractions
3. Introduction of pairs of unlike fractions and "matching" equivalent fractions with common denominators
4. Generalizing for the addition of unlike fractions, that is, that fractions must have common denominators

An example follows:

$$\frac{1}{2} + \frac{1}{3} = n$$
$$\frac{1}{2} = \frac{3}{6} \quad \text{and} \quad \frac{1}{3} = \frac{2}{6}$$

Therefore

$$\frac{1}{2} + \frac{1}{3} = \frac{3}{6} + \frac{2}{6} = \frac{5}{6}$$

Let us take a look at a primitive approach and consider the following example in the addition of unlike fractions.

$$\frac{1}{2} + \frac{1}{4} = n$$

Take a strip of paper of convenient size and let it be the unit. (Students will be expected to do the manipulation of the unit strip of paper.)

Fold the strip of paper in half. The strip is now divided into two equivalent sections.

Addition of Unlike Fractions 275

Take the same strip of paper and fold it in fourths. The paper strip will now have the following folds.

What will $\frac{1}{2} + \frac{1}{4}$ be on the strip of paper?

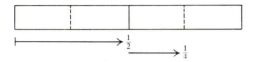

Do you see that $\frac{1}{2} + \frac{1}{4} = \frac{3}{4}$?

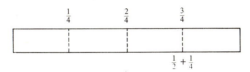

We note that $\frac{1}{2}$ is equivalent to $\frac{2}{4}$. Is $\frac{2}{4} + \frac{1}{4} = \frac{3}{4}$?

Let us look at the example problem discussed previously.

$\frac{1}{2} + \frac{1}{3} = n.$

Take a strip of paper of convenient size and let it be the unit.

Fold the strip of paper in half. The unit strip is now divided into two equivalent sections.

Note the fold and mark it $\frac{1}{2}$.

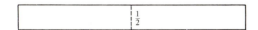

Take the same strip of paper and fold it into three equivalent sections.

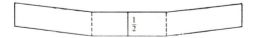

Mark the new folds $\frac{1}{3}$.

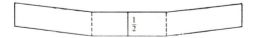

We can now relate the halves and the thirds on the paper strip. Can we determine $\frac{1}{2} + \frac{1}{3}$?

We know $\frac{1}{2}$.

276   Computations with Rational Numbers

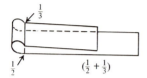

We also know $\frac{1}{3}$.

We can now mark off to the right of the $\frac{1}{2}$, a $\frac{1}{3}$ by folding the paper strip from left to right by setting the $\frac{1}{3}$ mark on the $\frac{1}{2}$ mark.

The strip of paper will now have the following markings:

Suppose we fold the $\frac{1}{3}$ section on the left of the paper strip into two equivalent sections. The paper strip will now have the following folds:

Will each of the folds be of equivalent size? How many equivalent sections is the unit strip of paper divided into?
Can we see that $\frac{1}{2} + \frac{1}{3} = \frac{5}{6}$?

We note that $\frac{1}{3}$ is equivalent to $\frac{2}{6}$ and $\frac{1}{2}$ is equivalent to $\frac{3}{6}$. Is $\frac{3}{6} + \frac{2}{6} = \frac{5}{6}$?
Let us examine $\frac{3}{4} + \frac{1}{6} = n$.

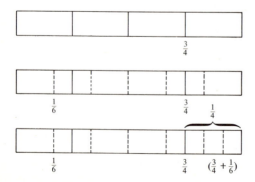

Addition of Unlike Fractions 277

Does it seem reasonable to fold each of the fourths into three equivalent sections?

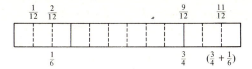

We conclude that $\frac{3}{4} + \frac{1}{6} = \frac{11}{12}$. We see that $\frac{1}{6}$ is equivalent to $\frac{2}{12}$ and $\frac{3}{4}$ is equivalent to $\frac{9}{12}$. Is $\frac{2}{12} + \frac{9}{12} = \frac{11}{12}$?

Let us look at another example to enhance discovery by the students. Specifically, let us look at

$$\frac{3}{8} + \frac{1}{3} = n.$$

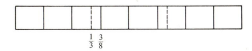

Paper strip is folded into eighths with $\frac{3}{8}$ noted.

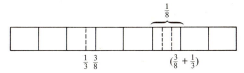

Paper strip now includes the fold in thirds.

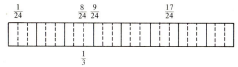

Paper strip shows $\frac{3}{8} + \frac{1}{3}$. Closer examination of the three folds at $\frac{3}{8} + \frac{1}{3}$ shows that they are equivalent, that is, the eighth section is divided into three equivalent folds. What would happen if the other eighth sections were folded into three equivalent sections each?

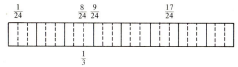

We note that the unit strip has twenty-four equivalent sections and $\frac{3}{8} + \frac{1}{3} = \frac{17}{24}$. We also note that $\frac{3}{8}$ is equivalent to $\frac{9}{24}$ and $\frac{1}{3}$ is equivalent to $\frac{8}{24}$.

Is $\frac{9}{24} + \frac{8}{24} = \frac{17}{24}$?

In summary:

$\frac{1}{2} + \frac{1}{4} = \frac{3}{4}$;   that is, $\frac{2}{4} + \frac{1}{4} = \frac{3}{4}$.

$\frac{1}{2} + \frac{1}{3} = \frac{5}{6}$;   that is, $\frac{3}{6} + \frac{2}{6} = \frac{5}{6}$.

$\frac{3}{4} + \frac{1}{6} = \frac{11}{12}$;   that is, $\frac{9}{12} + \frac{2}{12} = \frac{11}{12}$.

$\frac{3}{8} + \frac{1}{3} = \frac{17}{24}$;   that is, $\frac{9}{24} + \frac{8}{24} = \frac{17}{24}$.

278   **Computations with Rational Numbers**

We have established the intuitive basis for the addition of unlike fractions through the unit strip of paper. We have also noted the reference to equivalent fractions throughout the discussion. The need for equivalent fractions with common denominators in the addition of unlike fractions—the next step in the hierarchy of learning—becomes more meaningful.

REFERENCES FOR FURTHER READING

1. Allendoerfer, C. B., *Principles of Arithmetic and Geometry for Elementary School Teachers.* New York: The Macmillan Co., 1971. Chapter 23 has a more formal approach to computations with rational numbers than is given in this chapter.

2. Begle, E. G., *The Mathematics of the Elementary School.* New York: McGraw-Hill Inc., 1975. Chapters 15 and 17 contain additional material on computations with rational numbers that supplement nicely the discussion given in this chapter.

3. Heddons, J. W., and M. Hynes, "Division of fractional numbers," *AT*, **16** (1969), pp. 99–103. Gives a geometrical approach to the topic.

4. Henry, B., "Do we need separate rules to compute in decimal notation?" *AT*, **18** (1971), pp. 40–42. Points out that the rule for placing decimal points follows from the properties of common fractions.

5. Johnson, H. C., "Division of fractions—levels of meaning," *AT*, **12** (1966), pp. 362–368. Describes various ways of interpreting division of fractions.

6. Rappaport, D., "Percentage—noun or adjective?" *AT*, **8** (1961), pp. 25–26. Discusses confusion in regard to use of the words "percentage" and "percent."

7. Vance, I. E., "A natural way to teach division of rational numbers," *AT*, **16** (1969), pp. 91–93. A discussion of the division method described at the beginning of Section 10.5.

8. Winzenread, M. B., "Repeating decimals," *AT*, **20** (1973), pp. 678–682. Discusses arithmetic operations with repeating decimals.

9. Zweng, M., "The fourth operation is not fundamental—fractional numbers and problem solving," *AT*, **22** (1975), pp. 28–32. Argues that division of fractions does not deserve an important place in the primary school mathematics program.

Chapter 11

Measurement

11.1 INTRODUCTION

The kinds of measurement studied in the primary school include those of length, area, volume, angles, weight, time, and temperature, and the work progresses from simple measures of length in the early grades to measurement of areas and volumes in the upper grades.

Note, however, that there is a fundamental distinction between measurements of length, angles, time, temperature, and weight, and the "measurement" of area and volume. The latter two are usually calculated *indirectly* by means of formulas such as πr^2 for the area of a circle or a^3 for the volume of a cube, and the only actual measurements involved are those of length. (The use of measuring cups in cooking and a planimeter used by an engineer to measure area are exceptions to this but need not concern us here.)

The study of measurements of length, time, temperature, angles, and weight, in turn, can be divided into two parts. One part consists, so to speak, of the "theory": There are 24 hours in a day, 60 minutes in an hour; 90 degrees in a right angle; 12 inches in a foot, 100 centimeters in a meter*; 16 ounces in a pound, 1000 grams in a kilogram, etc. The other part consists of *actual* measurements of time, length, angles, weight, and temperature. It is the latter part that is all too frequently neglected. Children often can recite 12 inches equal 1 foot, 100 centimeters equal 1 meter, etc., and still be very clumsy in making actual measurements, as well as being exceedingly inept in the important skill of *estimation* of distances, areas, angle size, temperatures, weights, volumes, and time intervals.

---

* I am using the "er" endings (meter, liter, etc.) rather than the officially sanctioned "re" endings (metre, litre, etc.) simply because the former are still in common use on measuring devices and in the advertising of, for example, 1 liter bottles.

280   Measurement   **11.1**

What we are saying, then, is that work on measurement in the primary school should involve a good deal of actual measurement and also a good deal of practice in estimating measurements. Such activity not only is useful but will add interest to the study of measurement; children like activity.

At the present moment there is an additional problem in the teaching of measurement. Almost alone in the world, the U.S.A. uses the so-called English system of measurement (feet, pounds, etc.), whereas the rest of the world uses the metric system (centimeters, grams, etc.). Inevitably, the U.S.A. will go metric within the lifetime of the children in your classes. On the other hand, there will be, for some time, continued use of the English system. Thus you will be faced with the double task of teaching both the English system and the metric. Emphasis, however, should be on the metric system; and children should be given practice in estimating measurements *directly* in the metric system rather than thinking, for example, that a certain length is about 10 inches, and since 1 inch is about 2.5 centimeters, the length is about 25 centimeters.

One further aspect of measurement needs to be stressed. This is the fact that *all* measurements are approximations. Thus, for example, a measurement reported as 12 centimeters (cm) may be correct only to the nearest centimeter, to the nearest tenth of a centimeter, to the nearest hundredth of a centimeter, etc.

One very valuable aspect of the metric system is the ease with which the degree of accuracy of a measurement can be reported. For example, measurements reported as

$$12\text{ cm}, \qquad 12.0\text{ cm}, \qquad 12.00\text{ cm}$$

all mean something different. The 12 cm means the length is 12 centimeters to the nearest centimeter, and hence the length may actually be as much as 12.5 cm and as little as 11.5 cm; the 12.0 cm means that the length is 12 cm to the nearest tenth of a centimeter, and hence may actually be as much as 12.05 cm and as little as 11.95 cm; the 12.00 cm means that the length is 12 cm to the nearest hundredth of a centimeter, and hence may actually be as much as 12.005 cm and as little as 11.995 cm.

In turn, then, these limitations on the accuracy of direct measurements are reflected in the limitations in the accuracy of computed measurements. For example, if the diameter of a circle is measured as 1.2 cm (i.e., correct to the nearest tenth of a centimeter), it is silly to report the circumference as having a length of

$$(3.1416 \times 1.2)\text{ cm} = 3.76992\text{ cm}$$

A more reasonable statement would be that the circumference is 4 cm (correct to the nearest centimeter). Likewise, if the base of a rectangle is measured as 1.2 cm and the height as 5.2 cm (i.e., each correct to the nearest tenth of a centimeter), direct multiplication gives us

$$1.2 \times 5.2 = 6.24$$

for an area of 6.24 sq cm. A more informative reporting for the area, however, would be 6 sq cm (correct to the nearest square centimeter).

The details of how to report computations with measurements get rather complicated and are not generally part of a primary-school mathematics program. You can, and should, however, pay *some* attention to this important topic in your teaching, along the lines suggested above.

We now proceed to a discussion of measurement of length, volume, area, angles, time, temperature, and weight, in turn.

11.2 LENGTH

How long is the line segment shown in Fig. 11.1: $\frac{1}{3}$? 4? or 7? I hope the question makes no sense to you, and that you said "$\frac{1}{3}$, 4, or 7 *what*?"! Certainly the first task in assigning a measure to a length is to agree on a **unit** of length.

Figure 11.1

In the primary-school classroom, the study of length is best begun by emphasizing this need for an agreed upon unit of length. Thus, if each child measures the width of the classroom by using his steps as a unit of measure, many different numbers will be obtained as a measure of this width. Such an activity is shown in Fig. 11.2, taken from a third-grade text.

Jim measured the distance across the room by counting the number of steps he took to walk from one side to the other. His step was the "segment." He took 12 steps.

When Sue walked across the room, she took 16 steps.

Ann and Joe each made it in 14 steps.

Figure 11.2

282 Measurement   **11.2**

EXERCISES

1. Who took the most steps?

2. Who took the fewest steps?

3. Who took the longest steps?

4. Who took the shortest steps?

5. Which two children took about the same size steps?

6. Which two children can best compare the size of two different rooms by steps?

7. Cathy took shorter steps than anyone in the class. Will she take more steps or fewer steps than Sue to cross the room?

8. Tom took longer steps than any of the other children. Guess the number of steps Tom needs to get across the room.

Cathy's segment

Tom's segment

Figure 11.2 (continued)

As you know, there are many units of length in use today—and many more that were in common use in the past: inches, feet, centimeters, meters, hands (commonly used for measuring the height of horses), rods (in surveying), etc. Today, however, as we have said, the metric system is almost universally used, with the U.S.A. bound to adopt it in the very near future. Hence we will use the metric system almost exclusively throughout this chapter. For length, the commonly used metric units, with their approximate equivalents in the English system, are:

1 millimeter (mm) = 0.039 inches (in.)

10 mm = 1 centimeter (cm) = 0.39 in. (about the width of a popsicle stick!)

100 cm = 1 meter (m) = 39.37 in. (about a yard)

1000 m = 1 kilometer (km) = 0.62 miles (mi)

(There are other units such as the decimeter, which is 10 cm, and the dekameter, which is 10 m; but the ones listed here are the only ones in common use.)

Notice how simple it is to convert from one metric unit to another. All we have to do is to multiply or divide by a power of 10. Thus, for example,

2.63 m = (2.63 × 100) cm = 263 cm,

8243 m = (8243 ÷ 1000) km = 8.243 km

In contrast, to convert yards to feet, we multiply by 3; to convert feet to miles, we divide by 5280, etc.

Even after we have chosen to use the metric system, the best choice of a unit for a particular measurement is not automatically determined. Just as it would not be sensible (although it would be possible) to measure the length of a classroom by using a toothpick as a unit, or the length of a toothpick by using the width of a classroom as a unit, so it would not be sensible to measure the distance to the moon in centimeters, or the diameter of a penny in kilometers.

Clearly, an appropriate metric unit for measuring the line segment shown in Fig. 11.1 is the centimeter. To determine the measure of a line segment in centimeters (and similarly in terms of any unit of length), basically means to determine how many centimeter lengths, or parts of centimeter lengths, can be laid off end to end on the line segment, beginning at one end of the line segment and continuing until the other end is reached. An ordinary metric ruler, of course, simplifies this procedure by providing, side by side, copies of the unit centimeter, with the added refinement of including tenths of a centimeter (millimeters).

Measure the line segment of Fig. 11.1 with a metric ruler. If you measure carefully you will note that its length is between 7.2 cm and 7.3 cm but is closer to 7.2 cm. You might estimate an additional decimal place as, for example, 7.23 cm but, in any event, you can say that the length is 7.2 cm correct to the nearest tenth of a centimeter (or 72 mm correct to the nearest millimeter). Note that the convention described in the introduction says that 7.2 cm is to be understood as reporting a measurement correct to the nearest tenth of a centimeter, in contrast to 7.20 cm which would assert that the length was 7.2 cm correct to the nearest hundredth of a centimeter. For children, however, it is best to write out the degree of precision used. The important thing to emphasize is that *all* measurement is approximate.

Finding the **perimeters** of various polygons simply involves finding the sum of the lengths of the sides. Thus, if the lengths of the sides of the triangle of Fig. 11.3 are measured as shown to the nearest tenth of a centimeter, the perimeter of the triangle will be found to be (4.1 + 3.2 + 6.0) cm = 13.3 cm, to the nearest tenth of a centimeter.

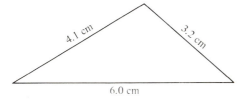

Figure 11.3

But what about the lengths of curves other than those made up of line segments as, for example, a circle, or the curve shown in Fig. 11.4? One way to attempt to find such lengths would be to fit, as well as possible, a piece of string to the curve, and then straighten the string and measure its length. (And certainly one would have children carry out activities of this kind.) Another way would be

Figure 11.4

to "approximate" the curve by line segments, and then find the sum of the lengths of these line segments, as suggested in Fig. 11.5. Better approximations could then be obtained by decreasing the lengths of the line segments involved.

Figure 11.5

Let's carry out some of this process to find a rough approximation to the perimeter (**circumference**) of a circle of **radius** r units. Figure 11.6 shows a regular hexagon inscribed in our circle. The circumference of the circle is then approximated (very roughly) by the perimeter of the hexagon. Since the triangles shown in Fig. 11.6 are all equilateral triangles whose sides are of length r units, this perimeter is $6r$ units, or, in terms of the diameter d ($= 2r$), $3d$. Thus our first approximation to the circumference of the circle is $6r$ units.

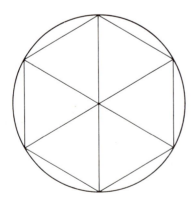

Figure 11.6

Using regular polygons with more sides (and some considerable algebra!) gives us the following table, where the perimeters are given correct to four decimal places.

Number of sides	Length of 1 side	Perimeter of polygon
6	r	$3d$
12	$0.51763r$	$3.1058d$
24	$0.26105r$	$3.1326d$
48	$0.13081r$	$3.1394d$
96	$0.065438r$	$3.1410d$
192	$0.032723r$	$3.1414d$
384	$0.016362r$	$3.1416d$
768	$0.0081812r$	$3.1416d$

This leads us to the famous formula for the circumference of a circle, $C = \pi d$, where π is an irrational number (i.e., a nonterminating and nonrepeating decimal—see Chapter 9) which, to 4 decimal places, is 3.1416, and d ($= 2r$) is the **diameter** of the circle. (Another simple and useful approximation to π is $3\frac{1}{7}$.)

Programed Lesson 11.2

1. 1 m = _____ cm.

   100

2. 2.5 m = _____ cm.

   250

3. 526 cm = _____ m.

   5.26

4. 1 km = _____ m.

   1000

5. 62 km = _____ m.

   62,000

6. 545 m = _____ km.

   0.545

7. 1 cm = _____ mm.

   10

8. 24 cm = _____ mm.

   240

286 Measurement   11.2

9. 15 mm = _____ cm.

    1.5

10. 4 yards = _____ feet.

    12

11. 5 feet = _____ yards.

    $1\frac{2}{3}$

12. 2 miles = _____ feet.

    10,560

13. 15 cm = _____ inches (approximately).

    5.85

14. 6 km = _____ miles (approximately).

    3.72

15. The perimeter of a circle is called its _____.

    circumference

Exercises 11.2

1. Measure the length in centimeters of the curve shown in Fig. 11.4 by using a string.

2. Measure the length in centimeters of the curve shown in Fig. 11.4 by adding the measures of the lengths of the line segments shown in Fig. 11.5. Compare your result with that obtained in Exercise 1.

3. Cut out a cardboard circle of diameter 10 cm and mark a point on its circumference. Place the circle with this mark on a line, and mark this position on the line as A in the figure below. Now roll the circle along the line until the mark comes back to the line, as shown in the figure. Mark this position on the line as B, and measure the distance AB. How is the distance AB related to π? Why?

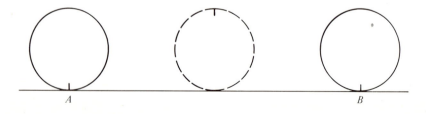

4. The diameter of a circle is measured as 2.3 cm, correct to the nearest tenth of a centimeter. Find its circumference, giving your answer to the nearest tenth of a centimeter.

5. Imagine a rope stretched tightly around the earth at the equator. How much more rope would you need to place the rope exactly 1 meter above the equator at every spot? (Assume that the earth is a perfect sphere of diameter 13,000,000 meters.) Now consider the same problem for an asteroid that is a perfect sphere of diameter 1000 meters.

11.3 AREA

Initial work on measurement of areas with children customarily involves first convincing them that the most useful unit of area is a square. Thus, for example, an attempt to use a circular region as the unit of area leads to a situation such as that shown in Fig. 11.7. Clearly there is too much open space left between the unit circular areas for us to be able to say with any accuracy how many of these unit areas are contained in the rectangle.

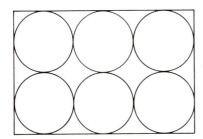

Figure 11.7

Suppose then that we take as our unit area, the area of a square of side 1 cm. Such a square is said to have an area of 1 **square centimeter** (cm²). A *direct* measurement of an area would then involve placing this unit square (and replicas thereof) on the area to be measured, as shown in Fig. 11.8. Of course, approximation may be involved if, as shown in Fig. 11.9, the squares do not fit the area exactly. Here the area of the rectangle is certainly greater than 10 cm². Since each of the eight areas $a, b, c, \ldots, h$ is clearly less than half our unit area, we can conclude that this area is less than $(10 + 8 \times \frac{1}{2})$ cm² = 14 cm². To obtain a better approximation we could (in theory, at least!) use as our unit square one which has a side of length 0.1 cm. Doing this would result in fitting in about 1380 of these squares, giving us an area of $\frac{1380}{100} = 13.8$ cm². (See the article at the end of the chapter for an activities approach to measurement of area for children.)

1	2	3	4	5	6	7
8	9	10	11	12	13	14
15	16	17	18	19	20	21

Figure 11.8

1	2	3	4	5	f
6	7	8	9	10	g
a	b	c	d	e	h

Figure 11.9

In practice, of course, we rely on various formulas for the determination of areas from **linear** (length) measurements of its dimensions. Thus, for example, the area, A, of a rectangle of length ℓ units and width w units is given by the formula $A = (\ell \times w)$ square units.

Three of the most useful formulas for the areas of polygonal regions are listed below, and ways of deriving some of them are suggested in programed lesson 11.3. (Children will use various manipulative devices such as the geoboard, or cutout cardboard figures, in developing these.)

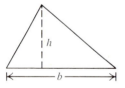

Parallelogram

$A = h \times b$

(h is the length of the **altitude** and b is the length of the **base** of the parallelogram.)

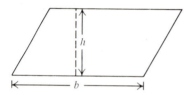

Triangle

$A = \tfrac{1}{2}h \times b$

(h is the length of the **altitude** and b is the length of the **base** of the triangle.)

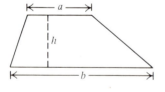

Trapezoid

$A = \tfrac{1}{2}h(a + b)$

(h is the length of the **altitude** and a and b the lengths of the **bases** of the trapezoid.)

A word about terminology: We often speak, for example, of the area of a triangle. Technically, however, the area of any triangle is 0, since a triangle is the union of three *line segments*, and line segments certainly have an area of 0! One should, then, really speak of the area of the *interior* of a triangle, or the area of a triangular *region*. Having said this, let us refer to the areas of triangles, rectangles, etc., knowing perfectly well what we mean!

11.3

A more difficult task, of course, is to determine the areas of figures whose sides are not all line segments. Again, approximations, in this case by rectangles, can be used, as suggested in Fig. 11.10. As the number of rectangles (all of the same width) is increased, the approximation is improved. (A development of this idea leads to the integral calculus.) Another way of estimating irregular-shaped areas is suggested in Exercise 10 of Exercises 11.3.

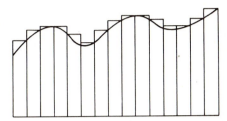

Figure 11.10

Such a procedure can be used to develop the formula $A = \pi r^2$ for a circle of radius r. Another approach, more suitable for children, is as follows: Consider a circle divided up, like a pie, into pieces (**sectors**) of equal size, as shown in Fig. 11.11, where there are 12 of these sectors. This figure also shows these sectors spread out to form an "almost" rectangle which is very close in area to the rectangle $ABCD$. Now what is the area of rectangle $ABCD$? First of all, its width is approximately r. Now the sum of all the 12 arc lengths shown in the "almost" rectangle is, of course, equal to the circumference of the circle, which is $2\pi r$. Hence the sum of the arc lengths of one side of the "almost" rectangle is $\frac{1}{2}(2\pi r) = \pi r$. Thus, an approximation to the area of rectangle $ABCD$ is $r \cdot \pi r = \pi r^2$ and hence πr^2 is also an approximation to the area of the "almost" rectangle (= area of the circle). By increasing the number of sectors, our approximation is bettered. This is indicated in Fig. 11.12, where we have drawn just one of the smaller sectors. As you can see, AB is now more closely equal to r and the length of $\overline{AE}$ more closely equal to the length of the arc $\widehat{AE}$ than they were in Fig. 11.11.

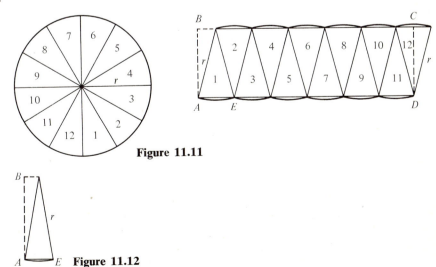

Figure 11.11

Figure 11.12

290 Measurement   **11.3**

What we have given, then, is not a *proof* that the area of a circle of radius r units is πr^2 square units, but a *plausibility* argument suitable for presentation to an upper primary class.

The commonly used units of area in the metric system are, in addition to the square centimeter, the **square meter** (m^2), which is the area of a square whose sides are 1 m in length, the **hectare** (ha), the area of a square whose sides are 100 meters in length, and the **square kilometer** (km^2), which is the area of a square whose sides are 1 kilometer in length. We have

$$1 \, m^2 = (100 \times 100) \, cm^2 = 10{,}000 \, cm^2,$$
$$1 \, ha = (100 \times 100) \, m^2 = 10{,}000 \, m^2,$$

and

$$1 \, km^2 = 100 \, ha$$

In the English system, the commonly used units of area are the **square inch** (sq in.), the **square foot** (sq ft), the **square yard** (sq yd), the **acre,** and the **square mile** (sq mi). We have

$$1 \, sq \, ft = (12 \times 12) \, sq \, in. = 144 \, sq \, in.,$$
$$1 \, sq \, yd = (3 \times 3) \, sq \, ft = 9 \, sq \, ft,$$
$$1 \, acre = 4840 \, sq \, yd,$$

and

$$1 \, sq \, mi = 640 \, acres$$

Again, note that, as is true for linear measure, conversion from one metric unit of measure to another is much easier to carry out than such conversions in English units.

The metric units of area measurement and the English units are related as follows (correct to 3 decimal places):

$$1 \, cm^2 = 0.155 \, sq \, in.$$
$$1 \, m^2 = 1.196 \, sq \, yd$$
$$1 \, ha = 2.471 \, acres$$
$$1 \, km^2 = 0.386 \, sq \, mi$$

Programed Lesson 11.3

1. In the figure on page 291 we see that the area of triangle $AB\check{C}$ is equal to the area of triangle *DEF*. Hence the area of parallelogram *ABEF* is equal to the area of rectangle _____ shown next. Since the area of this rectangle is _____, we can conclude that the area of parallelogram *ABEF* is also _____.

11.3 Area

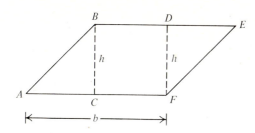 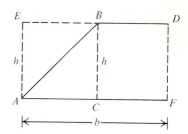

EAFD, h × b, h × b

2. In the figure below, the area of the right triangle *BAC* is one-half the area of the rectangle *ABDC*. Since the area of the rectangle *ABDC* is _____, we can conclude that the area of triangle *BAC* is _____.

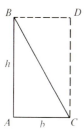

h × b, ½h × b

3. In the figure below, the triangle *ABC* can be broken up into the two right triangles, _____ and _____, whose areas, by Exercise 2, are _____ and _____, respectively.

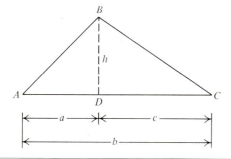

BDA, BDC, ½h × a, ½h × c (Different orders of letters possible)

4. Continuing from Exercise 3, we have that the area of triangle *ABC* = area of right triangle _____ + area of right triangle _____ = ½ _____ + ½ _____ = ½h × (___ + ___) = ½h × b.

BDA, BDC, h × a, h × c, a, c

292 Measurement 11.3

5. In the figure below, the area of triangle ABC = area of right triangle ___ − area of right triangle ___ = $\frac{1}{2}$___ − $\frac{1}{2}$___ = $\frac{1}{2}$___ × (___ − ___) = $\frac{1}{2}h$ × ___.

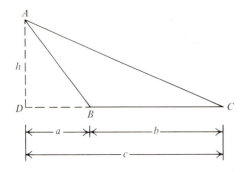

ADC, ADB, $h \times c$, $h \times a$, h, c, a, b

6. In the figure below, the area of the trapezoid $ABCD$ = area of right triangle ___ + area of rectangle ___ + area of right triangle ___ = $\frac{1}{2}$___ + ___ + $\frac{1}{2}$___.

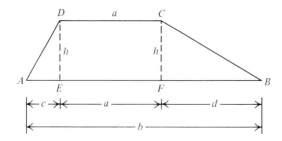

DEA, DEFC, CFB, $h \times c$, $h \times a$, $h \times d$

7. Continuing Exercise 6, we have that the area of trapezoid

$ABCD = \frac{1}{2}h \times c + h \times a + \frac{1}{2}h \times d$
$= \frac{1}{2}h \times c + (\frac{1}{2}h \times a + \frac{1}{2}h \times a) + \frac{1}{2}h \times d$
$= \frac{1}{2}$___ × $(c + a + a + d)$
$= \frac{1}{2}h \times [a + ($___ $+ c + d)]$
$= \frac{1}{2}h \times (a + $___$)$

h, a, b

8. To find the surface area of the right circular cylinder on page 293, we can first observe that the area of the top and bottom together is 2 × ___. Now we can think of the cylinder as being cut from top to bottom and unrolled as shown in the next figure (without the top or bottom). The rectangle formed has width h and length ___ and so has an area of ___.

11.3 Area

πr^2, $2\pi r$, $2\pi r \times h$

9. Continuing Exercise 8, we see that the surface area of the cylinder is $2\pi r^2 +$ _____ $= 2\pi r($ __ $+$ __ $)$.

$2\pi r \times h$, r, h

10. By counting squares and estimating parts of squares, find an approximation to the number of square units in the figure shown below.

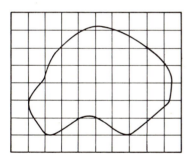

About 35 sq units

11. A rectangle is 34 cm wide and 2 m long. Its area is _____ cm² or _____ m².

6800, 0.68

12. To change the measurement of an area given in square centimeters to one in square meters, we _____ the area in square centimeters by _____. Conversely, to change the measurement of an area in square meters to one in square centimeters, we _____ the area in square meters by _____.

divide, 10,000, multiply, 10,000

13. One square yard = ___ square feet. Hence, to change the measurement of an area given in square yards to one in square feet, we _____ the area in square yards by ____. Conversely, to change the measurement of an area in square feet to one in square yards, we _____ the area in square feet by ____.

9, multiply, 9, divide, 9

294 Measurement 11.4

14. If the length of a square is doubled, its area is ———.

quadrupled

15. 1 ha = —— m^2.

10,000

16. 2.3 ha = —— m^2.

23,000

17. 12,000 m^2 = —— ha.

1.2

18. 1 km^2 = —— ha.

100

19. 4.6 km^2 = —— ha.

460

20. 35 ha = —— km^2.

0.35

21. 1 sq ft = —— sq in.

144

22. 5 sq ft = —— sq in.

720

23. 864 sq in. = —— sq ft.

6

24. 9 acre = —— sq yd.

43,560

11.4 VOLUME

Measurement of volume again involves the choice of a suitable unit of volume which, in this case, is a cube, as shown in Fig. 11.13. Then by way of blocks (or sugar cubes!), the basic formula for the volume of a rectangular parallelopiped (a box!) can be developed, as suggested by Fig. 11.14.

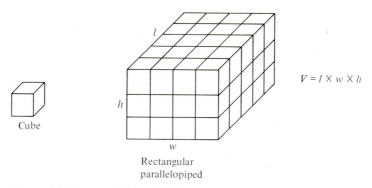

Figure 11.13   **Figure 11.14**

Developing formulas for volumes of other solids along the lines of the development of formulas for areas is not an easy task. Fortunately, an experimental procedure suitable for children is available to develop some of these formulas. This procedure involves the use of various-shaped containers that can hold water (and these are available from a number of companies dealing in primary-school supplies). For example, suppose we have, as shown in Fig. 11.15, a right circular cone and a right circular cylinder, with the same altitudes and the same radius of the bases. Filling the cone with water and emptying it into the cylinder can be done three times, leading us to the formula $\frac{1}{3}\pi r^2 h$ for the volume of a cone, if we assume the (reasonable) formula for the volume of the cylinder, base × height, or $\pi r^2 h$.

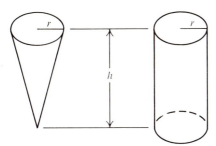

Figure 11.15

The most commonly used units of volume in the metric system are the **cubic centimeter** (cm^3 or cc), which is the volume of a cube the length of whose edges is one centimeter, the **liter** (ℓ) which is equal to 1000 cc, and the **milliliter** (ml), which is $\frac{1}{1000}$ of a liter and hence equal to the cubic centimeter.

In the English system there is a bewildering array of measures of volume. (See Exercise 1 of programed lesson 11.4.) They can all be related to the metric system by using the fact that one liter = 1.057 quarts (correct to 3 decimal places).

Programed Lesson 11.4

Use a dictionary or other source of information, if necessary, to do Exercises 1 and 2.

1. The number of fluid ounces in a teaspoon is ___; in a tablespoon is ___; in a dram, ___; in a cup, ___, in a pint, ___; in a quart, ___; in a gallon, ___; in a pony, ___; in a gill, ___. (This is to convince you of the merits of the metric system!)

   $\frac{1}{6}, \frac{1}{2}, \frac{1}{8}$, 8, 16, 32, 128, 1, 4 (But you may find different answers in dictionaries for teaspoon and tablespoon.)

2. The volume of a sphere of radius r is ___.

   $\frac{4}{3}\pi r^3$

3. 1 liter = ___ cubic centimeters. Hence, to change the measurement of a volume given in liters to one in cubic centimeters, we _____ the volume in liters by ___. Conversely, to change the measurement of a volume given in cubic centimeters to one given in liters, we _____ the volume in cubic centimeters by ___.

   1000, multiply, 1000, divide, 1000

4. 1 cubic yard = ___ cubic feet. Hence, to change the measurement of a volume given in cubic yards to one in cubic feet, we _____ the volume in cubic yards by ___. Conversely, to change the measurement of a volume given in cubic feet to one given in cubic yards, we _____ the volume in cubic feet by ___.

   27, multiply, 27, divide, 27

5. If the length of a cube is doubled, its volume will be multiplied by ___.

   eight

11.5 ANGLES

The most commonly used unit of angle measure is the **degree.** A right angle ABC, as shown in Fig. 11.16, has degree measure 90 and we write $m(\angle ABC) = 90°$.

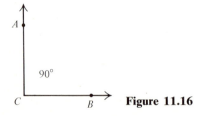

Figure 11.16

11.6 Other Measurements 297

To measure an angle we can use a **protractor.** Figure 11.17 shows a protractor being used to measure angles of 50° and 130°.

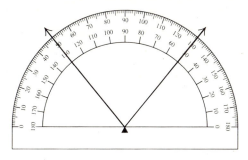

Figure 11.17

Subdivisions of a degree can be in tenths, hundredths, etc., as 13.2°, or in **minutes** and **seconds**, where 60 minutes (60′) is equal to 1 degree, and 60 seconds (60″) is equal to 1 minute.

Exercises 11.5

Measure with a protractor the angles in each of the two triangles shown below and compute the sum of the degree measures of the angles of each triangle. Conclusion?

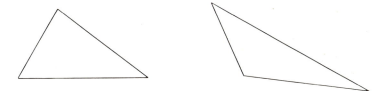

11.6 OTHER MEASUREMENTS

The other measurements commonly discussed in the primary school are, as mentioned previously, those of time, temperature, and weight.

About time we need say little here, except to point out that (1) the alternative ways of stating a time can be confusing to children, e.g., ten-forty, forty minutes past ten, forty minutes after ten, twenty minutes to eleven, twenty to eleven; (2) there is increasing use of the **twenty-four hour** clock, whereby, for example, 2 P.M becomes 14 hours.

Likewise, little needs to be said about the measurement of temperature here, except to note that, in the metric system, temperature is measured in degrees **Celsius** (also called **centigrade**), and in the English system, in degrees **Fahrenheit**. In the Celsius system the freezing point of water (at sea level) is taken as 0° and the boiling point (also at sea level) as 100°, whereas the comparable temperatures on the Fahrenheit scale are 32° and 212°, respectively.

The two measurements are related by the following formulas, where C is the temperature in degrees Celsius and F the temperature in degrees Fahrenheit:

$$C = \tfrac{5}{9}(F - 32) \quad \text{and} \quad F = \tfrac{9}{5}C + 32$$

Weight in the metric system is commonly measured in **grams** (g) and **kilograms** (kg) where 1000 g = 1 kg. (The weight of a U.S. nickel is about 5 grams.) Other metric units are the **milligram** (= 0.001 gram) and the metric ton (= 1000 kg)—also sometimes called a **long ton**. We have, correct to 2 decimal places,

$$1 \text{ kg} = 2.20 \text{ pounds (lb)}$$

In the English system we have 16 ounces (oz) = 1 pound and 2000 lb = 1 ton.

There is a nice relation between metric weight and metric volume, in that 1 gram is the weight of 1 cm^3 of water at 4°C. (Not that the temperature makes much difference except in the most precise scientific work—but water is at its densest at 4°C!)

Exercises 11.6

1. In an international airline schedule, the time of departure of a certain flight is listed as 20 hours. What is the time in terms of A.M. or P.M.?

2. What time is 5 A.M. on a 24-hour clock? 6 P.M.?

3. What Celsius temperature corresponds to a temperature of 41°F? to 55°F? to −10°F?

4. What Fahrenheit temperature corresponds to a temperature of 45°C? to 64°C? to −15°C?

5. A Celsius thermometer and a Fahrenheit thermometer both read the same number of degrees. What is the temperature?

6. Suppose steak costs £2 a kilogram in England (assume £1 = $1.86). What would be the cost in dollars of one pound?

7. What is the weight, in kilograms, of the water in the filled tank shown below?

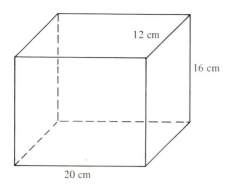

8. What is the weight, to the nearest hundredth of a kilogram, of the water in the filled cylindrical tank shown below? (Use 3.14 as the approximate value of π.)

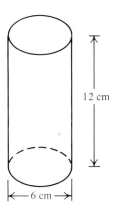

Chapter Test

In preparation for this test, you should review the following terms. (I assume that you are already familiar with the English system of units.)

Millimeter (mm) (282)
Centimeter (cm) (282)
Meter (m) (282)
Kilometer (km) (282)
Square centimeter (sq cm) (287)
Square meter (sq m) (290)
Hectare (ha) (290)
Square kilometer (sq km) (290)
Cubic centimeter (cc) (295)
Milliliter (ml) (295)
Liter (l) (295)
Degree (for angles) (296)
Minute (for angles) (297)
Second (for angles) (297)
Protractor (297)
Celsius (297)

Fahrenheit (297)
Gram (298)
Kilogram (298)
Circumference (284)
Radius (284)
Diameter (285)
Altitude of a triangle (288)
Altitude of a parallelogram (288)
Altitude of a trapezoid (288)
Base of a triangle (288)
Base of a parallelogram (288)
Base of a trapezoid (288)
Perimeter (283)
Area (287)
Volume (294)

You should also review how to:

1. Convert from one metric unit to another (282, 290, and 295);
2. Convert from one English unit to another (285, 290, and 296);
3. Convert metric measurements to English measurements (285, 290, and 296);
4. Convert from Celsius to Fahrenheit and vice versa (297–298);
5. Convert time on a 24-hour clock to time on a 12-hour clock, and vice versa (298);
6. Measure an angle with a protractor (297);
7. Find the perimeter of various geometric figures (283–284);
8. Find the area of various geometric figures (288–294);
9. Find the volume of various geometric solids (295–296).

300   Measurement

1. 392 cm = ____ m.

2. 12.2 m = ____ cm.

3. 1562 m = ____ km.

4. 3.5 yd = ____ ft.

5. 15 mi = ____ yd.

6. 2.5 ha = ____ sq mi.

7. 0.52 sq km = ____ ha.

8. 6.3 sq yd = ____ sq ft.

9. 368 cc = ____ ℓ.

10. 2.6 ℓ = ____ cc.

11. 262 cc = ____ ml.

12. 5.2 cu yd = ____ cu ft.

13. 5184 cu in. = ____ cu ft.

14. 6 pints = ____ quarts.

15. 15°F = ____ C.

16. 15°C = ____ F.

17. On a 24-hour clock, 3 P.M. = ____ hours.

In Problems 18 through 22, the measurements given are to be considered as correct to the nearest tenth of a unit, and the answers should be given to the nearest tenth of a unit.

18. A circle has a radius of length 2.3 cm. Find its circumference and its area. (Use 3.14 as an approximation to π.)

19. The length of the base of a parallelogram is 7.5 cm, and its altitude is of length 3.2 cm. Find its area.

20. The length of the base of a triangle is 3.7 cm and its altitude is of length 1.7 cm. Find its area.

21. The lengths of the bases of a trapezoid are 10.3 m and 7.4 m, respectively, and the length of its altitude is 2.6 m. Find its area.

22. Using 3.14 as an approximation to π, find the volume of a right circular cone with an altitude of length 3.7 cm and a base whose radius is of length 2.1 cm.

TEST ANSWERS

1. 3.92	2. 1220
3. 1.562	4. 10.5
5. 26,400	6. 0.00965 (approximately)
7. 52	8. 56.7

9. 0.368

10. 2600

11. 262

12. 140.4

13. 3

14. 3

15. $-9\frac{4}{9}$

16. 59

17. 15

18. 14.4 cm,   16.6 cm^2

19. 24.0 cm^2

20. 3.1 cm^2

21. 23.0 m^2

22. 17.1 cm^3

Grids, tiles, and area

by Kathryn Besic Strangman

Reprinted from *The Arithmetic Teacher*, Vol. 15 (1968), pp. 668–672.

The purpose of this article is to report a method for introducing area to fifth-grade pupils which is mathematically sound and readily teachable. Some programs begin their approach to area indirectly by making direct linear measurements and computing the area obtained from these. Some methods define the area of a rectangle as the product of the length and the width of the rectangle and then proceed to discover relationships between the areas of a rectangle and other polygons and, later on, other types of regions.

These indirect methods may mask the real meaning of area. Several questions arise here. What is area in general? Is area length times width? Are children learning key underlying concepts of area which will prove useful to them in their further study and encounters with the real world? What do we want to teach children about the concept of area? How can we go about doing this? We hope to answer these questions to some degree in the following discussion.

The approach reported here is direct, and it is analogous to the way in which linear measure is carried out. Thus it is a method that follows easily from the previous experience of the child. This, along with the fact that it is a direct method, provides an approach to area measure which is easy to teach with meaning and understanding.

In this approach area is defined as a number; namely, as a number assigned to a region. We usually say that it is a measure of the amount of surface contained in some plane region. Length is defined as a number assigned to a line segment. It is a measure of a distance along some straight line. How does one proceed to determine the length of a line segment?

Suppose we wish to measure the line segment shown in Fig. 1. The first step we take is to choose a unit of reference and agree to consider its measure as exactly 1 unit. Let us select the line segment in Fig. 2(*a*) as the unit.

Figure 1

Figure 2

We next attempt to compare the unit segment with the line segment under consideration. This comparison can be accomplished by laying off copies of the unit segment, end to end, on the line containing the segment to be measured, and counting the number of units needed to cover this segment. The units can be laid off on the line segment by using a compass with the radius equal to the length of the unit. (See Fig. 3.)

Figure 3

302

It is impossible to determine the exact number of units needed to cover the segment, but we can easily determine two points between which the end of the segment lies, and thus two numbers between which is the exact length of the segment. It is customary to choose the number associated with the closer point as the length of the segment. In this instance the length is 3 units to the nearest unit. Obviously an approximation is involved. However, the approximation can be made more exact, if desired, by choosing a smaller unit of measure. For example, using the measuring unit in Fig. 2(b), the length of the segment is 7 units, and the approximation is better. Figure 4 illustrates that the approximation was improved.

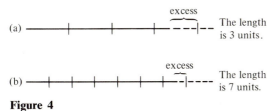

Figure 4

The procedure described above of laying off copies of the unit segment on a line amounts to the construction of an instrument for measuring length. This, of course, is a ruler, the instrument commonly used for measuring length.

It is possible to approach area measure in a similar, direct manner. The first step is to choose a measuring unit. For area the unit must be a certain amount of surface in a region. Although other units may be used, such as a triangle, rectangle, hexagon, or other region, it is convenient to choose some square region, which we will agree has an area of 1 unit. For illustration let us choose the unit shown in Fig. 5 and determine the area of the closed plane region shown in Fig. 6.

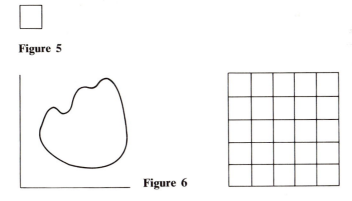

Figure 5

Figure 6   **Figure 7**

The next step is one of comparison. We must try to find how many square units completely cover the plane region we wish to measure, and no more. Next, place copies of the measuring unit (Fig. 5) side by side, row upon row, to obtain what we shall call a grid. (See Fig. 7.) Note that this works easily because of the shape of the measuring unit we chose. This grid is our measuring instrument. (Recall that in measuring length we placed copies of the unit line segment end-to-end to form a ruler.) As in working with length, the measurement problem now becomes that of determining how many of these measuring units are needed to completely cover the given region.

In the classroom one might proceed as follows. After discussing the term "amount of surface," choose a measuring unit of area (such as that in Fig. 5) and construct a grid. Duplicate the grid on translucent paper or plastic. Now place the grid over the region whose area is to be determined. (For illustration, consider the area of the region in Fig. 6.) In order that each pupil obtain the same results, use guide marks on the lower left of the figure. (Note guide marks in Fig. 6.) We are interested, of course, in the number of measuring units that completely cover the region. We give the following instructions to the pupils.

1. Completely shade every unit of which at least part covers some of the surface of the region. (See Fig. 8.)

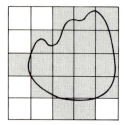

Figure 8

2. Count the number of units just shaded. (The answer is 18.) This number is called the *outer area* of the region. Note that the area of the region is less than this number, since more surface than the surface of the region itself was shaded.

3. Next, completely shade (using a different kind of shading or a new grid of the same unit) every unit that lies entirely inside the region. (See Fig. 9.)

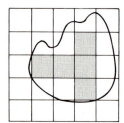

Figure 9

4. Count the number of units shaded this time. (The answer is 5.) This number is called the *inner area* of the region. Note that the area of the region is greater than this number, since 5 units are not enough to completely cover the region.

5. Now we put together the information we obtained in questions 2 and 4 and conclude the following:

   I. 5 < area of the region < 18; that is, the area of the region is a number between 5 and 18.

We have gained a lot of information about the area of the region. (Note that if the grid had been placed in a different position the limits on the area might have been different.) However, we can do better. Next we use a different grid. For classroom use it is convenient to choose a grid in which four of the square units completely cover one larger unit in the first grid. (See Fig. 10.) Use the same region and repeat the entire process—steps 1–5 above. (See Figs. 11 and 12.) This time we conclude:

II. 31 < area of the region < 59.

Figure 10

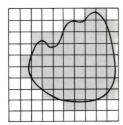

Figure 11   **Figure 12**

In order to compare the results from conclusions I and II we do the following. Consider covering the region with tiles that are the size of the large reference unit (Fig. 5). If each tile costs 8¢, how much will it cost to tile the entire region? Conclusion I tells us we need between 5 and 18 large tiles. Therefore, the cost will be between 40¢ and 144¢ ($1.44).

Next, suppose we use tiles the size of the smaller unit. Since each large tile covers as much of the region as 4 small tiles, each small tile would cost 2¢. Conclusion II tells us we need between 31 and 59 small tiles. Therefore, the cost will be between 62¢ and 118¢ ($1.18). In the second instance, we obtain an estimate of the cost that is better than the first one, since the interval 62–118 is smaller than the interval 40–144.

Following this, more examples and other grid sizes should be used. After the pupils have had considerable practice determining inner and outer areas of different regions and the costs of tiling such, we begin to generalize that *the smaller the grid size, the better the approximation for the area of the region.* Thus with a smaller size grid we have a better measure of the area.

Let us summarize the important concepts discussed so far.

1. Area is a number assigned to a given region.
2. Just as in linear measure, we cannot determine this number, i.e., the measure of a region, exactly, but we can make approximations to it.
3. Areas of different regions can be compared.
4. We determine limits for the area of a region by comparing the area of the region in question to the area of a measuring unit.
5. By choosing a smaller measuring unit it is possible to obtain a better approximation for the area.

Pupils should thoroughly comprehend these basic concepts so that they gain a clear understanding of area. Once this has been accomplished, the area of other special plane figures can be easily taught.

Consider a rectangle whose length and width measure 4 units and 3 units respectively—theoretically, of course. (See Fig. 13.) When we place the grid of Fig. 7 on the rectangular region of Fig. 13, lining up the grid with the two adjacent sides of the rectangle, we note a very interesting thing. The inner area is 12 and the outer area is 12 also! From what we have seen before, we must conclude that 12 ≤ area of the rectangular region ≤ 12. But this implies that the area of the rectangular region equals 12! It is further pointed out that since there are 3 rows of measuring units with 4 in each row, we have 3 sets with 4 in each set, and by the definition of multiplication (number of sets times the number in each set equals the total number of objects) we can obtain the area in this special case by multiplying 3 by 4 to obtain 12. The 3 rows correspond to the fact that the rectangle is 3 measuring units in width, whereas 4 in each row corresponds to the fact that the rectangle is 4 measuring units in length. After considering similar examples, we generalize that the area of a rectangular region can be found by multiplying the length of the rectangle by its width. This can be extended further when the pupils learn to multiply fractions.

Figure 13

One can proceed from rectangles to other special regions such as triangular regions and, later, circular regions. Standard measuring units, such as the square inch, can be introduced when the need for them is recognized by the pupils. These and other particular aspects and special cases of area should be presented to the pupils after a good conceptual foundation has been laid.

In summary, it is felt that the above approach to area can contribute a great deal to the understanding of what area really is. Area is measured directly, as length is, and the general region is considered first in an attempt to point out some of the basic general properties of area. The special cases, such as the area of a rectangular region, then fall into place very easily and simply. It is hoped that this approach will lead students to the fact that area means more than "length times width."

REFERENCES FOR FURTHER READING

1. Bailey, T. G., "Linear measurement in the elementary school," *AT,* **21** (1974), pp. 520–525.

2. Banks, J. H., "Concepts of Measurement," Chap. 7 of *Enrichment Mathematics for the Grades,* National Council of Teachers of Mathematics, Reston, Va., 1963.

3. Botts, T., "Linear measurement and imagination," *AT,* **9** (1962), pp. 376–382. Emphasizes the difference between actual measurement (approximate) and the concept of ideal points, lines, triangles, etc.

4. Bourne, H. N., "The concept of area," *AT,* **15** (1968), pp. 233–342. Describes the use of practical experiences by children to form a concept of area.

5. Bruni, J. V., and H. Silverman, "Developing the concept of linear measurement," *AT,* **21** (1974), pp. 570–577. Describes details of beginning the topic by using nonstandard units: "baby" feet, "giant" feet, etc.

6. Dubisch, R., "Some comments on teaching the metric system," *AT*, **23** (1976), pp. 106–107.

7. Kelley, J. L., and D. Rickert, *Elementary Mathematics for Teachers*, Chapter 8, Holden-Day, San Francisco, 1970.

8. Krause, E. L., "Elementary school metric geometry," *AT*, **15** (1968), pp. 673–681. Here "metric" simply refers to measurement and not to the metric system!

9. NCTM Implementation Committee. "Metric—not *if* but *how*," *AT*, **21** (1974), pp. 366–369. Contains an extensive bibliography in regard to teaching the metric system.

10. Payne, J. N., and R. C. Seber, "Measurement and Approximation," Chapter 5 of *Growth of Mathematical Ideas*. National Council of Teachers of Mathematics, Reston, Va., 1959.

11. Smart, J. R., and J. L. Marks, "Mathematics of measurement," *AT*, **13** (1966), pp. 283–287.

Chapter 12

Recapitulation

As should be clear from the material presented so far, the mathematics of the primary school is no longer confined to arithmetic and applications of arithmetic. In the concluding chapters of this book, we will consider other topics that are being included in contemporary programs of primary-school mathematics and that are likely to increase in importance in the years to come.

Nevertheless, the arithmetic and geometry that we have considered up to this point is now, and will undoubtedly continue to be, the core of the primary-school mathematics program. Hence it seems appropriate to pause here to see, in further detail, the relation of the mathematics we have discussed to a typical program of contemporary primary-school mathematics.

Most primary-school mathematics text series are provided with a "scope and sequence" chart, which states in considerable detail when various topics are introduced and how these topics are carried through the grades. The major headings vary somewhat, but a fairly typical chart might have as its major headings (or "strands"):

1. Sets
2. Number and number systems
3. Structure and properties
4. Numeration
5. Sentences
6. Elementary probability and statistics
7. Functions
8. Operations
9. Geometry
10. Reasoning and proof
11. Problem solving
12. Measurement

In thinking about these strands, it should first be observed that there is considerable interweaving among them. Sets, for example, are used in discussing number systems, in describing geometric objects, and in defining operations with numbers; numeration is used in writing numbers in a number system and also in discussing operations with numbers; and so on.

At this point in the text, we have given some consideration to all of these topics except probability, statistics, and functions. Chapter 13 will have more to say about operations; Chapter 14 will deal with functions; Chapter 15 will consider another aspect of geometry; Chapters 16 and 17 will be on probability and statistics, respectively; and the concluding Chapter 18 will include further work on reasoning and proof.

Now let us look briefly at all the "strands" except (6) and (7), from the point of view of what is covered in a typical primary-school mathematics program (Grades 1 through 6) today and also from the point of view of what you have been studying. It is not my purpose to describe in detail what aspects of each topic are discussed in each grade (there is considerable variation from one text book series to the next) but simply to give a general idea of a typical program.

SETS

The primary-school mathematics program begins much as we have done, with the idea of a set and of one-to-one correspondence, and then, in due time, utilizes the concept of a set and union and complements of sets to define addition and subtraction of whole numbers. Later the brace ({ }) notation is introduced, and geometrical figures are described as sets of points, etc. Naturally, as we have emphasized many times, there is much less formal notation and vocabulary— especially in the early grades—than has been given in this text.

NUMBER AND NUMBER SYSTEMS

The concept of whole numbers as a property of matching sets is developed in the first grade (again, of course, in a very informal way). Gradually, the concept of number is extended to fractions, with attention also being paid to such concepts as that of a prime number, the L.C.M., and the G.C.D. Also included in some texts are divisibility tests, which we will consider in Chapter 18 (along with other number-theoretic results).

STRUCTURE AND PROPERTIES

From the very beginning, contemporary primary-school mathematics programs call attention to the commutative and associative properties, first for addition of whole numbers, and later for multiplication. The special properties of 0 and 1 are emphasized, and the distributive property is introduced and extensively used. Later these properties are considered in relation to the set of fractions.

NUMERATION

Beginning with the numerals $1, 2, 3, \ldots, 9$ in the first grade, the concepts of place value and the role of 0 in numeration are gradually developed, culminating in decimal fractions. Common fractions are, of course, also introduced, beginning with such simple fractions as $\frac{1}{2}$ and $\frac{1}{3}$ in the third grade. Some series also include some work with bases other than ten, in grades five or six; others do not.

SENTENCES

Under this heading are included such simple sentences (equalities and inequalities) as $2 + 3 = 5$, $2 \times 5 = 10$, $2 + 3 < 8$, etc. Also included are what are commonly called "open sentences", such as $\square + 3 = 5$, leading, eventually, to such standard algebraic notation as $x + 3 = 5$. Finally, this topic includes formulas of various kinds, such as $A = b \times h$, as discussed in Chapter 11. Much of this work can be described as "pre-algebra," and it forms an excellent preparation for a formal course in algebra in the eighth or ninth grade.

OPERATIONS

Here, of course, we refer to the program of developing understanding of and skill in the operations of addition, subtraction, multiplication, and division, beginning with simple addition of whole numbers in the first grade and culminating in work with decimal fractions in the fifth and sixth grades. Obviously, the major part of our work here to date has been concerned with an understanding of the procedures for performing these operations. (I have assumed that you already have the skills!) A more general approach to the concept of operations is considered in Chapter 13.

GEOMETRY

In Section 1.6 and in Chapter 8 we have considered some of the geometry commonly dealt with in the primary school—with an emphasis on activities rather than on formal definitions. Additional work in geometry will be given in Chapter 15.

REASONING AND PROOF

At each stage of a contemporary mathematics program, the basic aim is that of understanding, before drill and memorization. Certainly this has been the aim in this book. Here, of course, we have been somewhat more formal in our discussions than would be the case in primary school; but I have also tried to indicate

312   **Recapitulation**

how the results we obtained can be made plausible to young children. (And "plausible" reasoning underlies formal reasoning even at advanced levels of mathematics.)

The concept of proof is certainly one of the most difficult ones in mathematics (whether "old" or "new" mathematics!); the development of an appreciation for, and ability to make, proofs is a slow process indeed. A good primary series moves slowly to achieve an understanding of proof, beginning with very informal ("plausibility") proofs, and only gradually becoming more formal.

Further work on both formal and informal reasoning is included in the remaining part of this book.

PROBLEM SOLVING

Our work here has been mainly concerned with the abstract ideas of mathematics, since these are the foundations on which the applications of mathematics rest. Some applications, however, have been included in the exercises. But any good primary mathematics program will have a *host* of applications, beginning with such problems as "John has 2 candy bars and Mary has 3. How many do they have together?" Similarly, problems involving measures of perimeters, area, volumes, etc., certainly help to make mathematics real to children and serve as motivation for its study.

MEASUREMENT

The material on this topic that is taught in the primary school has been surveyed in Chapter 11.

Summary Test for Chapters 7 through 12

1. −4 is called the _____ of four, or the _____ _____ of 4.
2. The set of positive integers is also called the set of _____ numbers.
3. Complete the following addition "rabbit".

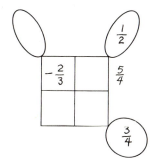

4. Complete the following multiplication "rabbit".

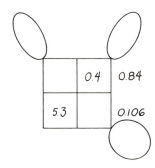

314   Summary Test for Chapters 7 through 12

5. If r, s, and t are any rational numbers such that $r < s$ and $t < 0$, then $r \times t \underline{\hspace{1cm}} s \times t$.

6. $\left| -\frac{7}{2} - \left(-\frac{3}{4} \right) \right| = \underline{\hspace{1cm}}$.

7. If a polygon is a regular polygon, all of its sides must be $\underline{\hspace{1.5cm}}$ and all of its angles must be $\underline{\hspace{1cm}}$.

8. A parallelogram is a quadrilateral that has its opposite sides $\underline{\hspace{1cm}}$.

9. A baseball is in the shape of a $\underline{\hspace{1cm}}$.

10. A matchbox is in the shape of a $\underline{\hspace{2cm}}$ $\underline{\hspace{2cm}}$.

11. A regular hexagon can be formed from six $\underline{\hspace{2cm}}$ triangles.

12. Which of the fractions $\frac{12}{5}, \frac{2}{9}, \frac{12}{4}, \frac{3}{6}$ are proper fractions? Which are written in lowest terms?

13. An irrational number is one that can be written as a $\underline{\hspace{2cm}}$ and $\underline{\hspace{2cm}}$ decimal.

14. Express $\frac{2}{13}$ as a repeating decimal.

15. We can write 0.023 as the repeating decimal $\underline{\hspace{2cm}}$.

16. Write $1.3\overline{27}$ as a common fraction in lowest terms.

17. Joe's salary of $550 a month was increased to $588.50. By what percent was his salary increased?

18. Do the following addition by using the fundamental principle, and write your answer as a mixed fraction.

$$\frac{2}{3} + \frac{8}{9} + \frac{5}{12}$$

19. Draw a figure, as in the text, to illustrate $\frac{4}{3} \times \frac{2}{5}$.

20. Show how the subtraction, $3.01 - 1.42$, can be carried out on an abacus.

21. Do the division, $\frac{4}{3} \div \frac{5}{6}$, by the "divide numerator and denominator" method, and check your answer by using the "invert and multiply" rule. Express your answer in lowest terms.

22. Do the division, $0.004494 \div 0.014$, by each of the three formats given on page 264.

23. The ratio of men to women on a certain committee having seven members is $3:4$. If 6 more men are added to the committee, how many women must be added if the ratio of $3:4$ is to be maintained?

Summary Test for Chapters 7 through 12   **315**

24. $1 \text{ m} = \underline{\hspace{1cm}} \text{cm}$; $1 \text{ ha} = \underline{\hspace{1cm}} \text{m}^2$; $1 \, \ell = \underline{\hspace{1cm}} \text{cc}$; $1 \text{ kg} = \underline{\hspace{1cm}} \text{g}$.

25. $41°\text{F} = \underline{\hspace{1cm}} \text{C}$; $40°\text{C} = \underline{\hspace{1cm}} \text{F}$.

26. A circle has a radius of 21 cm. Using $3\frac{1}{7}$ as an approximation to π, find its circumference and its area.

27. The length of the base of a triangle is 2.6 cm and the length of its altitude is 1.7 cm. Find its area.

28. Using 3.14 as an approximation to π, find, to the nearest tenth of a cubic centimeter, the volume of a cylinder whose diameter is 4.2 cm and whose height is 5 cm.

29. Write 0.32_{six} as a common fraction in lowest terms in base ten.

30. $(\frac{4}{5})_{\text{six}} \times (\frac{11}{3})_{\text{six}} = (\underline{\hspace{1cm}})_{\text{six}}$.

Chapter 13

Binary Operations

13.1 EXAMPLES AND DEFINITIONS

We have considered many sets in our work thus far and, in particular, the set

$$W = \{0, 1, 2, 3, \ldots\}$$

of whole numbers, the set

$$N = \{1, 2, 3, 4, \ldots\}$$

of natural (or counting) numbers, the set

$$I = \{\ldots, -3, -2, -1, 0, 1, 2, 3, \ldots\}$$

of integers, and the set R of rational numbers. On these sets we have also considered **operations**, which take two members of the set under consideration and produce a third member of the set. Thus from 2 and 3 in N we get, by means of the operation of addition, the number $5 \in N$ and, by the operation of multiplication, the number $6 \in N$. Since two members of the set are involved in producing a third, we speak of **binary** operations (from the Latin prefix *bi*, referring to two) of addition and multiplication on N.

More generally, a binary operation on a set S is any rule that assigns to each pair of elements of S in a given order (i.e., an ordered pair) another element of S. When the binary operation is not an operation such as addition or multiplication for which there are standard symbols, we often use a symbol such as $*$ (read "star") to indicate the operation. Thus, in symbols, if $a * b$ represents the result of applying $*$ to a and b, we say that $*$ is a binary operation on a set S if for all a and b in S, $a * b \in S$.

Suppose $S = W$. Is subtraction a binary operation on S? No, since, for example, $3 \in W$ and $4 \in W$ but $3 - 4 \notin W$. (It is true that $4 - 3$, $5 - 2$, $7 - 6$, etc., are all in W but we require that $a - b$ be in W for *all* a and b in W.)

317

318   Binary Operations   **13.1**

If, however, we take $S = I$, then subtraction is a binary operation on I, since, if $a \in I$ and $b \in I$, then $a - b \in I$. (See Section 7.3.) (Note, here, the importance of order in defining this binary operation since, for example, $4 - 3 \neq 3 - 4$.)

Here are some other examples.

1. Let E be the set of even natural numbers, $\{0, 2, 4, 6, \ldots\}$. Then "$+$" and "$\times$" are both binary operations on E, since both the sum and product of two even numbers are even numbers.

2. Let O be the set of odd natural numbers, $\{1, 3, 5, 7, \ldots\}$. Then "$\times$" is a binary operation on O since the product of two odd numbers is an odd number. But "$+$" is not a binary operation on O since the sum of two odd numbers is not an odd number.

3. Let S be the set of subsets of the set $\{a, b, c, d\}$. Then "$\cup$" and "$\cap$" are both binary operations on S since, if A and B are subsets of S, so are $A \cup B$ and $A \cap B$. On the other hand, "$\times$" (Cartesian product) is not a binary operation on S since, for example, $\{a\} \in S$ and $\{b\} \in S$ but $\{a\} \times \{b\} = \{(a, b)\} \notin S$.

4. Let $S = N$ and, for all $a \in S$ and $b \in S$, define $a * b = (a, b)$ (the G.C.D. of a and b). Then "$*$" is a binary operation on S, since the G.C.D. of any two natural numbers is a natural number.

5. As in Example 4, now with $a * b = [a, b]$ (the L.C.M. of a and b). Again "$*$" is a binary operation on S since the L.C.M. of any two natural numbers is a natural number.

6. Let $S = W$ and define $a * b = 2a + b$ for all a and b in S. Thus

$$3 * 4 = (2 \times 3) + 4 = 10, \qquad 6 * 0 = 2 \times 6 + 0 = 12, \qquad \text{etc.}$$

Then "$*$" is a binary operation on S. (For later work it is important to note that the choice of letters in defining "$*$" is unimportant. Thus,

$$b * a = 2b + a, \qquad b * c = 2b + c,$$
$$(a * b) * c = (2a + b) * c = 2(2a + b) + c, \qquad \text{etc.}$$

Exercises 13.1

Give reasons for your answers in these exercises.

1. Is division a binary operation on N? on W? on I? on R? on the set of positive rational numbers?

2. Let S be the set of prime numbers, $\{2, 3, 5, 7, 11, \ldots\}$. Is "$+$" a binary operation on S? Is "$\times$"?

3. Let $S = \{x : x = n^2 \text{ for } n \text{ a whole number}\} = \{0, 1, 4, 9, 16, \ldots\}$. Is "$+$" a binary operation on S? Is "$\times$"?

13.2   **Properties of Binary Operations**   **319**

4. Let $S = \{x: \quad x = 3m \quad$ for m a whole number$\} = \{0, 3, 6, 9, \ldots\}$. Is "+" a binary operation on S? Is "×"?

5. Let $S = W$ and define $a * b = (a \times b) + 1$ for all a and b in S. Is "∗" a binary operation on S?

6. Let $S = N$ and define $a * b = a^b$ for all a and b in S. Is "∗" a binary operation on S?

13.2 PROPERTIES OF BINARY OPERATIONS

The main reasons for introducing the concept of binary operations here (and in some primary-school programs) are (1) to provide a unifying concept for the various operations of arithmetic and (2) to furnish us with some examples of operations other than subtraction and division that are not commutative or not associative. For example, consider the binary operation on W of Example 6 above, namely

$$a * b = 2a + b$$

We have already noted that $3 * 4 = 10$. On the other hand, $4 * 3 = (2 \times 4) + 3 = 11$. Thus $3 * 4 \neq 4 * 3$, so that this binary operation is not a commutative one. Similarly,

$$(3 * 4) * 5 = 10 * 5 = (2 \times 10) + 5 = 25$$

whereas

$$3 * (4 * 5) = 3 * [(2 \times 4) + 5] = 3 * 13 = (2 \times 3) + 13 = 19$$

Since $25 \neq 19$, we conclude that this binary operation is also not associative.

Exercises 13.2

Determine whether the following binary operations on the given set are commutative operations, and then determine whether they are associative operations. If your answer is in the negative, give an example to justify your conclusion.

1. The binary operations of "×" and "+" of Example 1 of Section 13.1.

2. The binary operation of "×" of Example 2 of Section 13.1.

3. The binary operations of "∪" and "∩" of Example 3 of Section 13.1.

4. The binary operation of Example 4 of Section 13.1.

5. The binary operation of Example 5 of Section 13.1.

6. The binary operation of "×" of Exercise 3 of Exercises 13.1.

7. The binary operations of "+" and "×" of Exercise 4 of Exercises 13.1.

8. The binary operation of Exercise 5 of Exercises 13.1.

9. The binary operation of Exercise 6 of Exercises 13.1.

10. The binary operation "*" on $S = N$ where $a * a = a$ and, if $b \neq a$, $a * b =$ greater of a and b. (For example, $2 * 2 = 2$, $2 * 3 = 3$, and $5 * 4 = 5$.)

11. Do the following exercise from a fifth-grade text. (Note that the single word "operation" here is used for our "binary operation"; that the set S is not explicitly stated but could be any one of W, N, I, or R; and that "grouping principle" means "associative property".)

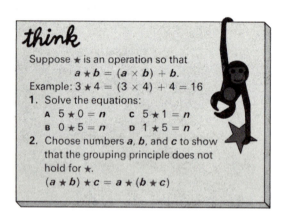

13.3 CLOCK ARITHMETIC

The binary operations of addition and multiplication on the set of whole numbers can be modified to apply to what is called **clock arithmetic** in many primary-school mathematics texts. Clock arithmetics (or, in more technical language, **modular arithmetics**) provide us with mathematical systems that possess many of the properties of ordinary arithmetic and thus serve as excellent tools for reviewing important arithmetic concepts in a new setting.

Consider, then, how we do some problems involving time on an ordinary 12-hour clock. If the time is now 9 A.M., then 2 hours later, the time will be (9 + 2) A.M. = 11 A.M. Here only ordinary addition is involved. But suppose we ask what the time will be 5 hours from 9 A.M. The answer, of course, would be 2 P.M. (It would be reported as 14 hours on the 24-hour clock used, for example, in the armed services and by international airlines, but here we are assuming a 12-hour clock.) Thus the correct "addition" involved can be symbolized by

$$9 \oplus 5 = 2$$

where the plus sign with a circle around it has been used to indicate that this is not ordinary addition but, rather, addition in 12-clock arithmetic (or, in more technical terms, addition **modulo** 12). Clearly, to do addition of whole numbers

13.3 Clock Arithmetic

modulo 12, we compute a sum as in ordinary addition and then find the remainder when that sum is divided by 12. Thus

$5 + 2 = 7$, $7 \div 12 = 0$ r 7,   and so $5 \oplus 2 = 7$
$5 + 10 = 15$, $15 \div 12 = 1$ r 3,   and so $5 \oplus 10 = 3$

Figure 13.1 shows the introduction of clock arithmetic in a fifth-grade text.

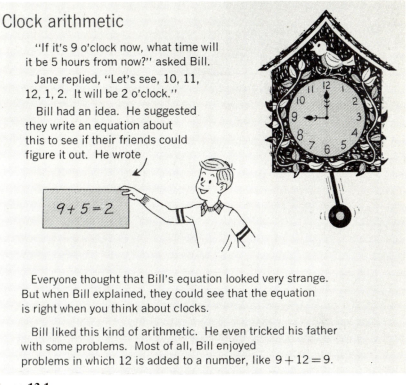

Figure 13.1

After this motivation for clock arithmetic is presented, it is customary to consider in more detail n-clock arithmetic for smaller values of n such as 3, 4, 5, or 6. Here, of course, we compute a sum of whole numbers in n-clock arithmetic by finding the ordinary arithmetic sum and then finding the remainder when that sum is divided by n.

If $n = 5$, an addition table modulo 5 will, as you should check, look like this:

$\oplus$	0	1	2	3	4
0	0	1	2	3	4
1	1	2	3	4	0
2	2	3	4	0	1
3	3	4	0	1	2
4	4	0	1	2	3

322   Binary Operations   **13.3**

What about multiplication in 5-clock arithmetic? We can use the additive definition of multiplication as in ordinary arithmetic. Thus, for example, just as in ordinary arithmetic

$$3 \times 4 = (4 + 4) + 4 = 8 + 4 = 12$$

so, in 5-clock arithmetic,

$$3 \otimes 4 = (4 \oplus 4) \oplus 4 = 3 \oplus 4 = 2$$

On the other hand, we can simply perform the ordinary multiplication of whole numbers and then find the remainder when this product is divided by n. Thus in 5-clock arithmetic we have

$$2 \otimes 2 = 4 \qquad \text{because } 2 \times 2 = 4 \quad \text{and} \quad 4 \div 5 = 0 \text{ r } 4$$

and

$$4 \otimes 4 = 1 \qquad \text{because } 4 \times 4 = 16 \quad \text{and} \quad 16 \div 5 = 3 \text{ r } 1$$

Using either procedure, we can construct a multiplication table modulo 5 as shown below (which, again, you should check):

$\otimes$	0	1	2	3	4
0	0	0	0	0	0
1	0	1	2	3	4
2	0	2	4	1	3
3	0	3	1	4	2
4	0	4	3	2	1

We now have a system, commonly denoted by J_5 (12-clock arithmetic is J_{12}, 4-clock is J_4, etc.) on which are defined two binary operations, "$\oplus$" and "$\otimes$". It "looks" a good deal like the system of whole numbers with the binary operations of "$+$" and "$\times$" being replaced by "$\oplus$" and "$\otimes$", respectively, except that, of course, it is a finite system rather than an infinite one. We will show, however, that it really more closely resembles the system of rational numbers (including negative rational numbers) under the operations of "$+$" and "$\times$"! More specifically, we will see that J_5, like the system of rational numbers, R, is an example of a field.

Fields were defined in Section 10.9. In terms of "$\oplus$" and "$\otimes$" this definition becomes:

J_5 is a field if, for all r, s, and t in J_5, it is true that

Addition   **Multiplication**

1. Closure properties

$$r \oplus s \in J_5 \qquad\qquad\qquad r \otimes s \in J_5$$

2. Commutative properties

$$r \oplus s = s \oplus r \qquad\qquad\qquad r \otimes s = s \otimes r$$

13.3 **Clock Arithmetic** **323**

3. Associative properties

$$r \oplus (s \oplus t) = (r \oplus s) \oplus t \qquad\qquad r \otimes (s \otimes t) = (r \otimes s) \otimes t$$

4. Existence of identities

$$r \oplus 0 = 0 \oplus r = r \qquad\qquad r \otimes 1 = 1 \otimes r = r$$

5. Existence of inverses (to be discussed later)

Distributive Property

$$6. \quad r \otimes (s \oplus t) = (r \otimes s) \oplus (r \otimes t)$$

Let us now see whether J_5 has all of the properties listed.

1. Since only 0, 1, 2, 3, and 4 appear in our addition and multiplication tables, properties 1 hold in J_5.

2. As is true for ordinary addition and multiplication of whole numbers, the symmetry of the addition and multiplication tables about the upper-left-to-lower-right diagonal tells us that both "$\oplus$" and "$\otimes$" are commutative binary operations.

3. Associativity is not as easy to see as commutativity. We can, however, check some cases to convince ourselves that both "$\oplus$" and "$\otimes$" are associative operations. Here is one case of each; others are asked for in the exercises:

$$2 \oplus (3 \oplus 4) = 2 \oplus 2 = 4 \qquad\qquad 2 \otimes (3 \otimes 4) = 2 \otimes 2 = 4$$

and also   and also

$$(2 \oplus 3) \oplus 4 = 0 \oplus 4 = 4 \qquad\qquad (2 \otimes 3) \otimes 4 = 1 \cdot \otimes 4 = 4$$

4. Our tables of addition and multiplication in J_5 show clearly that properties 4 hold.

6. Again, the distributive property is not as easy to verify as some of the other properties. Here is one example illustrating that the distributive property holds; others are asked for in the exercises:

$$2 \otimes (3 \oplus 4) = 2 \otimes 2 = 4$$

and also

$$(2 \otimes 3) \oplus (2 \otimes 4) = 1 \oplus 3 = 4$$

Finally we come to properties 5 which, in the listing of properties of the rational numbers given in Section 10.9, read:

Addition   **Multiplication**

$$r + (-r) = (-r) + r = 0 \qquad\qquad \text{If } r = \tfrac{a}{b} \neq 0, \text{ then } \tfrac{a}{b} \times \tfrac{b}{a} = 1$$

Clearly these properties do not hold exactly as stated above in J_5; $J_5 = \{0, 1, 2, 3, 4\}$, and so we don't have $-1, -2, -3, -4, \tfrac{2}{3}, \tfrac{3}{4}$, etc., as elements of J_5.

324   Binary Operations   **13.3**

However, properties 5 can be restated in the following form, which applies to both the system of rational numbers and J_5, and, in fact, to any field F. This restatement runs as follows:

Addition	**Multiplication**
For any $r \in F$, there exists an $s \in F$ such that $r + s = s + r = 0$. (Existence of additive inverses)	For any $r \in F$ with $r \neq 0$, there exists a $t \in F$ such that $r \times t = t \times r = 1$. (Existence of multiplicative inverses)

Now for r a rational number we have $s = -r$ and, for $r = \frac{a}{b} \neq 0$, $t = \frac{b}{a}$. For J_5 we note that:

If $r = 0$, then $s = 0$, because $0 \oplus 0 = 0$;

thus 0 is the additive inverse of 0;

If $r = 1$, then $s = 4$, because $1 \oplus 4 = 4 \oplus 1 = 0$;

thus 4 is the additive inverse of 1;

If $r = 2$, then $s = 3$, because $2 \oplus 3 = 3 \oplus 2 = 0$;

thus 3 is the additive inverse of 2;

If $r = 3$, then $s = 2$, because $3 \oplus 2 = 2 \oplus 3 = 0$;

thus 2 is the additive inverse of 3; and

If $r = 4$, then $s = 1$, because $4 \oplus 1 = 1 \oplus 4 = 0$;

thus 1 is the additive inverse of 4. Similarly we note that:

If $r = 1$, then $t = 1$, because $1 \otimes 1 = 1$;

thus 1 is the multiplicative inverse of 1;

If $r = 2$, then $t = 3$, because $2 \otimes 3 = 3 \otimes 2 = 1$;

thus 3 is the multiplicative inverse of 2;

If $r = 3$, then $t = 2$, because $3 \otimes 2 = 2 \otimes 3 = 1$;

thus 2 is the multiplicative inverse of 3; and

If $r = 4$, then $t = 4$, because $4 \otimes 4 = 1$;

thus 4 is the multiplicative inverse of 4.

Thus every element r in J_5, like every element in R, has an additive inverse and, if $r \neq 0$, a multiplicative inverse. The inverses of elements of R are also elements of R and the inverses of elements of J_5 are also elements of J_5.

Hence we conclude that J_5 is a field under the operations of "$\oplus$" and "$\otimes$".

Chapter Test 325

Exercises 13.3

1. Further verify the associative property of addition in J_5 by checking that the following equalities hold.
   a) $1 \oplus (2 \oplus 3) = (1 \oplus 2) \oplus 3$ **b)** $2 \oplus (1 \oplus 4) = (2 \oplus 1) \oplus 4$

2. Do as in Exercise 1 for the associative property of multiplication in J_5.
   a) $2 \otimes (3 \otimes 3) = (2 \otimes 3) \otimes 3$ **b)** $4 \otimes (2 \otimes 3) = (4 \otimes 2) \otimes 3$

3. Do as in Exercise 1 for the distributive property in J_5.
   a) $4 \otimes (2 \oplus 3) = (4 \otimes 2) \oplus (4 \otimes 3)$ **b)** $3 \otimes (2 \oplus 4) = (3 \otimes 2) \oplus (3 \otimes 4)$

4. Construct addition and multiplication tables for J_6.

5. Investigate which of the field properties hold in J_6.

6. Make a conjecture as to conditions on n for J_n to be a field.

Chapter Test

In preparation for this test you should review the meaning of binary operation (317) and clock (modular) arithmetic (320).

    You should also review how to:

1. Test whether a given operation on a set is a binary operation (317–318);
2. Determine whether a given binary operation on a set is commutative or associative (319);
3. Perform additions and multiplications in a clock arithmetic (320–322).

1. $*$ is said to be a binary operation on a set S if, whenever $a \in S$ and $b \in S$, then $a * b$ _____ .

2. If $*$ is a binary operation on a set S, then $*$ is a commutative operation if, for all a and b in S, _____ .

3. If $*$ is a binary operation on a set S, then $*$ is an associative operation if, for all a, b, and c in S, _____ .

4. Let $S = W$ and define, for a and b in S, $a * b = a + b - 1$. Why is $*$ not a binary operation on S?

5. Let $S = I$ and define, for a and b in S, $a * b = a + b - 1$. Show that $*$ is a binary operation on S. Is $*$ a commutative operation on S? An associative operation?

6. Construct addition and multiplication tables for J_3 and determine whether or not J_3 is a field.

7. Do as in Exercise 6 for J_4.

326 Binary Operations

TEST ANSWERS

1. $\in S$
2. $a * b = b * a$
3. $a * (b * c) = (a * b) * c$
4. Because if $a = b = 0$, $a * b = 0 + 0 - 1 = -1 \notin S$.
5. If $a \in I$ and $b \in I$, then $a + b - 1 \in I$ so that $*$ is a binary operation on $S = I$. Since $a * b = a + b - 1$ for all integers a and b and $b * a = b + a - 1$, it follows that $*$ is a commutative operation on S since $a + b - 1 = b + a - 1$ for all integers a and b. Now $a * (b * c) = a * (b + c - 1) = [a + (b + c - 1)] - 1 = (a + b + c) - 2$. Also $(a * b) * c = (a + b - 1) * c = [(a + b - 1) + c] - 1 = (a + b + c) - 2$. Hence $a * (b * c) = (a * b) * c$ for all integers a, b, and c, so that $*$ is also an associative operation.

6.

$\oplus$	0	1	2
0	0	1	2
1	1	2	0
2	2	0	1

$\otimes$	0	1	2
0	0	0	0
1	0	1	2
2	0	2	1

J_3 is a field. The commutative, associative, and distributive properties can be checked as in the text for J_5. The existence of additive and multiplicative inverses is shown by

$0 \oplus 0 = 0$ $1 \otimes 1 = 1$

$1 \oplus 2 = 0$ $2 \otimes 2 = 1$

$2 \oplus 1 = 0$

7.

$\oplus$	0	1	2	3
0	0	1	2	3
1	1	2	3	0
2	2	3	0	1
3	3	0	1	2

$\otimes$	0	1	2	3
0	0	0	0	0
1	0	1	2	3
2	0	2	0	2
3	0	3	2	1

All the properties of a field hold in J_4 except for the existence of multiplicative inverses; there is no element b of J_4 such that $2 \otimes b = 1$. Hence J_4 is not a field.

An application of modular number systems

by Julia Adkins

Reprinted from *The Arithmetic Teacher*, Vol. 15 (1968), pp. 713–714.

Some mathematics books for Grades 4–6 include a section on modular number systems (also called modular arithmetic or clock arithmetic). A simple application of this type of material is described below.

As a child, I often played games in which we "counted off" or recited rhymes to determine who would be chosen "It." Let us examine this method of selection to determine how one might "rig the game" so as to predetermine the chosen person.

Example 1. We used to say the sentence, "My mother told me to take this one." The person pointed to as the last word was pronounced was the person chosen. The sentence we recited has eight words. Assume that there are three children in the group.

$$8 \equiv 2 \quad (\text{mod } 3)$$

Therefore, make certain that the child you wish to select is in the second position for counting.

0	0	0
My	Mother	told
me	to	take
this	one	

Example 2. A rhyme we often used was the following one:

Eenie, meenie, mynie, moe.
Catch a big one by his toe.
If he hollers,
Let him go.
Eenie, meenie, mynie, moe.

Again, the person pointed to as the last word was spoken was the chosen person.

This verse has twenty-one words. Assume that there are five children in the group.

$$21 \equiv 1 \quad (\text{mod } 5)$$

Then, make certain that the child you wish to select is in the first position (or is the child with whom you begin the counting).

0	0	0	0	0
Eenie	meenie	mynie	moe	Catch
a	big	one	by	his
toe	If	he	hollers	let
him	go	Eenie	meenie	mynie
moe				

328 Binary Operations

Another variation that can be changed often enough that the children might not "catch on" would be the following one.

Example 3. Assume that there are seven children who have volunteered to perform a certain task, but you would prefer that Jim do it for you. You must make the selection in a way that "seems fair" to all the children. Have the children stand in a row before you. Announce that you will count to a certain number (and tell the students what the number is), and the child you point to as you say that number will be the chosen child. Assume that Jim is in the fourth position in the row. Then choose a number congruent to four in the modulo 7 system, and you will be certain that Jim is chosen.

$$4 \equiv 11 \equiv 18 \equiv 25 \qquad (\text{mod } 7)$$

0	0	0	Jim	0	0	0
1	2	3	4	5	6	7
8	9	10	11	12	13	14
15	16	17	18	19	20	21
22	23	24	25	and so on		

A variation of the technique described in Example 3 could be used as a game to play with the students.

Example 4. Allow the class to select the number of students to stand in a row before you and to determine which child will be the last one "pointed to" when you complete your counting. You reserve the right to select the number for the "counting off" (tell the students the number you have chosen). Assume that the class decides to have a row of eight students and that Sally is to be the "selected" one. Also assume that Sally is in the third position in the row. You should choose a number that is congruent to 3 (mod 8).

$$3 \equiv 11 \equiv 19 \equiv 27 \equiv 35 \qquad (\text{mod } 8)$$

0	0	Sally	0	0	0	0	0
1	2	3	4	5	6	7	8
9	10	11	12	13	14	15	16
17	18	19	20	21	22	23	24
25	26	27	28	29	30	31	32
33	34	35	and so on				

You may wish to reveal the above technique to your students as an application of modular number systems—or you may wish to keep it as *your* secret for your own use!

REFERENCES FOR FURTHER READING

1. Ginther, John L., "Some activities with operation tables," *AT,* **15** (1968), pp. 715–717. Describes work with the L.C.M. and G.C.D. as binary operations on *N*.

2. Muller, Francis J., "Modular Arithmetic," Chapter 4 of *Enrichment Mathematics for the Grades*, National Council of Teachers of Mathematics, Reston, Va. (1963).

3. Oosse, William J., "Properties of operation: A meaningful study," *AT,* **16** (1969), pp. 271–275. Gives examples of various binary operations and their properties.

Chapter 14

Functions

14.1 THE "GUESS MY RULE" GAME

I have emphasized throughout this book the importance of mathematical games, puzzles, and activities in a good primary-school mathematics program. However, one very useful and stimulating activity, the "guess my rule" game, I have deliberately postponed until now, since it fits in so closely with the topic of this chapter.

We can think of all of the various mathematical operations in terms of "rules." For example, the "add 1 rule" would be an instruction to add 1 to any given number. Once children understand the idea of a rule, they are ready to play the "guess my rule" game. For example, you may have in mind the "add 3 rule." Now you invite your pupils in, say, a first-grade class to give you numbers. Suppose one child gives you 4. You say "seven." Another child gives you 8. You say "eleven." Very soon some child will say "You arc using the 'add three rule'."

Sometimes, as shown in Fig. 14.1, taken from a fourth grade book, the "guess my rule" game is called "the function game."

Investigating the Ideas

Kay and Paul were playing the function game.
When Kay said 2, Paul answered 5.
When Kay said 3, Paul answered 7.
When Kay said 5, Paul answered 11.

When Kay said 6, what do you think Paul answered?
What is Paul's rule?

Figure 14.1

330 **Functions** 14.2

Children do enjoy such activity at all grade levels, and obviously get a lot of practice in arithmetic in trying to guess the rule. The exercises below will clearly indicate how the activity can be made quite a challenge to anyone.

Exercises 14.1

Imagine that you are a pupil and that your teacher responds as shown to the numbers that you supply. Guess the rule that the teacher has in mind.

1. You	Teacher		2. You	Teacher		3. You	Teacher
3	10		2	6		2	5
5	16		0	2		3	10
8	25		5	12		8	65
0	1		3	8		4	17

4. You	Teacher		5. You	Teacher		6. You	Teacher
1	4		-2	-7		0	1
3	14		-1	0		2	2
5	24		1	2		3	$2\frac{1}{2}$
4	19		2	9		4	3
			3	28		5	$3\frac{1}{2}$

7. You	Teacher		8. You	Teacher		9. You	Teacher
1	7		2	8		1	12
3	19		5	35		0	6
0	4		0	0		4	42
6	52		3	15		3	30

10. You	Teacher		11. You	Teacher		12. You	Teacher
5	60		1	0		1	1
3	32		2	1		2	0
1	12		5	0		3	0
7	96		8	1		4	1
			9	0		5	0
						6	1

14.2 FUNCTION "MACHINES"

The concept of a "rule" underlies the concept of a function, as we will discuss in the next section. Another activity preparatory to a formal presentation of this basic mathematical concept is by way of function "machines." For example, the "add three" rule might be thought of as the machine shown in Fig. 14.2. Whatever number we put into the machine, the machine adds 3 to it and puts the result out at the other end.

Our inputs and/or outputs need not be numbers. For example, we might consider a machine, as shown in Fig. 14.3, into which we insert the name of a U.S. president, and receive as output, the date of his birth.

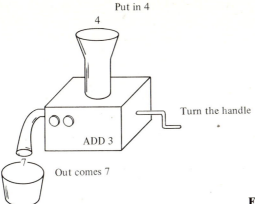

Figure 14.2

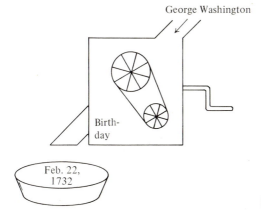

Figure 14.3

It is always necessary that the function machine operate unambiguously. For example, suppose we were to try to construct a function machine which, when fed a number x, produces a number y such that $y^2 = x$. Then, given the number 4, the machine would not know whether to produce 2 or -2 since $2^2 = (-2)^2 = 4$. No such indecision is allowed for a function machine! (By saying that the machine is to produce a *positive* number y such that $y^2 = x$ we can, of course, make it possible to have the machine be a function machine.)

Note that, on the other hand, it is *not* necessary that we always get a different output for every distinct input. For example, the "squaring machine" will have 4 as an output for the input -2 as well as for the input 2.

Figure 14.4, from a third-grade book, shows a more sophisticated function machine than those shown previously. (Personally, I prefer the cruder looking ones, as I think they show more vividly the idea of input versus output!)

The following exercises are taken from a fourth-grade text. They illustrate the concept of **composition** of functions, where the output of one function machine becomes the input of another.

332   **Functions**   14.2

The function machine

Study the pictures to see how the function machine works. A record of the operations of the function machine is shown below. Find the missing numbers.

Figure 14.4

Exercises 14.2

The function machine

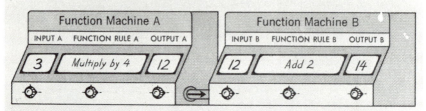

Function machine A is connected to function machine B. The output number from machine A becomes the input number for machine B. We put in 3. Machine A operates. Machine B operates. We get 14.

EXERCISES

Think about connected function machines and give the numbers you think should go in the gray spaces.

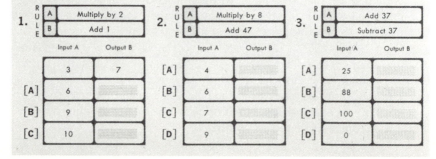

14.3 **Definition of a Function** **333**

4.
5.
6.

Input A	Output B		Input A	Output B		Input A	Output B
4	4		4	6		3	10
9	9		6	7		5	26
[B] 6		[B]	2		[B]	4	
[C] 8		[C]	10		[C]	8	

14.3 DEFINITION OF A FUNCTION

There are many places in mathematics where rules occur by which one number is determined by another. (Indeed, we will also see examples of "rules" in geometry in Chapter 15, whereby one geometric figure determines another.) Mathematicians refer to such rules as "functions." Before we consider a formal definition of a function, however, let us look more closely at the "add 3" rule or, equivalently, the "add 3" machine described in the previous two sections. Note that it was not stated explicitly what numbers could be given by the children or put into the machine. But since the activity is a simple one, it is reasonable to suppose that it might be carried out in first grade so that only whole numbers would be involved. In any event, the set of numbers (or geometric figures, names, etc.) which we agree to use as inputs is called the **domain** of the function, and the resulting set of outputs is called the **range** of the function.

With this background, we can now define a function:

A **function** f consists of a rule, a set D (the domain of f), and a set R (the range of f) such that when $x \in D$ is given, a unique element $y \in R$ is determined by the rule. Furthermore, every $y \in R$ is the result of applying this rule to one or more elements of D.

In terms of the machine analogy, the function f consists of a rule (which corresponds to what happens inside the machine), the domain (the set of things that are put into the machine), and the range (the set of things that come out of the machine).

Given f and $x \in D$, it is customary to refer to the element $y \in R$ determined by x as $f(x)$. (This is read "f of x" and is *not* a symbol for "f times x".) For example, for our "add 3" function, we have $f(x) = x + 3$, so that

$$f(1) = 1 + 3 = 4, \qquad f(2) = 2 + 3 = 5, \qquad \text{etc.,}$$

and for our "birthday" function, we have f(George Washington) = Feb. 22, 1732 and f(Abraham Lincoln) = Feb. 12, 1809, etc.

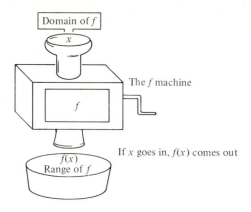

Figure 14.5

Using our machine analogy once more, we can picture our definition and terminology as shown in Fig. 14.5. Note that, using the set-builder notation, we can write: range of $f = \{f(x): x \in D\}$. As a convenient "shorthand" for "f is a function such that $f(x) = x + 3$", we sometimes write

$$f: f(x) = x + 3$$

Although it is common to use the letter f for a function and x for an element of the domain, other letters can be, and are, used. Thus $f: f(x) = x + 3$ could just as well be described by

$$g: g(x) = x + 3, \qquad f: f(t) = t + 3, \quad \text{or} \quad h: h(y) = y + 3;$$

—all these notations refer to the same "add 3" rule. Similarly,

$$f: f(x) = x^2, \qquad g: g(x) = x^2, \quad \text{and} \quad h: h(t) = t^2$$

all describe the "squaring" function.

Notice the distinction between a function f (a machine with its inputs and outputs) and $f(x)$ (an output). f is the function; $f(x)$ is a **functional value.**

In most of elementary mathematics, the domain of a function is some set of real numbers and $f(x)$ is given as some equation such as $f(x) = x + 2$, $x^2 - 2x + 3$, sin x (in trigonometry), log x (in the study of logarithms), etc. It is important to note, however, that this is not a requirement for a function. For example, we might take as our domain the set of natural numbers and define a function p by $p(x) =$ number of primes less than or equal to x. Then $p(1) = 0$, $p(2) = 1$, $p(3) = 2$, $p(4) = 2$, and $p(5) = 3$, since the first three primes are 2, 3, and 5. Here the range of p is the set of whole numbers.

Very often the domain of a function is not given explicitly. For example, it is quite common to see statements such as "consider the function f where $f(x) = 2x + 1$" with no domain specified. What is ordinarily assumed, then, is that the domain is taken to be the largest possible set of real numbers for which $f(x)$ is defined. In the case where $f(x) = 2x + 1$, this would be the entire set of real numbers. On the other hand, if $f(x) = 1/x$, it would be the set of nonzero real numbers, since $1/0$ is not defined, and if $f(x) = \sqrt{x}$, it would be the set of

14.3　　　　　　　　　　　　　　　　　　　**Definition of a Function**　　**335**

nonnegative real numbers, since the square root of a negative number is not a real number.

Exercises 14.3

1. If $f(x) = 2x + 3$, what is $f(0)$? $f(-1)$? $f(\sqrt{2})$?

2. If $f(x) = x^2$, what is $f(0)$? $f(1)$? $f(-1)$? $f(\sqrt{2})$?

3. The domain of the function f is $\{1, 2, 3\}$ and $f(x) = x^2 + 1$. What is the range of f?

4. The domain of the function g is $\{0, 1, 4\}$ and $g(x) = 3\sqrt{x}$. What is the range of g?

5. The function h has domain the set of real numbers and $h(t) = 0$ if t is a rational number, whereas $h(t) = 1$ if t is an irrational number. What is $h(2)$? $h(\sqrt{2})$? $h(\frac{1}{2})$? $h(\pi)$? $h(-2.14)$? What is the range of h?

6. A function f has domain the set of real numbers and range $\{1\}$. Write an equation for $f(x)$. (f is an example of a constant function.)

7. The domain of $f: f(x) = 2x$ is the set of whole numbers. What is the range of f?

8. The domain of $g: g(x) = |x|$ is the set of rational numbers. What is the range of g?

9. If $f(x) = 2x - 3$ and $g(x) = 3x + 4$, for what value of x is $f(x) = g(x)$?

10. If $g(t) = 3t + 5$ and $h(t) = t - 2$, for what value of t is $g(t) = h(t)$?

11. Do the following exercises taken from a sixth-grade text.

● *Let's explore a special function.*

Investigating the Ideas

Tim thought of a special function. Since he could not use addition, subtraction, multiplication, or division to describe the function rule, Tim used the symbol $\overset{\rightarrow}{\underset{\leftarrow}{n}}$. Study the tables below to understand Tim's function.

THE FUNCTION MACHINE

FUNCTION RULE $\overset{\rightarrow}{\underset{\leftarrow}{n}}$

n	$f(n)$
$5\frac{1}{4}$	5

A Function Rule $\overset{\rightarrow}{\underset{\leftarrow}{n}}$

n	$f(n)$
5	5
$5\frac{1}{8}$	5
$5\frac{3}{8}$	5
$5\frac{5}{8}$	6
$5\frac{7}{8}$	6

B Function Rule $\overset{\rightarrow}{\underset{\leftarrow}{n}}$

n	$f(n)$
$8\frac{1}{4}$	8
$8\frac{1}{2}$	9
$8\frac{3}{4}$	9
9	9
$9\frac{1}{4}$	9

C Function Rule $\overset{\rightarrow}{\underset{\leftarrow}{n}}$

n	$f(n)$
$\frac{8}{3}$	3
$\frac{7}{2}$	4
$\frac{1}{4}$	0
$\frac{17}{2}$	9
$23\frac{5}{6}$	24

336   Functions   14.4

 Can you discover Tim's function rule and use the rule to make a table of your own?

Discussing the Ideas

1. Describe Tim's rule in your own words.
2. Explain how to find the missing numbers.

   A $\overrightarrow{3} = \text{||||}$ B $\overrightarrow{3\frac{1}{4}} = \text{||||}$ C $\overrightarrow{3\frac{1}{2}} = \text{||||}$ D $\overrightarrow{3\frac{3}{4}} = \text{||||}$ E $\overrightarrow{4\frac{7}{16}} = \text{||||}$

3. Explain how to find the sums.

   A $\overrightarrow{1\frac{1}{3}} + \overrightarrow{2\frac{1}{3}} = \text{||||}$ B $\overrightarrow{2\frac{3}{4}} + \overrightarrow{3\frac{1}{2}} = \text{||||}$

4. Can you explain why the two sums below are not the same?

   A $\overrightarrow{5\frac{3}{8}} + \overrightarrow{4\frac{3}{8}} = 9$ B $\overrightarrow{5\frac{3}{8}} + \overrightarrow{4\frac{3}{8}} = 10$

14.4 FUNCTIONS AS MAPPINGS

Functions are often referred to as **mappings** and illustrated by a diagram such as that shown in Fig. 14.6. We say that f **maps** x onto $f(x)$. For example, consider the function $f: f(x) = 2x + 1$ with domain $\{0, 1, 2\}$. We can diagram this function as shown in Fig. 14.7.

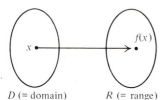

Figure 14.6   Figure 14.7

Such diagrams illustrate very clearly that to every $x \in D$ there is paired off an $f(x) \in R$. Note, in particular, that in a diagram of a function, we will never find two arrows from a single point in D to two different points in R. Thus the diagram shown in Fig. 14.8 cannot be the diagram of a function. On the other hand, it is perfectly possible to have two or more arrows from D terminating at the same point in R. This is true both for $g: g(x) = x^2$ and the "prime" function p of Section 14.3 as is illustrated in Fig. 14.9.

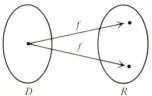

Figure 14.8

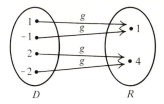
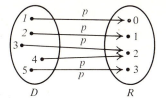

Figure 14.9

If only one arrow from D terminates at each point of R in the diagram of a function, we say that the function is **one-to-one,** whereas if there is at least one point in R such that two or more arrows from D terminate at this point, we say that the function is **many-to-one.** For example, the function $f: f(x) = 2x + 1$ is one-to-one, whereas the function $g: g(x) = x^2$ and the "prime" function p are both many-to-one. (In the case of g, two-to-one.)

The range of a function is determined by the domain, but, as Fig. 14.10 shows, there can be more than one function for a given domain and range.

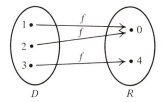
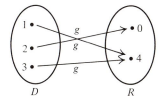

Figure 14.10

The concept of the composition of two functions was mentioned at the end of Section 14.2. Figure 14.11 illustrates the composition of the functions $f: f(x) = x + 1$ and $g: g(x) = x - 1$, where the domain of f is $\{1, 2, 3\}$ and the domain of g is the range of f. Note that the effect of first applying the mapping f and then applying the mapping g is to send 1 onto 1, 2 onto 2, and 3 onto 3, i.e., to send each element of $\{1, 2, 3\}$, the domain of f, onto itself. When any two functions f and g have the property, as in our example, that

$$x \xrightarrow{f} y \xrightarrow{g} x$$

for all x in the domain of f, we say that the function g is the **inverse** of the function f.

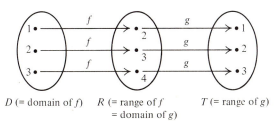

D (= domain of f) R (= range of f T (= range of g)
                        = domain of g)

Figure 14.11

Exercises 14.4

1. The domain of a function h is $\{1, 2\}$ and the range is $\{0, 3\}$. How many functions can you find having this domain and range? Illustrate each with a diagram. Are they all one-to-one?

2. The same as Exercise 1, except that the domain is $\{1, 2, 3\}$, the range is $\{0, 4\}$, and it is given that 1 maps onto 0.

3. Consider the statement: "The domain of a certain function has 9 members and the range has 10 members." Can this possibly be true? What happens if you try to draw a diagram for such a function?

4. A function f has $\{1, 2, 3, 4\}$ as both domain and range. Draw a diagram for such an f such that no member of the domain maps onto itself.

5. Draw a diagram like that of Fig. 14.11 to show the composition of the function f: $f(x) = x^2$ and g: $g(x) = 2x$. The domain of f is $\{0, 1, -1\}$ and the domain of g is the range of f.

6. Do as in Exercise 5, but with f: $f(x) = x - 1$, $g(x) = x^2 - 1$, and the domain of $f = \{1, 2, 3\}$.

7. Which, if either, of the two diagrams below is the diagram of a function?

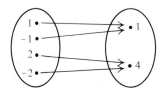 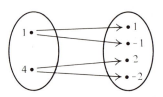

8. Show by a diagram like that of Fig. 14.11 that if f: $f(x) = 2x$ with domain $\{4, 7, 10\}$, then g: $g(x) = \frac{1}{2}x$ with domain the range of f is the inverse of f.

9. Show by a diagram like that of Fig. 14.11 that if h: $h(x) = \sqrt{x}$ with domain $\{1, 4, 9\}$, then k: $k(x) = x^2$ with domain the range of h is the inverse of h.

10. Show by a diagram like that of Fig. 14.11 that if s: $s(x) = x^2$ with domain $\{-1, 1, 2, -2\}$, then r: $r(x) = \sqrt{x}$ with domain the range of s is *not* the inverse of s.

14.5 FUNCTIONS AS SETS OF ORDERED PAIRS

Since with each x in the domain D of a function f is associated a unique element, $f(x)$, of its range, we see that every $x \in D$ determines an ordered pair $(x, f(x))$. For example, if $D = \{1, 2, 3\}$ and $f(x) = 2x + 1$, we get the set of ordered pairs $\{(1, 3), (2, 5), (3, 7)\}$—one ordered pair for each element of D. In general, for any function f with domain D we get a set of ordered pairs $\{(x, y): x$ in D and $y = f(x)\}$.

Such a set of ordered pairs gives us the same information as the "rule", domain, and range defining the function, and so it is just as satisfactory to *define* the function as being a set of ordered pairs as it is to define it in terms of a rule. Indeed, many mathematicians take this point of view of a function. Thus, for example, instead of talking about the function f with domain $\{1, 2, 3\}$ such that $f(x) = 2x + 1$, they write

$$f = \{(1, 3), (2, 5), (3, 7)\}$$

Similarly, $f: f(x) = x^2$ with domain the set of real numbers is written

$$f = \{(x, y): x \text{ a real number and } y = x^2\}$$

or as

$$f = \{(x, x^2): x \text{ a real number}\}$$

Although every function can be regarded as a set of ordered pairs, not every set of ordered pairs is a function. For example, consider the set $\{(1, 2), (2, 3), (1, 5)\}$. Figure 14.12 shows a diagram of this set like the diagrams of Section 14.4. But since there are two arrows from 1, we know that this is not a diagram of a function.

Any set of ordered pairs, however, defines what is called a **relation**; $\{(1, 2), (2, 3), (1, 5)\}$ is a relation, and Fig. 14.12 is a diagram of a relation. Functions, then, are special kinds of relations.

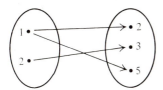

Figure 14.12

Exercises 14.5

1. Which of the following relations are functions?
    a) $\{(1, 2), (2, 3), (3, 4), (4, 5), (5, 1)\}$
    b) $\{(0, 1), (1, 2), (2, 2), (1, 3), (0, 4)\}$
    c) $\{(2, 0), (1, 0), (3, 0), (-1, 0)\}$
    d) $\{(x, x^2 + 1): x \text{ a real number}\}$
    e) $\{(x, y): x \text{ a real number and } x + y = 7\}$
    f) $\{(x, y): x \text{ a real number and } x^2 + y^2 = 25\}$
    g) $\{(x, y): x \text{ a real number and } x^2 + y = 0\}$
    h) $\{(x, y): x \text{ a real number and } x + y^2 = 0\}$
    i) $\{(x, y): x \text{ a real number and } x^3 + y^3 = 100\}$

340 Functions 14.6

2. Write the following functions as sets of ordered pairs using set-builder notation if necessary.
   a) $f: f(x) = x + 3,\quad D = \{4, 5, 6\}$
   b) $g: g(x) = x^2 - 1,\quad D = \{0, 1, -1\}$
   c) $h: h(x) = x^3,\quad D = $ real numbers
   d) $j: j(x) = \sqrt{x},\quad D = $ nonnegative real numbers

3. Write the functions f and g of Exercise 8 of Exercises 14.4 as sets of ordered pairs.

4. Write the functions h and k of Exercise 9 of Exercises 14.4 as sets of ordered pairs.

5. Study the results of Exercises 3 and 4 and recall that g is the inverse of f and k is the inverse of h. Use the results of your study to write the inverses of the following functions given as sets of ordered pairs.
   a) $\{(2, 1), (3, 0), (5, 4)\}$
   b) $\{(-1, 2), (0, 1), (2, 3)\}$
   c) $\{(1, 2), (3, 4), (5, 6), (6, 7)\}$
   d) $\{(-1, 1), (2, -2), (3, -3), (4, -4)\}$

6. Thinking of inverses of functions in terms of sets of ordered pairs, consider the function $\{(1, 1), (-1, 1), (2, 4), (-2, 4)\}$. Why does this function not have an inverse? (Compare your conclusion with that reached in Exercise 10 of Exercises 14.4.)

14.6 GRAPHS OF FUNCTIONS

From your previous work in mathematics you should be familiar with what is called a **coordinate** or **Cartesian plane** as a device for showing ordered pairs of real numbers. Such a plane is shown in Fig. 14.13 with various points corresponding to ordered pairs of real numbers shown on it.

Since we can consider a function as a set of ordered pairs of numbers, we can consider the graph of a function as the set of points corresponding to this set of

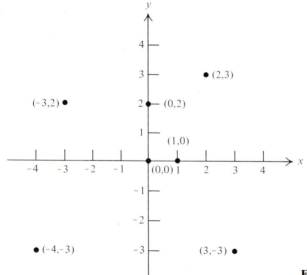

Figure 14.13

ordered pairs. Thus, for example, the graph of the function $f = \{(1, 3), (2, 5), (3, 7)\}$ is as shown in Fig. 14.14.

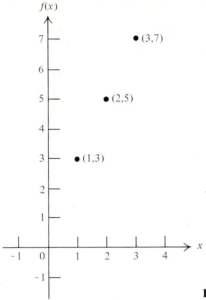

Figure 14.14

Now let us consider the graph of the function $g: g(x) = 2x + 1$. Since $g(1) = 3$, $g(2) = 5$, and $g(3) = 7$, we obtain the ordered pairs $(1, 3)$, $(2, 5)$, $(3, 7)$ of the function f whose graph was shown in Fig. 14.14. Note that the three points corresponding to these ordered pairs seem to lie on a straight line, as shown in Fig. 14.15. If we now calculate some additional ordered pairs of g as given in the

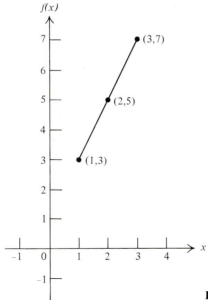

Figure 14.15

table below and plot the points corresponding to these ordered pairs as shown in Fig. 14.16, they, too, seem to lie on this line. It can be shown that all the points that correspond to $(x, 2x + 1)$ for any real number x, do indeed lie on this line.

x	0	−1	−2	−3
$g(x)$	1	−1	−3	−5

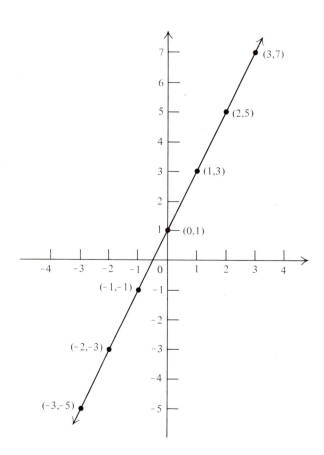

Figure 14.16

Our interest in graphs here is not from the standpoint of studying how graphs of various functions can be drawn but rather in a general characteristic of graphs of functions. Now any set of points in the plane is a set of points corresponding to a *relation*, i.e., a set of ordered pairs. But just as not every set of ordered pairs is a function, so not every set of points in a coordinate plane is the graph of a function. Consider for example the set of points shown in Fig. 14.17. They correspond to the set $\{(-2, 3), (1, 1), (1, 2)\}$ which is not a function (why?) but is a relation.

Now consider Fig. 14.18, which is the graph of a circle of radius 5. It includes the points corresponding to the ordered pairs $(3, 4)$ and $(3, -4)$, and hence it cannot be the graph of a function. Again, however, it is the graph of a relation.

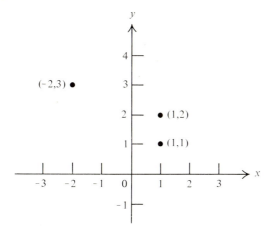

Figure 14.17

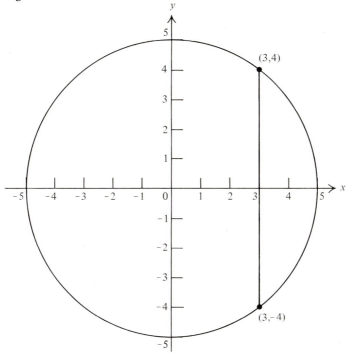

Figure 14.18

344 **Functions** **14.6**

On the other hand, the graph of the half-circle shown in Fig. 14.19 is the graph of a function: every value of x determines a unique value of y. Geometrically, this means that no vertical line intersects the graph in more than one point, whereas, in Figs. 14.17 and 14.18, there exist vertical lines that intersect the graph in more than one point.

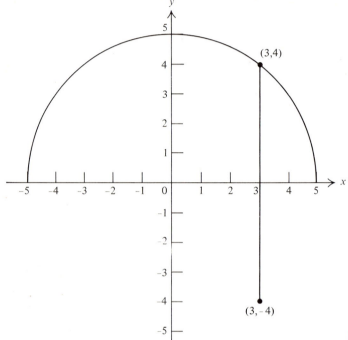

Figure 14.19

Exercises 14.6

1. Determine whether or not the given graph is the graph of a function. If it is, determine the domain and range of the function.

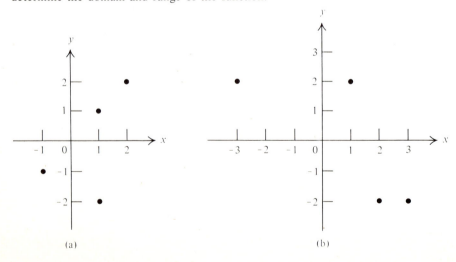

14.6 Graphs of Functions 345

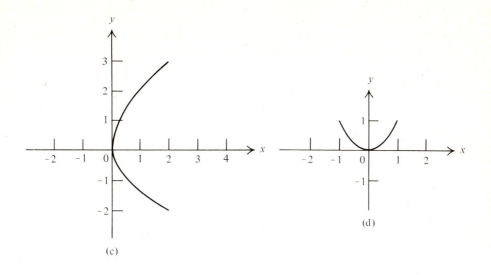

(c)

(d)

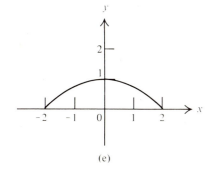

(e)

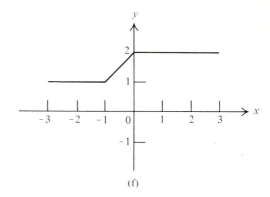

(f)

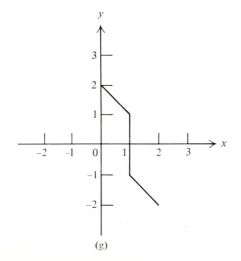

(g)

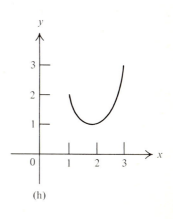

(h)

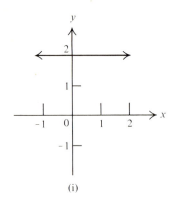
(i)

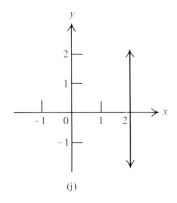
(j)

2. Draw the straight lines corresponding to the following functions. The domain in each case is the set of real numbers.
   a) $f: f(x) = 3x - 2$
   b) $g: g(x) = 2x + 1$
   c) $h: h(x) = x$
   d) $i: i(x) = 3$

Chapter Test

In preparation for this test you should review the meaning of the following terms and symbols:

Function (333)
Domain (333)
Range (333)
Functional value (334)
$f(x)$ (333)
Mapping (336)

One-to-one function (337)
Many-to-one function (337)
Inverse of a function (337)
Relation (339)
Graph of a function (341)

You should also review how to:

1. Invent and play "guess my rule" games (329);
2. Compute functional values (333);
3. Show a function by a mapping diagram (336);
4. Show a function as a set of ordered pairs (338–339);
5. Show a function as a graph (340–343);
6. Determine whether a given relation is a function (339);
7. Determine whether a given graph is a graph of a function (342–344).

1. Guess the rule implied by the following pupil–teacher exchange:
   Pupil   0   2   3   5
   Teacher   5   9   11 15

2. A function f consists of a _____, a set D (called the _____ of the function), and a set R (called the _____ of the function) such that when $x \in D$ is given, a _____ element $y \in R$ is determined by the rule. Furthermore, every $y \in R$ is the result of applying this rule to one or more elements of _____.

3. $f(x)$ is read _____.

Chapter Test 347

4. In set-builder notation, range of f = _____.
5. If $f(x) = x^2 + 2x - 1$, what is $f(0)$? $f(2)$? $f(-1)$?
6. If g is a function with domain $\{2, 0, -1\}$ and $g(x) = 3x - 1$, what is the range of g?
7. Draw a diagram illustrating the function of Exercise 6.
8. Show the function in Exercise 6 as a set of ordered pairs.
9. Which, if either, of the following two relations is a function:

$\{(1, -1), (-1, -1), (2, 0), (-2, 0)\}$ or $\{(1, -1), (2, 0), (1, 1)\}$?

10. Is $\{(x, y): x^4 + y^4 = 100\}$ a function?
11. Which, if either, of the following two graphs is the graph of a function?

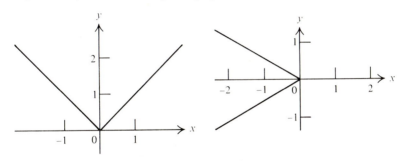

12. Do the following exercises taken from a fifth-grade text.

The example below will help you see how we can graph the number pairs from a function machine table.

function rule
$n + 1$

n	f(n)
0	1
1	2
2	
3	
	5
	6

Coordinates:
(0, 1)
(1, 2)
(2, ▓)
(3, ▓)
(▓, 5)
(▓, 6)

$f(n)$
6 . . . . .
5 . . . M . .
4 . . . . .
3 . N . . . .
2 . . . . .
1 . . . . .
0 └────────→ n
   0 1 2 3 4 5 6

DISCUSSION EXERCISES ─────────────────────────
1. Give the missing coordinates in the example above.
2. Give the coordinate for points M and N in the graph.
3. What do you notice about the set of points in this graph?
4. Using the function rule $f(n) = n + 1$, give the second coordinate for each of the following.
   $(1\frac{1}{2}, \text{▓})$ $(2\frac{1}{2}, \text{▓})$ $(3\frac{1}{2}, \text{▓})$ $(4\frac{1}{2}, \text{▓})$

TEST ANSWERS

1. $x \to 2x + 5$
2. Rule, domain, range, unique, D
3. f of x
4. $\{f(x): x \in D\}$
5. $f(0) = -1, \quad f(2) = 7, \quad f(-1) = -2$
6. $\{5, -1, -4\}$
7.

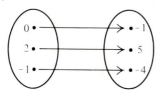

8. $\{(0, -1), (2, 5), (-1, -4)\}$
9. The first one
10. No
11. The lefthand one
12. (1) 3, 4, 4, 5   (2) (4, 5), (2, 3)
    (3) They all lie on a straight line.
    (4) $2\frac{1}{2}, 3\frac{1}{2}, 4\frac{1}{2}, 5\frac{1}{2}$

Let's consider the function!

by Rosemary C. Anderson

Reprinted from *The Arithmetic Teacher*, Vol. 16 (1969), pp. 280–284.

In 1963 the Cambridge Conference on School Mathematics recommended the introduction of the ideas of set and function in the early elementary grades. Other groups have made similar recommendations. Set concepts have received much attention in the last ten years, and these ideas are fairly clear-cut to many elementary school teachers today. But what is really meant by the concept of function, and how is this concept to be implemented in the elementary-school mathematics program? This article will attempt to clarify the meaning of function and point out what is being done and what could be done to emphasize this concept.

What is a Function?

It has been said that the function concept is one of the most far-reaching and important ideas in mathematics. We are concerned with more than just identifying and quantifying; we are concerned also with the relationship of things. A special sort of relationship is called a function.

Unfortunately the word "function" is used in several different ways. Basically, the term does not yet have a definition that is universally accepted. Webster's Third New International Dictionary gives as one definition "any quality, trait, or fact so related to another that it is dependent upon and varies with it." To some mathematicians a function is a graph on the Cartesian plane such that no vertical line cuts it more than once. Others describe a function as a rule that prescribes a unique value of y for a given value of x. Still others would call it a mapping of one set on another in which no two elements of the second set come from the same element of the first. The first set is called the domain, the second the range.

For the purposes of this discussion we will define a function as a set of ordered pairs, no two of which have the same first component. A common awareness of this function relationship is frequently indicated by expressions such as "it depends upon..." For example, we would say the distance traveled by a spaceship depends upon, and thereby is a function of, time and speed of travel. The yield of a corn crop "depends upon" the amount of rainfall. The amount of postage for sending a package by first-class mail "depends upon" the weight of the package. The function concept is an abstraction from many everyday situations in which two changing quantities are related.

The function relation may be expressed by the equation, the formula, and the graph. An equation $x^2 = y$ could be shown by the mapping illustration in Fig. 1.

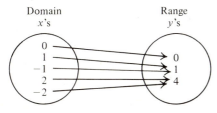

Figure 1

Functions

A formula is a specialized form of the equation. The formula for the area of a rectangle, $A = \ell w$, is a statement of a functional relationship. The area depends upon the measurement of the length and width of the rectangle and varies as the length and width vary.

Graphs are used to express functional relationships also. The following example (Table 1 and Fig. 2) makes use of tabulated data and a graph to describe the function relationship between time and temperature.

Table 1. Hourly Temperatures Measured at a Weather Station

Time in hours—A.M.	1:00	2:00	3:00	4:00	5:00	6:00
Temperature in degrees F.	68	67	65	61	60	62

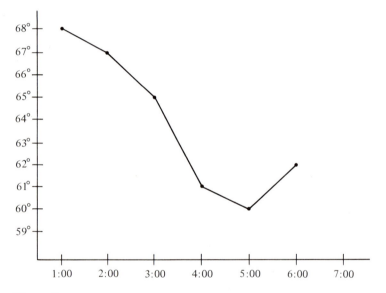

Figure 2

The idea of quantities being functionally related is basic in problem solving. It is these dependencies, because they show the connections between the known and the unknown, that make solutions possible.

Recommendations

In the report of the Cambridge Conference on School Mathematics, the following specific recommendations are made:

> The ideas of set and function should be introduced as soon as possible (K–2):
>
> 1. Number as a property of finite sets.

Let's Consider the Function! **351**

2. The comparison of cardinals of finite sets with emphasis on the fact that the result is independent of which mapping function is used.

3. Numerical functions determined by very simple formulas.

They continue, to suggest the following applications:

Visual display of data on Cartesian coordinates, such as recording growth of seedlings by daily measurement of height, or graphs of time and temperatures for hourly readings of a thermometer.

These suggestions assume a general pattern of pre-mathematics to introduce each topic, to be followed later by as much formal study as may be appropriate.

1. Study of sets, relations and functions.

2. Graphs of relations and functions, both discrete and continuous.

3. Graphs of empirically determined functions.

At this point it seems to be necessary to clarify for the reader the distinctions between relations and functions.

A *relation* is a set of ordered pairs. It could look something like this:

$$\{(1, 2), \quad (5, -3), \quad (1, -3), \quad (5, 1)\}$$

The graph of this relation has no intended pattern or order; the relation is arbitrary and planless. We are familiar with many kinds of relations. For example: Paul is the brother of Jim; Chicago is west of New York.

A *function* is a special sort of relation in which the ordered pairs form a pattern. The set

$$\{(0, 5), \quad (1, 4), \quad (2, 3), \quad (3, 2), \quad (4, 1), \quad (5, 0)\}$$

has an obvious pattern. This is the set of all pairs of positive integers whose sum is 5.

A function is therefore a relation, but a relation is not necessarily a function. A function is a relation in which no two different ordered pairs have the same first member.

The Minnesota Elementary Curriculum Project in Grades K–3 lists as the first important goal of its program to teach the use of a coordinate system. Those who worked on the project feel that as soon as the pupil understands this, he will have a geometrical algorism for graphing the function. Dr. Rosenbloom defines a function as a collection of ordered pairs usually having a definite relationship. For example, an illustration of a function consisting of three pairs such that the second element is twice as large as the first is the following:

$$\{(1, 2), \quad (3, 6), \quad (5, 10)\}$$

Dr. David A. Page of the University of Illinois Arithmetic Project discusses functions in his book, *Number Lines, Functions, and Fundamental Topics.* He distinguishes between a function and a relation that is not a function in the following manner.

The main ideas of a function are: starting numbers, landing numbers, and a rule for finding what landing number corresponds to each starting number. We should also say that one function is not permitted to have two different landing points corresponding to one starting point.

352 Functions

Page recommends that we teach number-line-rule notation to elementary school children and avoid the word "function" completely. He also presents a simplified notational system.

| *Example* | **Page's Notation** | **Conventional Notation** |

Rule for the function:

$$\Box' \to 2\,\Box + 5 \qquad\qquad f(x) = 2x + 5$$

The Madison Project, under the leadership of Dr. Robert B. Davis, specifically includes the following topics:

1. The concept of a function

2. The graph of a function

Dr. Davis states, however, that the Madison Project materials provide a supplementary program, not a substitute for the standard program.

A report of the advisory committee on mathematics of the California Curriculum Commission recommends that, at every elementary-school level, textbooks and teachers should utilize every opportunity to provide a background of readiness experiences upon which the pupil can develop a correct understanding of the function concept. This report suggests these readiness activities for the elementary grades:

1. Experiences in pairing that will develop an awareness that different sets of objects may be paired with the same number but that a finite set of objects is paired with one and only one number.

2. Ordered one-to-one pairings in which each single object is paired uniquely with a single object.

3. Ordered many-to-one pairings in which several objects are paired uniquely with the same object. These one-to-one and many-to-one pairings lead to the concept of a function.

4. Pairing of geometric figures with numbers, which occurs in measurement and mensuration.

5. The introduction of simple formulas in the arithmetic program.

6. Graphs and tables used to present functions.

These intuitive experiences will pave the way toward recognizing a function as a set of ordered pairs in which no two pairs have the same first element.

One current textbook series in arithmetic specifically introduces and names the function concept in its texts. The fourth-grade text uses the idea of a function machine.

Example

	Function Machine	
Input	Function Rule	Output
15	Input number −7	8
○	○	○

Figure 3

The teacher's manual explains:

> One vital feature of these pairs (from the function machine) is that for each first number there is only one second number.
>
> In summary, we have a set and a rule for each function. When we apply the rule to an element of the set, we get just *one* answer. Thus for each function we have a set of ordered pairs, no two of which have the same first number.

The first lesson dealing with functions (fifth grade) presents a notation $f(2) = 9$; f means "apply the function rule to the number in parentheses."

Example. Simple addition facts

$$Rule: n + 2$$
$$f(1) = 3$$

The function rule can be as complex or as simple as one likes so long as it assigns exactly one number to each number in the reference set.

The last lesson in Grade 5 deals with graphs of mathematical sentences or functions. The children are assisted in making a table of number pairs associated with a given function rule and then in graphing these sets of number pairs. Attention is paid to certain patterns that are found in graphs of functions.

In a review section in the fifth-grade text the following problem is given as a review of the function concept.

Give the missing numbers.

Function Rule
$(2 \times n) + 9$

n	$f(n)$
6	21
—	29
37	—

Other series develop the concept of function on an intuitive basis, but few make any reference to this concept in the teacher's guide or use the term "function."

Summary

The idea of a mathematical function has its roots in the basic experience of pairing, found in most primary programs today. Because the function concept has to be abstracted out of a wide spectrum of experience, the student is not ready to generalize or formulate the idea until he has experienced and identified it in many different settings.

Today's mathematics curriculum in the elementary schools contains topics that could easily be adapted to emphasize the function concept. The student should be led to realize that functions can be presented or described by statements, formulas, equations, tabulated data, and graphs. Students should practice discovering and stating relationships in given problem situations. This practice should lead to the development of a better understanding of the function concept and to an awareness of the dependence of one variable on another in the environment. The student is then on his way toward mastery of an important concept that will open the door to many mathematical ideas of far-reaching application.

354   Functions

REFERENCES FOR FURTHER READING

1. Johnson, David C., and Cohen, Louis S., "Functions," *AT*, **17** (1970), pp. 305–315. Discusses the role of the concept of a function in the primary-school curriculum.

2. Karlin, Marvin, "Machines," *AT*, **12** (1965), pp. 327–334. Discusses function machines and their uses.

3. May, Kenneth D., and Van Engen, Henry, "Relations and Functions," Chap. 3 of *Growth of Mathematical Ideas*, National Council of Teachers of Mathematics, Reston, Va., 1959.

Chapter 15

Transformation Geometry

15.1 MOVING FIGURES IN A PLANE

I have previously suggested that geometry is all too often taught as a static subject with emphasis on definitions. One of the easiest ways to make geometry a dynamic subject is to develop the ideas of **transformation** geometry, or, to use a simpler term, **motion** geometry. Such an approach is very common in many European countries. It is also used in some U.S. primary-school programs and is likely to be used by many more in the years ahead.

Consider Fig. 15.1, where we can imagine moving rectangle *A* to the position of rectangle *B*. Or, to involve actual motion, we can make a copy of rectangle *A* by tracing it on a sheet of thin paper and then move the copy until it fits exactly on rectangle *B*.

Figure 15.1

There are three basic ways in which we can move a figure onto a new figure of the same size and shape (a congruent figure). We will discuss each of these ways in turn.

Note that from the point of view of functions as discussed in Chapter 14, we will be considering functions with domain and range the set of all geometric figures in a plane (points, line segments, triangles, etc.). The particular kind of functions that we call transformations in this chapter map any figure onto a figure congruent to it. Other kinds of transformations are considered in connection with other aspects of geometry. Thus, for example, we might consider all the functions that map a figure onto a similar figure (same shape but not necessarily the same size). Our set of congruency transformations, then, would be a subset of this set of similarity transformations.

15.2 TRANSLATIONS

Consider △*ABC* and the line *L* shown in Fig. 15.2. To perform a **parallel movement** or **translation** of △*ABC* in the **direction** of *L* and the **distance** *PP′*, we make a copy of Fig. 15.2 on thin paper and place the copy on the figure so that it fits exactly, as shown in Fig. 15.3. Then we slide the copy until point *P* on the copy coincides with point *P′* on the figure underneath. With a pin we mark the new positions of *A*, *B*, and *C* on the figure and, upon removing the copy, label the points marked by the pin as *A′*, *B′*, and *C′*, respectively. (See Fig. 15.4.)

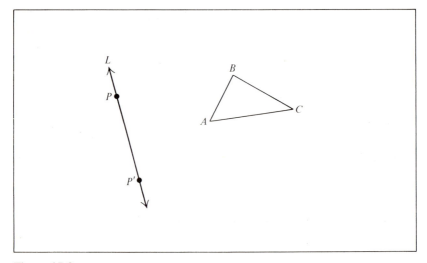

Figure 15.2

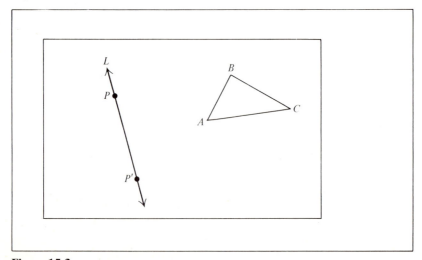

Figure 15.3

15.2 Translations

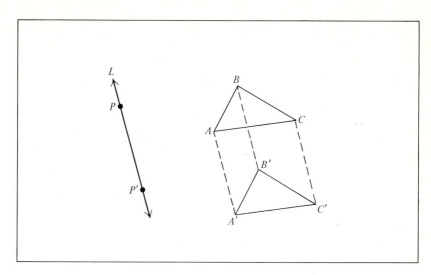

Figure 15.4

Clearly, $A'B' = AB$, $B'C' = BC$, $A'C' = AC$, and indeed, $\triangle A'B'C'$ is congruent to $\triangle ABC$. We say that $\triangle ABC$ has been moved (**transformed**) onto $\triangle A'B'C'$ by a parallel movement or by a translation (the former term probably is more suitable for children).

We can imagine every point in the plane being moved onto other points while $\triangle ABC$ is being moved onto $\triangle A'B'C'$. Furthermore, if we know where any point is moved to, we can determine where any other point is moved to. If a point A is moved onto point A' we say that the movement **assigns** A' to A, and that A' is the **image** of A under the movement. Our experiment indicates that

1. A translation in a plane assigns to each point A in the plane exactly one point A', its image under the translation.

2. If A' is the image of A, and B' the image of B under a translation, then $\overline{AB}$ is congruent to its image $\overline{A'B'}$.

If A' is the image of A, we refer to A as the **pre-image** of A'.

Exercises 15.2

1. Trace two copies of the next figure on thin paper and label both copies as indicated. Place one copy on top of the other so that the two sets of figures coincide. Now slide the top copy along the line $\overleftrightarrow{AA'}$ until point A on the top copy coincides with point A' on the bottom copy. With a pin, mark the new positions of P, C, D, E, F, and G on the bottom copy, and, upon removing the top copy, label these points on the bottom copy P', C', D', E', F', and G', respectively, and draw $\overline{C'D'}$ and $\triangle E'F'G'$. Then P' is the image of P under this translation, $\overline{C'D'}$ is the image of $\overline{CD}$, and $\triangle E'F'G'$ is the image of $\triangle EFG$.

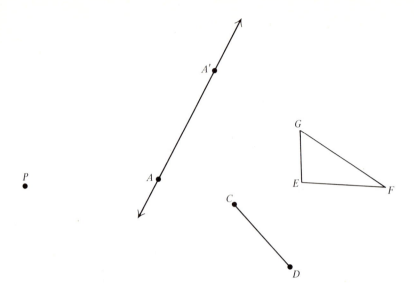

2. Here, as shown in the figure below, a translation parallel to L takes P onto P'. Proceed as in Exercise 1 to find the image of $\overline{AB}$ under this translation, and also find, by taking P' onto P, the line segment $\overline{CD}$ that is the pre-image of $\overline{C'D'}$ (i.e., the line segment that is moved onto $\overline{C'D'}$ under this translation).

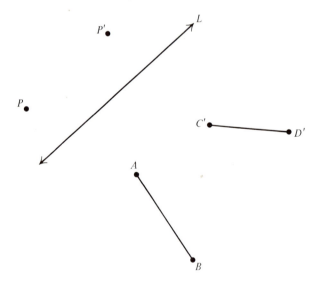

15.3 ROTATIONS

Another kind of movement is one in which the points of a figure move along circles in the plane with the same center. Such movements are called **rotations,** or, again using a term perhaps more suitable for children, **turning movements.** The common center is called the **center of rotation** or **turning point** of the movement.

15.3

We can carry out a rotation in a way similar to the way we carried out a translation in Section 15.2. Figure 15.5 shows $\triangle ABC$ and a center of rotation O. The rotation about O is to take P onto P'.

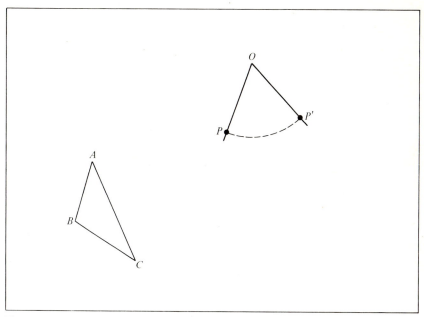

Figure 15.5

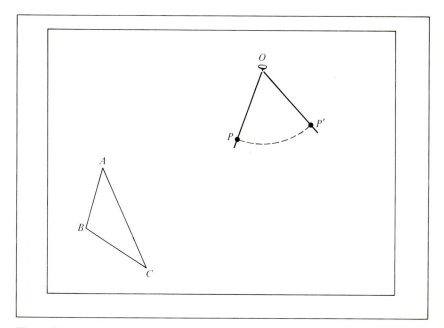

Figure 15.6

We make a copy of Fig. 15.5 on thin paper and place the copy on the figure so it fits exactly, as shown in Fig. 15.6. Then, placing the figure and the copy on a piece of cardboard, we put a pin or thumbtack through O. Now we rotate the copy until P on the copy coincides with P' underneath. With a pin we mark the new positions of A, B, and C on the figure and, upon removing the copy, label the points marked by the pins as A', B', and C', respectively. The result is shown in Fig. 15.7.

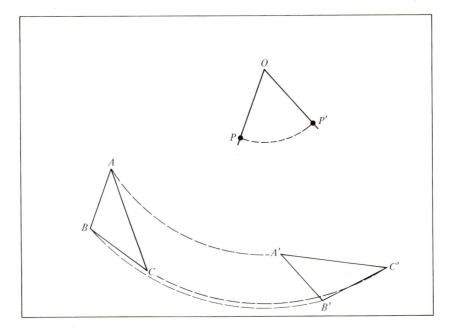

Figure 15.7

The same rotation about O takes each point except O into another point. Furthermore, if we know where any one point (other than O) is moved under the rotation about O, we can determine where any other point is moved to. As was the case for translation, if a point A is moved onto point A', we say that the movement assigns A' to A, that A' is the image of A, and that A is the pre-image of A' under the movement. (Note that the center of rotation is its own image and pre-image.)

Our experiment indicates that:

1. A rotation in a plane assigns to each point in the plane exactly one point A', its image under the rotation.

2. If A' is the image of A and B' the image of B under a rotation, then $\overline{AB}$ is congruent to its image $\overline{A'B'}$.

Exercises 15.3

1. Trace two copies of the figure below on thin paper and label both copies as indicated. Place one copy on top of the other so that the two sets of figures coincide, and use a pin or thumbtack through O to fasten both to a piece of cardboard. Now rotate the top copy around O until point A coincides with point A' on the bottom copy. With a pin, mark the new positions of P, C, D, E, F, and G on the bottom copy and, upon removing the top copy, label these points on the bottom copy P', C', D', E', F', and G', respectively, and draw $\overline{C'D'}$ and $\triangle E'F'G'$. Then P' is the image of P under this rotation, $\overline{C'D'}$ is the image of $\overline{CD}$, and $\triangle E'F'G'$ is the image of $\triangle EFG$. What is the image of O?

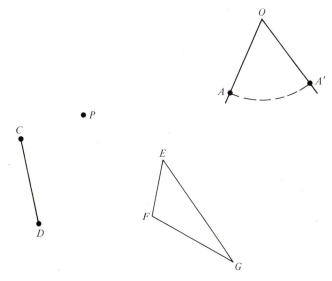

2. Here, as shown below, a rotation about O takes P onto P'. Proceed as in Exercise 1 to find the image of $\overline{AB}$ under this rotation, and also find, by taking P' onto P, the line segment $\overline{CD}$ that is the pre-image of $\overline{C'D'}$.

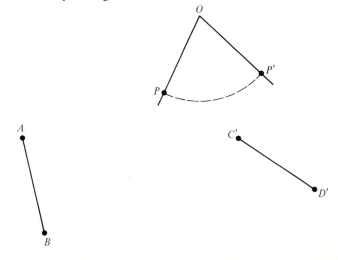

362 Transformation Geometry 15.4

3. In the figure below, A' is the image of A under a certain rotation. One possible center of this rotation is O; P and Q are also possible centers of this rotation. Find other possible centers of rotation. How many are there and how are they located?

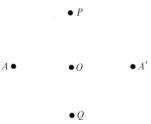

4. Use your answer to Exercise 3 to find the center of rotation that moves $\overline{AB}$ onto $\overline{A'B'}$ in the figure below.

15.4 REFLECTIONS

The third, and final, type of motion is called a **reflection.** It is called a reflection because the result of the motion to be described is mirrorlike: If a line L is thought of as showing the position of a mirror, then in Fig. 15.8 the mirror reflection of point A will appear to be at A'—the image of A under the reflection—and the mirror reflection of the face F will appear to be the face F'—the image of F under the reflection.

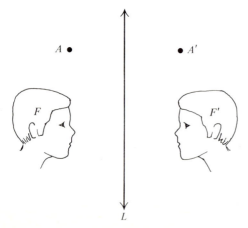

Figure 15.8

15.4 Reflections

To carry out a reflection of $\triangle ABC$ about a line L shown in Fig. 15.9, we again make a labeled copy of the figure on thin paper, fold the paper along L, and, with a pin, mark the positions of A, B, and C as they appear through the folded paper. Unfolding the paper, we label the points marked by the pin as A', B', and C', respectively. Now, placing the copy on the original figure, we mark with our pin the positions of A', B', and C' on the original figure and also label these positions A', B', and C', respectively. (Of course, if the original figure is on thin paper that can be folded, rather than a figure in a book, the last step is not needed!) The result of this activity would be as shown in Fig. 15.10.

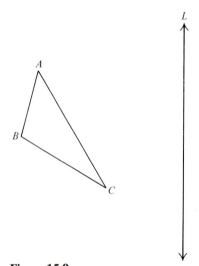

Figure 15.9

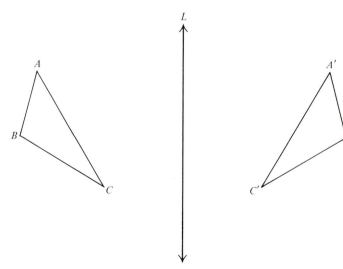

Figure 15.10

364   Transformation Geometry   15.4

As was the case for translation and rotation, if a point A is moved onto point A' by a reflection, we say that the movement assigns A' to A, that A' is the image of A, and that A is the pre-image of A' under the movement. Note that any point of L is its own image and pre-image and that the image of any point is also the pre-image of that point. Thus in Fig. 15.8, A' is both the image of A and the pre-image of A.

Our experiment indicates that:

1. A reflection about a line in a plane assigns to each point in the plane exactly one point A', its image under the reflection.

2. If A' is the image of A and B' the image of B under a reflection, then $\overline{AB}$ is congruent to its image $\overline{A'B'}$.

Exercises 15.4

1. Trace a copy of the figure below on thin paper and label the copy as indicated. Fold the copy along line L, and with a pin, mark the positions of P, C, D, E, F, and G as they appear through the folded paper. Unfold the paper and label the points marked by the pin as P', C', D', E', F', and G', respectively, and draw $\overline{C'D'}$ and $\triangle E'F'G'$. Then P' is the image of P under this reflection, $\overline{C'D'}$ is the image of $\overline{CD}$, and $\triangle E'F'G'$ is the image of $\triangle EFG$. What is the image of any point on L?

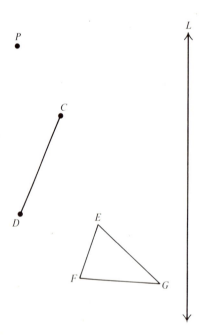

2. Proceed as in Exercise 1 with the next figure to find the image of $\overline{AB}$ under reflection in the line L and also find the line segment $\overline{CD}$ that is the pre-image of $\overline{C'D'}$.

15.4 **Reflections** **365**

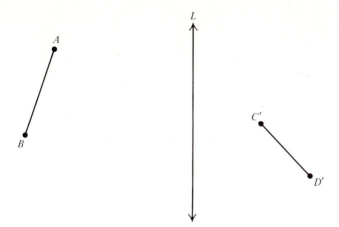

3. In the figure below are shown some shapes and their images under reflection in the same line. Trace the figure on a sheet of thin paper and draw the line of reflection.

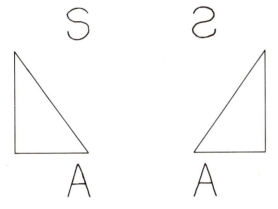

4. Answer the questions in the following exercise taken from a fourth-grade text.

ACTIVITY CARD 5

When you fold along a line of symmetry, one half exactly matches the other half.

The letter **A** has 1 line of symmetry.
The letter **H** has 2 lines of symmetry.

Which letters can you cut from old newspaper headlines and show, by folding, that they

have just 1 line of symmetry?

have exactly 2 lines of symmetry?

5. Study the figure on the left below to find an alternate to the folding technique to perform a reflection. Then use this technique to reflect $\triangle ABC$ in the line L in the righthand figure.

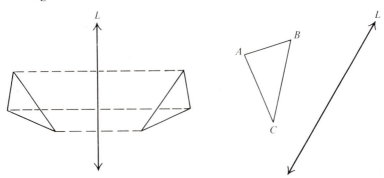

15.5 COMBINING MOVEMENTS

We have considered three types of movements (transformations) in a plane: translations (parallel movements), rotations (turning movements), and reflections. We can combine these movements, as shown in Fig. 15.11, where we first perform a translation which takes triangle A onto triangle A' and then a reflection which takes A' onto A''. Since A is congruent to A' and A' is congruent to A'', it follows that A is congruent to A'' (transitive property of congruence—see Section 8.3).

Not only does a translation followed by a reflection send any figure onto a congruent figure; any combination of the three basic transformations, called a **rigid motion,** sends any figure onto a congruent figure.

A systematic study of these transformations can lead to an alternative approach to the traditional Euclidean approach to geometry in secondary school,

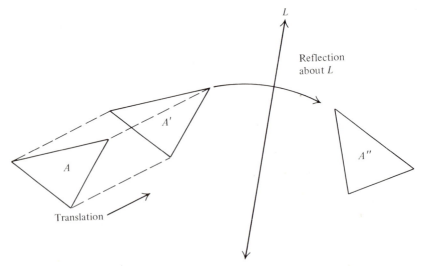

Figure 15.11

15.5 Combining Movements 367

and, indeed, this alternative approach is widely used today outside of the U.S.A. and is recommended by many people for use in the U.S.A. The following exercises will give some idea of how this transformation approach can be developed.

Exercises 15.5

1. Draw a figure showing the translation of a line segment $\overline{AB}$ onto $\overline{A'B'}$ followed by another translation (in a different direction) of $\overline{A'B'}$ onto $\overline{A''B''}$. Now show a single translation that takes $\overline{AB}$ directly onto $\overline{A''B''}$. (That is, show that the "product" of two translations is another translation.)

2. Draw a figure showing a rotation about a point P that takes a line segment $\overline{AB}$ onto $\overline{A'B'}$, followed by another rotation about the same point P that takes $\overline{A'B'}$ onto $\overline{A''B''}$. Now show a single rotation about point P that takes $\overline{AB}$ directly onto $\overline{A''B''}$. (That is, show that the "product" of two rotations about the same point is another rotation about that point.)

3. Draw a figure showing the reflection in a line L of a line segment $\overline{AB}$ onto $\overline{A'B'}$, followed by a reflection of $\overline{A'B'}$ in a line K, parallel to L, that takes $\overline{A'B'}$ onto $\overline{A''B''}$. Now show a translation that takes $\overline{AB}$ directly onto $\overline{A''B''}$. (That is, show that the "product" of two reflections about parallel lines is a translation and not another reflection.)

4. Draw a figure showing the reflection in a line L of a line segment $\overline{AB}$ onto $\overline{A'B'}$, followed by a reflection of $\overline{A'B'}$ in a line K, not parallel to L, that takes $\overline{A'B'}$ onto $\overline{A''B''}$. Now show that a rotation with center at $L \cap K$ takes $\overline{AB}$ directly onto $\overline{A''B''}$. (That is, show that the "product" of two reflections about nonparallel lines is a rotation and not another reflection.)

5. In the figure below find where $\triangle CDE$ is carried when you first perform a translation that carries A onto A', and then make a reflection about L.

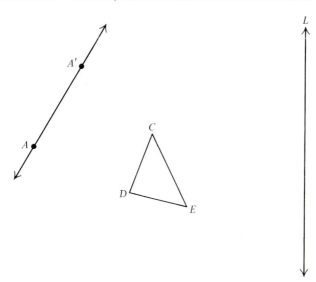

6. Do as in Exercise 5 except that you first perform the reflection and then the translation. Are the results the same?

7. Find where △CDE in the figure below is carried when you first perform a translation that carries A onto A', then a reflection about L, and finally, a rotation about O that carries P onto P'.

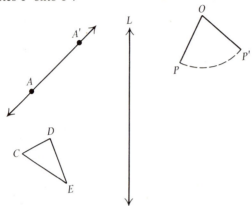

8. Do as in Exercise 7, except first do the reflection, then the rotation, and then the translation. Compare your result with that of Exercise 7.

15.6 SYMMETRIES OF FIGURES IN A PLANE

In our discussion of transformation geometry, we have said that any combination of rigid motions carries any figure onto a congruent figure. We did not, however, define "congruent" except to say that congruent figures are of the "same size and shape"—a rather vague definition, even if intuitively meaningful. Actually, in a treatment of geometry via transformations, we *define* congruence by saying that one figure is congruent to another if it is the image of that figure under some combination of rigid motions.

There are some figures that, under various rigid motions, are transformed onto themselves. For example, if we make a reflection of the rectangle *ABCD* shown in Fig. 15.12 about the line *L* that bisects $\overline{AD}$ and $\overline{BC}$, each point of the rectangle will be moved onto a point of the same rectangle. In particular, this transformation takes *A* to *D*, *D* to *A*, *B* to *C*, and *C* to *B*. We say that this reflection of the rectangle in *L* is a **symmetry** or, more precisely, a **reflection symmetry** of the rectangle. (Since the existence of reflection symmetries can be easily investigated by the folding of paper cutouts of figures, we can also call them **folding symmetries,** a term more appropriate for use with young children.)

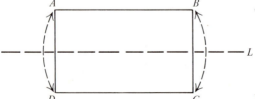

Figure 15.12

15.6   Symmetries of Figures in a Plane

Other examples: A square has four reflection symmetries, one about each of the dashed lines shown in Fig. 15.13 which also indicates that an isosceles triangle has one reflection symmetry, and that an equilateral triangle has three.

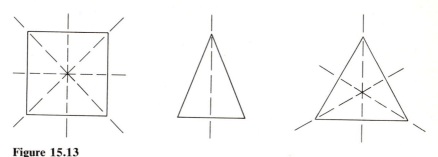

Figure 15.13

A figure may also possess **rotational (turning) symmetry.** Thus we can rotate a square about its center until it looks (except for any labeling of the vertices) exactly as it did before the rotation. Figure 15.14 shows such a clockwise rotation of 90°. We obtain similar results with clockwise rotations of 180°, 270°, and 360°. We could, of course, continue to rotate through 450°, 540°, etc., or we could rotate in a counterclockwise direction. If we agree, however, to limit all rotations to angles of degree measure less than or equal to 360° and to rotate in one direction only (clockwise or counterclockwise), we can say that a square has exactly four rotational symmetries.

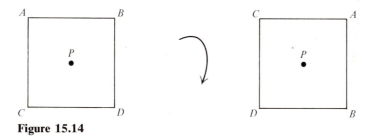

Figure 15.14

Under this restriction we can say that if a triangle has exactly three rotational symmetries, then it is an equilateral triangle, and that a circle has an unlimited number of rotational symmetries.

Exercises 15.6

1. Cut out and fold paper figures, if necessary, to answer the following questions.
   a) A rectangle that is not a square has how many reflection symmetries?
   b) If a quadrilateral has exactly two reflection symmetries, must it be a rectangle?
   c) Can a triangle have exactly two reflection symmetries? Two rotational symmetries?

d) How many reflection symmetries does a circle have?

e) Does any plane figure other than a circle have an unlimited number of rotational symmetries?

2. Decide how many rotational and reflection symmetries each of the figures below has.

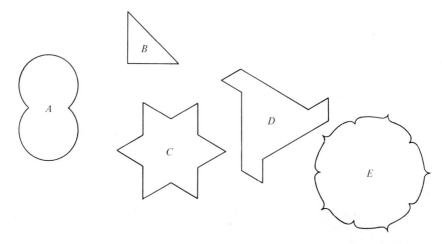

3. Which of the letters A, B, ..., Z have reflection symmetries? Which have rotational symmetries (other than the 360° one)?

4. How many reflection symmetries do each of the following regular polygons have? How many rotational symmetries?
   a) A regular pentagon
   b) A regular hexagon
   c) A regular octagon
   Use your results to suggest a general rule for determining the number of reflection symmetries and the number of rotational symmetries of any regular polygon.

Chapter Test

In preparation for this test, you should review the meaning of the following terms:

Image of a point (357)
Pre-image of a point (357)
Parallel movement (translation) (356)
Rotation (turning movement) (358)
Center of rotation (turning point) (358)

Reflection (362)
Rigid motion (366)
Reflection (folding) symmetry (368)
Rotational (turning) symmetry (369)

You should also review how to:

1. Perform translations (356), rotations (359–360), and reflections (363–364) of figures to determine images and pre-images;
2. Determine what reflection and rotational symmetries a given figure has (368–369).

In Problems 1 through 3, use the tracing procedures described in Exercises 15.2 through 15.4 to answer the questions.

1. In the figure below, find the image of rectangle ABCD under the translation that takes P onto P' and label it A'B'C'D'. Also find the pre-image of $\triangle E'F'G'$ and label it EFG.

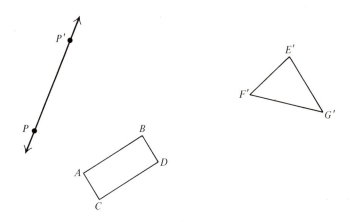

2. In the figure below, find the image of rectangle ABCD under the rotation about O that takes P onto P' and label it A'B'C'D'. Also find the pre-image of $\triangle E'F'G'$ and label it EFG.

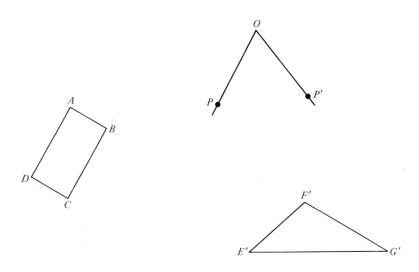

372   Transformation Geometry

3. In the figure below, find the image of rectangle *ABCD* about the line *L* and label it *A'B'C'D'*. What is the pre-image of *ABCD*?

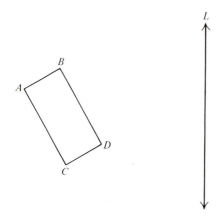

4. If a triangle has exactly one reflection symmetry, then it must be an _____ triangle; if it has exactly three reflection symmetries, then it must be an _____ triangle; if it has no reflection symmetries, then it is neither an _____ triangle nor an _____ triangle.

5. How many rotation symmetries and how many reflection symmetries does a rhombus have if the rhombus is not a square?

6. Do the following exercise taken from a fifth-grade text. ($\frac{1}{4}$ rotation means through 90°; $\frac{1}{2}$ rotation through 180°; and $\frac{3}{4}$ rotation through 270°.)

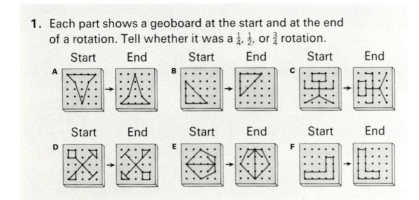

Chapter Test

TEST ANSWERS

1.

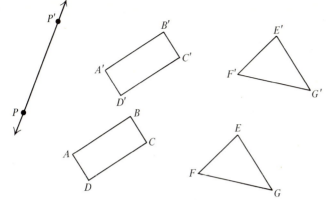

Rectangle $A'B'C'D'$ is both the image and the pre-image of rectangle $ABCD$.

2.

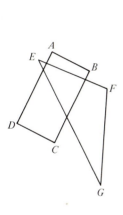

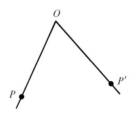

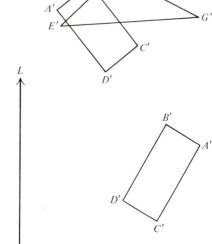

3.

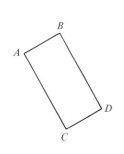

374   Transformation Geometry

4. Isosceles, equilateral, isosceles, equilateral
5. Two of each
6. A $\frac{1}{2}$ B $\frac{1}{4}$ C $\frac{3}{4}$ D $\frac{1}{2}$ E $\frac{3}{4}$ F $\frac{1}{4}$

Informal geometry through symmetry

by J. Richard Dennis

Reprinted from *The Arithmetic Teacher*, **16** (1969), pp. 433–436.

The geometry component of the elementary school program is the basis of much discussion today. Many of the efforts in this area of mathematics have approached elementary-school geometry from the point of view of Euclid's postulates. Another approach to geometric topics is based on the idea of symmetry. A unit using this approach would involve five to six weeks of class time and would do much to augment the elementary school geometry program.

The first idea that we need to make clear is the meaning of the phrase "exactly alike" as it is used in geometry. Students already have opinions on the meaning of this phrase, but their opinions frequently differ. Agreement is needed on an experimental test, the results of which will be acceptable in cases of differing opinions.

A pair of figures that appear to be exactly alike are shown in Fig. 1. How can we tell for sure? The test that we agree upon is to trace one figure and then try to match the tracing with the other figure. This match need not be achieved in any particular position or orientation. It may be possible to achieve a matching in several positions. It must be possible to achieve a matching in at least one position (Fig. 2).

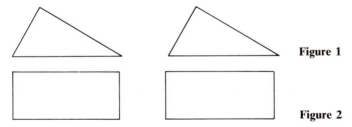

Figure 1

Figure 2

When it is possible to exactly match a tracing of one figure with another figure we say that the figures are *congruent*, and the various matchings are called *congruences*.

Teachers should create many opportunities for children to experiment with the "trace and try to match" process described above. It is from such experiments that basic intuitions about congruence are derived. These intuitions will become a foundation for the discovery and exploration of more complicated properties of geometric figures. For example, by experimenting with a tracing, congruences can be immediately separated into two types:

1. *Face-down* congruences, for which the tracing must be turned over to make it match (Fig. 3).

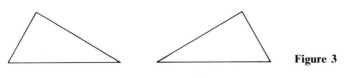

Figure 3

2. *Face-up* congruences for which the tracing is not turned over, just moved around to make it match (Fig. 4).

Figure 4

Having distinguished between these two types of congruences, we concentrate on the effects of each of them. We ask questions like the following:

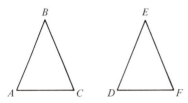

Figure 5

1. For the face-up congruence of these figures, with what part of triangle *DEF* does the tracing of segment *AB* match?
2. For the face-down matching, with what part of triangle *DEF* does the tracing of angle *ACB* match?
3. Again for the face-down matching, with what part of triangle *DEF* does the tracing of segment *AC* match?

Some important facts to observe are these:

1. For both matchings, the tracing of point *B* matches point *E*.
2. For both matchings, the tracing of segment *AC* matches segment *DF* but the tracings of individual points of the segments do not match in the same way.

Work of this nature gives an introduction to the notion of corresponding parts for a congruence, and, of course, since the tracing is used to match these parts, corresponding parts of congruent figures are congruent.

The next step in this development is to apply the notions of congruence and corresponding parts to a single figure rather than to two figures. Specifically we study *self-congruences* of a figure. Again these are of two types—face up and face down.

Figures with face-down self-congruences have a very important property. There is a line each of whose points corresponds with itself for that face-down matching. For example, consider the triangle in Fig. 6.

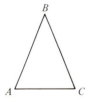

Figure 6

For the face-down self-congruence of this triangle, point B corresponds with itself. There is also a point of segment AC that corresponds with itself for this matching. In fact, if we draw a line through these two points, each point of that line corresponds with itself for the face-down congruence. It is such lines that we shall call *lines of symmetry*.

Through this definition, each line of symmetry is associated with a face-down congruence of a figure. So, when given an exercise such as to find all lines of symmetry for a particular figure, the student need only count the face-down matchings of a tracing. After a little practice, students easily move to the stage of just thinking about the tracing. It is important to note, however, that when all else fails, a *tracing* will make answers to questions quite obvious.

We are now ready to begin our study of triangles. It is assumed that many of the figures used in the previous work have been triangles. Students should see and experiment with triangles with no lines of symmetry, triangles with one line of symmetry, and triangles with three lines of symmetry. An important exercise is to have students try to sketch a triangle with exactly two lines of symmetry. Another important exercise is to try to sketch a triangle with a symmetry line that does not go through a vertex.

You will recognize the triangles with one (or more) lines of symmetry as those usually called *isosceles* triangles. Those with three lines of symmetry are usually called *equilateral* triangles. Having looked at the possible line symmetries for triangles, students are in a position to find properties of each type of triangle. For example, consider a triangle with one line of symmetry, i.e., one face-down self-congruence (Fig. 7).

Figure 7

1. There is a pair of congruent sides, because for the face-down self-congruence the tracing of one side of the triangle matches another side of the triangle.
2. There is a pair of congruent angles, because for the face-down self-congruence the tracing of one angle matches another angle of the triangle.
3. The symmetry line goes through the middle point of one side, because for the face-down self-congruence the tracing of one part of this side matches the other part of this side.
4. The symmetry line bisects one angle (for similar reasons).
5. The symmetry line "divides" the triangle into two congruent regions (for similar reasons).

At this stage some other important questions should be considered:

1. Could a triangle have a pair of congruent sides without having a line of symmetry?
2. Could a triangle have a pair of congruent angles without having a line of symmetry?

For each of these questions, evidence is easily gathered from an experimental sketch and a piece of tracing paper. The properties of triangles with three lines of symmetry are presented in a like manner.

Before classifying quadrilaterals it is convenient to introduce the notions of perpendicular and parallel lines. For perpendicular lines we look at pairs of lines and ask: In what cases is one line a line of symmetry for the other? For example, here are a dashed line and a solid line (Fig. 8). The dashed line is a line of symmetry for the solid line. We can also show this with a tracing.

Figure 8

Here it is important to have made clear the idea that, at best, pictures of lines leave much to be desired. For figures like triangles, lines of symmetry appear to cut the picture in half. For lines, it is no longer possible to judge lines of symmetry by checking to see if the picture is cut in half. An important observation is that when one line is a line of symmetry for another, the two lines make square corners with each other. We say that two lines are *perpendicular* whenever one is a line of symmetry for the other.

For parallel lines we again look at pairs of lines, but this time we ask if the lines have a line of symmetry in common. Again we can use a tracing (Fig. 9). Parallel lines are those lines which do have a line of symmetry in common.

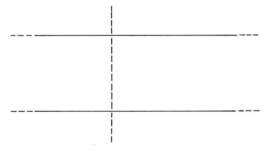

Figure 9

The face-down self-congruences gave us lines of symmetry. One type of face-up self-congruence is particularly important for the study of quadrilaterals. Sometimes a figure has a face-up self-congruence for a half-turn of a tracing. When this happens there is a point which corresponds with itself. Such a point is called a *center* or *point of symmetry* (Fig. 10). Again it is important to examine several figures for points of symmetry, and to look for corresponding parts under these half-turn self-congruences.

Figure 10

As we begin the study of quadrilaterals, we notice one important feature not found among triangles. Quadrilaterals may have lines of symmetry that do go through vertices or lines of symmetry that do not go through vertices (Fig. 11). So we introduce the phrase *diagonal symmetry line* for those that do go through vertices, and the phrase *nondiagonal symmetry line* for those that do not go through vertices.

Figure 11

When we classify quadrilaterals, we find those with:

1. No lines of symmetry.

2. One line of symmetry.

3. Two lines of symmetry.

4. Four lines of symmetry.

5. A point of symmetry.

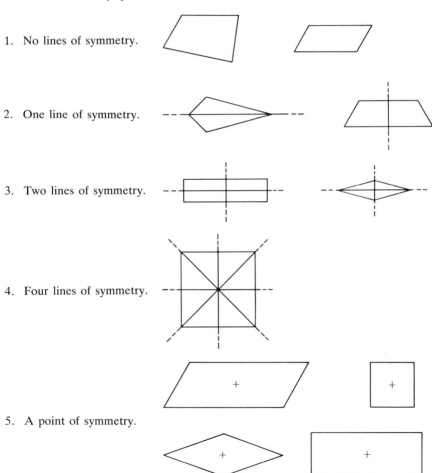

Notice that when quadrilaterals have two or more lines of symmetry they also have a point of symmetry. There are no quadrilaterals with exactly three lines of symmetry.

380 Transformation Geometry

As was the case for triangles, the usual properties about congruent and parallel sides, congruent angles, bisecting diagonals, perpendicular diagonals, etc., follow from the corresponding parts for the various self-congruences.

This has been a brief discussion of the topics involved, an appropriate sequence for these topics, and some sample questions at the various points of the development. Probably the most important word of caution to the teacher is to allow plenty of time for making experimental sketches and for conducting tracing experiments. Eventually many students reach a point where they can answer questions by merely conducting a thought experiment, but very few of them begin at this level.

REFERENCES FOR FURTHER READING

1. Eccles, Frank M., *An Introduction to Transformational Geometry*, Addison-Wesley, Reading, Mass., 1971.

2. Forseth, Sonia D., and Adams, Patricia A., "Symmetry," *AT*, **17** (1970), pp. 119–121.

3. Walter, Marion, "An example of informal geometry: mirror cards," *AT*, **13** (1966), pp. 448–452.

4. Walter, Marion, "Some mathematical ideas involved in the mirror cards," *AT*, **14** (1967), pp. 115–125.

Chapter 16

Probability

16.1 INTRODUCTION

Simple statistics (to be discussed in Chapter 17) in the form of construction of graphs of various kinds has long been a part of the primary-school program. More recently, some elementary ideas of probability have been appearing as early as Grade 5 and may very well begin to appear considerably earlier in the future. Both statistics and probability involve problems of interest to children, and work on these topics affords a good opportunity to practice arithmetic skills.

At a more advanced level, of course, there are numerous uses for probability and statistics. Both disciplines are involved in developing preventives for diseases such as polio, in the study of the relation of smoking to lung cancer, in the conducting of opinion polls, in economic forecasting, and so on. A background in probability or statistics (especially the latter) is a prerequisite for many courses in such areas of study as biology, business administration, economics, and education. Furthermore, the average citizen is bound to meet statistical data in reading newspapers and magazines (and, as will be pointed out in the next chapter, sometimes in a misleading form).

16.2 EXPERIMENTAL PROBABILITY

It is possible, of course, to just "think" about probability. Thus with an ordinary coin we can certainly assume, without actual experimentation, that it is equally likely that it will fall heads or tails. With children, however, it is very important to perform actual experiments with coin tossing, drawing different colored beads from a bag, etc., to check theoretically expected results against actual results. Also, in many applications of probability, the only way to determine the probabilities is to experiment. This is illustrated in Exercise 3 below, as well as in the bottom part of Fig. 16.1 taken from a sixth-grade book. (Note that a cube with its faces labeled $A, B, C, \ldots$, is used, rather than a die. More than one state has textbook adoption procedures that prevent the adoption of a primary textbook series involving dice because of their connotation with gambling!)

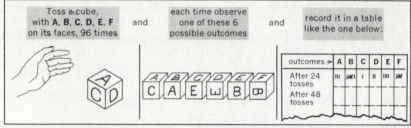

A Cube Experiment

DISCUSSION EXERCISES

1. Predict, as closely as you can, the number of times that each letter on the cube will appear after:

   [A] 24 tosses.   [B] 48 tosses.   [C] 72 tosses.   [D] 96 tosses.

   [E] your classmates each toss it once and the results are combined.

   Give reasons, if any, for your predictions.

2. Perform the experiment, tabulate your results and those of your class, and compare the results with your predictions. Discuss these results.

3. Would you have made the same predictions for the experiment if this figure had been tossed instead of the cube? Explain.

Figure 16.1

Indeed, even when it is theoretically possible to analyze the probability of an event, the analysis may be so complicated that it is easier to proceed on an experimental basis by using a high-speed computer to simulate the event. This technique is known as the **Monte Carlo** method. This Monte Carlo method can also be used for problems in other branches of mathematics, as is illustrated by Exercise 4 below.

Exercises 16.2

1. Toss a coin 50 times and record the number of heads obtained.

2. How do your results compare with your expectations of 50% heads?

16.3   **The Sample Space of an Experiment**   **383**

3. Toss a thumbtack. When it lands, it will either have its head flat ($\perp$) or it will fall on its side ($\wedge$). Toss it 50 times and keep a record of the number of times it falls flat. From your experiments do you conclude that the two possibilities "falling flat" and "falling on a side" are equally likely outcomes?

4. Cut out a paper square whose sides are two inches in length and, in the middle of the square, draw a circle of radius $\frac{1}{2}$ inch. Since you are going to throw darts at this square, either find a wall that will not be harmed by darts or fasten the square to a fairly large and heavy sheet of cardboard. In any event, wind up with the square on a wall at about eye-level.

    Now stand 8 to 10 feet away from the square and throw darts at it. Keep throwing darts until you have hit the square 25 times. (This will take a while unless you are an expert with darts! Do not, however, move closer to the wall to improve your chances of hitting the square.) Count the number of times, x, that you hit the circle, and then calculate the ratio of x to 25. This ratio should have something to do with the area of the circle! (*Hint:* the larger the circle, the more likely it will be that when the square is hit, the dart will also be inside the circle.)

16.3 THE SAMPLE SPACE OF AN EXPERIMENT

We use the word "experiment" here to refer to carrying out an activity, such as tossing a coin or throwing a die, where the outcome of the activity is uncertain. Here are some examples of experiments that will lead us into a discussion of probability.

Example 1. Suppose we toss a coin. If we assume that the coin will never land on its edge, we have two possible outcomes: "heads" (H) and "tails" (T). The set {H, T}, then, is the set of all possible outcomes. We write $S = \{H, T\}$ and call S the **sample space** of the experiment.

Example 2. Suppose we throw a die. Then the numeral showing on the upper side of the die will be 1, 2, 3, 4, 5, or 6, and we take as the sample space of the experiment, $S = \{1, 2, 3, 4, 5, 6\}$.

Example 3. Suppose we buy one lottery ticket out of a set of 1000 tickets numbered $1, 2, 3, \ldots, 1000$. Then we take $S = \{1, 2, 3, \ldots, 1000\}$ as the sample space of the experiment.

Example 4. Suppose we toss a coin and then toss it again. The possible outcomes are given by the following table:

First toss	H	H	T	T
Second toss	H	T	H	T

    If we write HH to stand for the first toss resulting in heads and the second toss also resulting in heads, HT to stand for the first toss resulting in heads and the second in tails, etc., we can write the sample space of the experiment as $S = \{HH, HT, TH, TT\}$.

384   Probability   **16.3**

Example 5.   Suppose we ask a person to tell us his month of birth. There are twelve possible outcomes, and S = {Jan., Feb., Mar., ... , Dec.} is the sample space of the experiment.

Example 6.   Suppose we ask two persons each to give us his month of birth. We can display the possible outcomes of this experiment in the chart shown in Fig. 16.2. Here each box in the chart represents a possible outcome. For example, the box marked with * corresponds to the first person having March as his birth month, and the second person having February as his birth month; the box marked with √ corresponds to the first person having May as his birth month and the second person having July as his birth month. (What do the boxes marked with × have in common?) So we take

$$S = \{(x, y): x \text{ and } y \text{ any of Jan., Feb., ... , Dec.}\}.$$

Then $n(S) = 12 \times 12 = 144$.

To summarize: Each time we consider an experiment we can define a sample space for the experiment—the set of all possible outcomes for the experiment.

Second person's birth month

	Jan.	Feb.	Mar.	Apr.	May	June	July	Aug.	Sept.	Oct.	Nov.	Dec.
Jan.	×											
Feb.		×										
Mar.		*	×									
Apr.				×								
May					×	√						
June						×						
July							×					
Aug.								×				
Sept.									×			
Oct.										×		
Nov.											×	
Dec.												×

First person's birth month

Figure 16.2

16.4   **Events**   **385**

Exercises 16.3

For each of the following experiments, define a suitable sample space S, using set-builder notation if convenient, and determine $n(S)$, the number of elements of S.

1. We ask someone to give us the name of a day of the week.

2. From a hat containing ten slips of paper numbered $1, 2, \ldots, 10$, we draw one slip of paper.

3. From the hat of Exercise 2, we draw one slip of paper, return it to the hat, and then again draw a slip of paper.

4. From the hat of Exercise 2, we draw one slip of paper and then draw another slip without returning the first to the hat.

16.4 EVENTS

The sample space of an experiment has as members *all* the possible outcomes of an experiment. In any particular experiment, however, we may be interested in only some of the outcomes—as, for example, drawing a winning lottery ticket or rolling a seven with two dice. We are thus interested in subspaces of S, which we call **events**. Here are some examples of experiments and events.

Experiment	**Sample space S**	**Event E (described in words and then as a subset of S)**
1. Throw a die	$S = \{1, 2, 3, 4, 5, 6\}$	Die shows an odd number: $E = \{1, 3, 5\}$.
2. Buy one lottery ticket out of 1000 with 10 winning tickets.	$S = \{1, 2, 3, \ldots, 1000\}$	You win: $E = \{w_1, w_2, \ldots, w_{10}\}$ with $w_1, w_2, \ldots, w_{10}$ the winning numbers.
3. Ask two persons to tell you their birth month.	$S = $ set containing 144 members, as given in Fig. 16.2	Both persons have the same birth month: $E = $ set containing those 12 members of S marked with $\times$ in Fig. 16.2.
4. Toss a coin and toss it again.	$S = \{HH, HT, TH, TT\}$	One head and one tail comes up: $E = \{HT, TH\}$

Consider now all the events that can be found if the sample space is $S = \{\text{win}, \text{lose}, \text{tie}\}$, as in a game of football. The possible events are:

       The subset with no members: $\{\ \ \}$
       The subsets with one member: $\{\text{win}\}, \{\text{lose}\}, \{\text{tie}\}$
       The subsets with two members: $\{\text{win}, \text{lose}\}, \{\text{win}, \text{tie}\}, \{\text{lose}, \text{tie}\}$
       The subset with three members: $\{\text{win}, \text{lose}, \text{tie}\}$.

386   Probability   **16.4**

In probability, any event represented by the empty set is called an **impossible event**. Thus the event in a game of football "neither a win, a loss, nor a tie" is an impossible event and is represented by the empty set. At the other extreme, the event represented by the entire sample space S is called a **sure event**. Thus the event in a game of football "either win, lose, or tie" is a sure event and is represented by the set $S = \{win, lose, tie\}$.

Note that we have used the word "event" here in a technical sense. In ordinary usage the word is taken to mean "a happening" rather than a set, as when we speak of "coming events," "the next event on the program", etc. In this common usage of the word "event," then, we might speak of the event that a game of football ends in a tie—meaning "a happening". But in our technical use of the word "event" we would consider the set {tie} as the event. Similarly, the "event" that a game of football ends in a win or a tie corresponds to the set {win, tie}.

In our formal discussions about events we will always consider them as sets, but in informal discussions we may speak, for example, of the event "game of football ends in a tie" (in quotation marks). Another notation for this situation (used in the following exercises) is:

> *Experiment:* Play a game of chess.
> *Event:* Game results in a draw (i.e., neither player wins).

Here the sample space is $\{win, lose, draw\}$ and the event is {draw}.

Exercises 16.4

For each of the following experiments, the sample space has been determined in Exercises 16.3. Write the event for which the verbal description is given and determine $n(E)$, the number of elements of E, for each event.

1. *Experiment:* We ask someone to give us the name of a day of the week.
    Events: a)   The name begins with "T".
                 b)   The name ends in "z".
                 c)   The name ends in "y".

2. *Experiment:* From a hat containing 10 slips of paper numbered $1, 2, \ldots, 10$, we draw one slip of paper.
    Events: a)   Number on the slip is even.
                 b)   Number on the slip is odd.
                 c)   Number on the slip is even or odd.
                 d)   Number on the slip is greater than 10.
                 e)   Number on the slip is less than 5.

3. *Experiment:* From the hat of Exercise 2, we draw one slip of paper, return it to the hat, and then again draw a slip of paper.
    Events: a)   Number on the first slip is 1.
                 b)   Number on the second slip is 1.
                 c)   Number on both slips is 1.
                 d)   Number on first slip is even and the number on the second slip is 2.

4. *Experiment:* From the hat of Exercise 2, we draw one slip of paper and then draw another slip without returning the first to the hat.

Events: a) Number on the first slip is 1.
b) Number on the second slip is 1.
c) Number on both slips is 1.
d) Number on the first slip is even and the number on the second slip is 2.

16.5 DEFINITION OF PROBABILITY

Consider the experiment of drawing a lottery ticket from 1000 tickets of which ten are winners. Our intuitive notion of probability certainly suggests that the chances of holding a winning ticket are 10 in 1000. In probability terminology we say that the probability of holding a winning ticket is 10/1000. Now our sample space for the experiment is $S = \{1, 2, 3, \ldots, 1000\}$, and the event "the ticket is a winner" is $E = \{w_1, w_2, \ldots, w_{10}\}$, where $w_1, w_2, \ldots, w_{10}$ are the winning numbers. Thus $n(S) = 1000$ and $n(E) = 10$, so that the probability of winning, $P(E)$, can be written as

$$P(E) = \frac{n(E)}{n(S)} = \frac{10}{1000} = \frac{1}{100}$$

As a second example, consider the toss of a die where $S = \{1, 2, 3, 4, 5, 6\}$ and let the event be "an odd prime number comes up." Then $E = \{3, 5\}$ and the formula

$$P(E) = \frac{n(E)}{n(S)} = \frac{2}{6} = \frac{1}{3}$$

again gives us a probability corresponding to our intuition.

As a third example, consider the experiment in which we ask two persons their birth month and ask what the probability is that they have the same birth month. There are (see Fig. 16.2) 144 elements in our sample space, and

$$E = \{(\text{Jan., Jan.}), (\text{Feb., Feb.}), \ldots, (\text{Dec., Dec.})\},$$

so that $n(E) = 12$. Thus, assuming that all of the outcomes (Jan., Jan.), ..., (Dec., Dec.) are equally likely (which would, of course, not be the case if the two persons were twins!) we have

$$P(E) = \frac{n(E)}{n(S)} = \frac{12}{144} = \frac{1}{12}$$

As a final example, consider the experiment in which we toss a green die and then a red one. Here $S = \{(x, y): x \text{ and } y \text{ any of } 1, 2, \ldots, 6\}$, so that $n(S) = 36$. If our event is "the sum of the two numbers on the dice is 7," then

$$E = \{(1, 6), (6, 1), (2, 5), (5, 2), (3, 4), (4, 3)\}.$$

388 Probability 16.5

Thus $n(E) = 6$, and so

$$P(E) = \frac{6}{36} = \frac{1}{6}$$

On the other hand, if our event is "the sum of the two numbers on the dice is 12", then $E = \{(6, 6)\}$. Thus, in this case, $n(E) = 1$ and

$$P(E) = \frac{1}{36}$$

These examples lead us to the definition of the probability of an event E,

$$P(E) = \frac{n(E)}{n(S)} \tag{1}$$

if all outcomes of the experiment are equally likely.

It is very important to note the qualification in (1) that all outcomes are equally likely. If, for example, I were to play a game of chess against the world champion, I would have the possibility of winning, losing, or drawing, i.e., $S = \{\text{win, lose, draw}\}$. Then $E = \{\text{win}\}$ corresponds to my winning the game. But I assure you that

$$P(E) \neq \frac{n(E)}{n(S)} = \frac{1}{3}!$$

Exercises 16.5

The first four exercises are those for which $n(S)$ and $n(E)$ have already been calculated in Exercises 16.3 and 16.4. Find the probability of each event.

1. *Experiment:* We ask someone to give the name of a day of the week.
   Events: a) The name begins with "T".
               b) The name ends in "z".
               c) The name ends in "y".

2. *Experiment:* From a hat containing ten slips of paper numbered $1, 2, \ldots, 10$, we draw one slip of paper.
   Events: a) Number on the slip is even.
               b) Number on the slip is odd.
               c) Number on the slip is even or odd.
               d) Number on the slip is greater than 10.
               e) Number on the slip is less than 5.

3. *Experiment:* From the hat of Exercise 2, we draw one slip of paper, return it to the hat, and then again draw a slip of paper.
   Events: a) Number on the first slip is 1.
               b) Number on the second slip is 1.
               c) Number on both slips is 1.
               d) Number on the first slip is even and the number on the second slip is 2.

16.5   **Definition of Probability**   **389**

4. *Experiment:* From the hat of Exercise 2 we draw one slip of paper and then draw another slip without returning the first to the hat.
   Events: a)   Number on the first slip is 1.
                 b)   Number on the second slip is 1.
                 c)   Number on both slips is 1.
                 d)   Number on the first slip is even and the number on the second slip is 2.

In Exercises 5 through 9, begin by finding a suitable sample space and write the event as a subspace of this sample space. Then, assuming equally likely outcomes, calculate the probability of the event using formula (1).

5. *Experiment:* We roll a green die and then a red one.
   Events: a)   Both dice show a "6".
                 b)   The sum of the numbers shown on the two dice is 8.
                 c)   The number shown on the green die is less than the number shown on the red die.
                 d)   The sum of the numbers shown on the two dice is less than 13.
                 e)   The sum of the two numbers shown on the two dice is greater than 12.

6. *Experiment:* We toss a coin, then toss it again, and then toss it a third time.
   Events: a)   All three tosses result in heads.
                 b)   Exactly two tosses result in heads.
                 c)   At least two tosses result in heads.
                 d)   At most two tosses result in heads.

7. *Experiment:* A survey of families with two children is made, and the sexes of the children in order of age (older child first) are recorded. (Assume that the ratio of male births to female births is $1:1$.)
   Events: a)   Family has two boys.
                 b)   Family has one boy and one girl.
                 c)   Family has at least one girl.
                 d)   Family has at most one girl.

8. *Experiment:* A survey of families with three children is made, and the sexes of the children (in order of age, oldest child first) are recorded. (Assume that the ratio of male births to female births is $1:1$.)
   Events: a)   Family has three girls.
                 b)   Family has at least two girls.
                 c)   Family has exactly two girls.
                 d)   Family has at most one girl.
                 e)   Family has at least one girl.
                 f)   Family has more girls than boys.

9. *Experiment:* We ask three persons to tell us their birth month.
   Events: a)   All three have the same birth month.
                 b)   All three have birth months beginning with "J".

10. *Experiment:* A coin is tossed until a head appears twice in a row, and then tossed again.
    Event: This last toss is a head.

390 Probability

16.6 SOME PROPERTIES OF $P(E)$

Property 1. If E is an impossible event, then $P(E) = 0$.
  Proof: If E is an impossible event, then $E = \emptyset$ and so $n(E) = 0$. Hence

$$P(E) = \frac{n(E)}{n(S)} = \frac{0}{n(S)} = 0$$

Property 2. If E is a sure event, then $P(E) = 1$.
  Proof: If E is a sure event, then $E = S$ and so

$$P(E) = \frac{n(E)}{n(S)} = \frac{n(S)}{n(S)} = 1$$

Property 3. If E is any event, then $0 \le P(E) \le 1$.
  Proof: We note that $n(E) \ge 0$ and $n(E) \le n(S)$. Thus

$$0 = \frac{0}{n(S)} \le \frac{n(E)}{n(S)} = P(E)$$

and

$$P(E) = \frac{n(E)}{n(S)} \le \frac{n(S)}{n(S)} = 1$$

This gives us the result, $0 \le P(E) \le 1$.

In Section 2.5 we defined the relative complement of two sets A and B with $B \subseteq A$ as $A \backslash B = \{x : x \in A \text{ and } x \notin B\}$. Since $E \subseteq S$ we can consider here the set $E' = S \backslash E$ and establish

Property 4. If E is any event, then $P(E) = 1 - P(E')$.
  Proof: Since $E \cup E' = S$ and $E \cap E' = \emptyset$, it follows that $n(S) = n(E) + n(E')$. (See Section 2.3.) Thus

$$\frac{n(S)}{n(S)} = \frac{n(E)}{n(S)} + \frac{n(E')}{n(S)} = P(E) + P(E')$$

Hence,

$$1 = P(E) + P(E') \qquad \text{and so} \qquad P(E) = 1 - P(E')$$

As you will see later, it can happen that it is easier to calculate $P(E')$ than to calculate $P(E)$. In such a situation, then, we calculate $P(E')$ and then find $P(E)$ easily by using Property 4. To do this, however, we need to know how to determine the "happening" E' given the "happening" E. For example, suppose that our experiment is the tossing of two coins and that E is the event "at least one toss is not a head". Then E' is the event "not E", i.e., E' is the event "both tosses result in a head". So, since $S = \{HH, HT, TH, TT\}$, $n(S) = 4$, $E' = \{HH\}$,

16.6   Some Properties of *P(E)*   **391**

and $n(E') = 1$, it follows that $P(E') = \frac{1}{4}$ and thus

$$P(E) = 1 - P(E') = 1 - \tfrac{1}{4} = \tfrac{3}{4}$$

Here, of course, it is not at all difficult to calculate $P(E)$ directly, since $E = \{HT, TH, TT\}$ and thus we have

$$P(E) = \frac{n(E)}{n(S)} = \frac{3}{4}$$

Consider now the experiment of asking three persons to give their birth month and letting E be the event "at least two of the three persons have the same birth month." In this case it is not an easy matter to calculate $n(E)$ directly. It is, however, as we will see in the next section, not too difficult to calculate $P(E')$ where E' is the event "not E", i.e., E' is the event "all three persons have a different birth month". Then, having $P(E')$, we can calculate $P(E)$ by Property 4.

As a final example, consider the experiment of tossing an unfair coin. If it is known that the probability of heads for this coin is 0.6, what is the probability of tails?

Here we take E as "coin comes up tails", and since we are given that $P(E') = 0.6$, we have

$$P(E) = 1 - P(E') = 1 - 0.6 = 0.4$$

(Note that Property 4 applies even when the outcomes of an experiment are not equally likely.)

Exercises 16.6

In Exercises 1 through 4, an experiment and an event E are given. You are to: (a) Describe E' as a "happening"; (b) write E' as a set; (c) calculate $P(E')$; and (d) find $P(E)$ by using Property 4. (See the tossing-of-two-coins experiment given in the text.)

1. *Experiment:* Toss a green die and then a red one.
   Event: The sum of the two numbers on the dice is less than 11.

2. *Experiment:* From a hat containing 7 slips of paper numbered 1, 2, 4, 6, 8, 10, 12, we draw one slip of paper, return it to the hat, and then again draw a slip of paper.
   Event: At most one of the two slips of paper drawn has a prime number written on it.

3. *Experiment:* From the hat of Exercise 2, we draw one slip of paper and then draw another slip without returning the first to the hat.
   Event: Neither of the two slips of paper drawn has a prime number written on it.

4. *Experiment:* A survey of families with three children is made, and the sexes of the children (in order of age, oldest child first) are recorded. (Assume that the ratio of male births to female births is $1:1$.)
   Event: Family has at least one girl.

392   Probability   **16.7**

In Exercises 5 through 8, an experiment and an event E are given and you are to describe E' as a "happening".

5. *Experiment:* You ask four persons to give you their birth month.
    Event: At least two of the four persons have the same birth month.

6. *Experiment:* One card is drawn from a deck of cards.
    Event: The card is not a face card (i.e., not a jack, a queen, or a king).

7. *Experiment:* A green die, a red die, and a yellow die are thrown.
    Event: None of the numbers shown on any of the dice is a prime number.

8. *Experiment:* Same as in Exercise 7.
    Event: At most one of the numbers shown on any of the three dice is a prime number.

16.7 A COUNTING PRINCIPLE

In our examples of probability so far, it has been a fairly simple matter to calculate $n(S)$ and $n(E)$ directly and, hence, $P(E)$. This is not always the case. For example, suppose we want to know the probability of being assigned the seven-digit phone number 543–1719, under the condition that all the phone numbers have the first digit either 5 or 6, the second digit 3 or 4, and that all possible seven digit numbers will be assigned. Clearly it would be an enormous task to write down all possible such phone numbers.

Before tackling this problem let us do a simpler but similar problem to see the principle involved. Suppose we want to find all three-digit numerals where the first digit is in the set $\{1, 2, 3\}$, the second is in the set $\{4, 5\}$, and the third is in the set $\{6, 7, 8\}$. We can find the number of such three-digit numerals by forming what is called a **tree-diagram**, as shown in Fig. 16.3. Direct counting, then, shows us that there are 18 such numerals possible. On the other hand, we can see that 2 "branches" of the "tree" lead from each of the first three branches, giving us $3 \times 2 = 6$ branches, and then, from each of the 6 branches, 3 more branches lead off, to give us a total of $6 \times 3 = 18$ branches.

That is, without direct counting, we can calculate the number of these three-digit numerals by simply noting that there are three ways of choosing the first digit, two ways of choosing the second digit, and three ways of choosing the third digit. These tasks can then be done in $3 \times 2 \times 3 = 18$ different ways.

More generally, we have the following **fundamental counting principle:**

If one task can be done in T_1 ways, a second task can then be done in T_2 ways, ..., the nth task can then be done in T_n ways, then the total number of ways that all n tasks can be done in that order is $T_1 \times T_2 \times \cdots \times T_n$.

Let us now return to the problem involving the telephone numbers. Since the first two tasks involve two choices (5 or 6, and then 3 or 4), we have $T_1 = T_2 = 2$. Then $T_3 = T_4 = T_5 = T_6 = T_7 = 10$ (any one of $0, 1, \ldots, 9$ can be chosen), so that $n(S) = 2 \times 2 \times 10 \times 10 \times 10 \times 10 \times 10$. Our event is "the number is 543–1719", so that $E = \{543\text{–}1719\}$ and $n(E) = 1$. Hence

$$P(E) = \frac{1}{2 \times 2 \times 10 \times 10 \times 10 \times 10 \times 10} = \frac{1}{400,000}$$

A Counting Principle

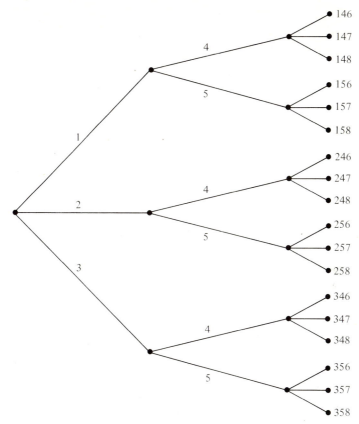

Figure 16.3

As a second example, suppose that a four-volume set of books, all of the same size, tumble down off a shelf, and we replace them without looking at the volume numbers. What is the probability that we will have replaced them in the correct order? There are four choices for the first book, leaving three for the second, then two for the third, and finally, only one for the fourth. Hence

$$n(S) = 4 \times 3 \times 2 \times 1 = 24$$

Since there is only one correct order, we have $n(E) = 1$, and so

$$P(E) = \frac{n(E)}{n(S)} = \frac{1}{24}$$

Now let us return to the birth-month problem mentioned in the previous section. That is, we want to calculate the probability that if we ask three persons (no twins or triplets!) their birth month, at least two of the three will have the same birth month. Since each person may be born in any one of 12 months, there are, by the fundamental counting principle, $12 \times 12 \times 12 = 1728$ possible outcomes to our experiment of asking three persons their birth month. However, it is

394   Probability   16.7

not easy to calculate $n(E)$ directly if E is the event "at least two of the three persons will have the same birth month". We can, however, calculate $n(E')$, where E' is the event "not E" (i.e., "all three persons have different birth months"), for $n(E')$ can be calculated by the fundamental counting principle, since there are 12 possibilities for the birth month of the first person, 11 for the second, and 10 for the third. Hence

$$n(E') = 12 \times 11 \times 10 \quad \text{and} \quad P(E') = \frac{12 \times 11 \times 10}{12 \times 12 \times 12} = \frac{55}{72}$$

Then, by Property 4,

$$P(E) = 1 - P(E') = 1 - \frac{55}{72} = \frac{17}{72} = 0.24 \quad \text{(correct to 2 decimal places)}$$

That is, there is about a 24 to 100 (or about a 1 to 4) chance that at least two of any three persons will have the same birth month.

The same method can be used to find the probability that at least two of four persons have the same birth month, that at least two of five persons have the same birth month, and so on. The following table gives the results of some of these calculations, where the decimals are given correct to two decimal places.

Number of persons	Probability that at least two have the same birth month
2	$\dfrac{1}{12} = 0.08$
3	$\dfrac{17}{72} = 0.24$
4	$\dfrac{41}{96} = 0.43$
5	$\dfrac{89}{144} = 0.62$
6	$\dfrac{1343}{1728} = 0.78$

Thus it appears that there is a considerably better than a 50% chance (62%) that among five persons at least two will have the same birth month. (Try this experiment!) Of course, in arriving at this result we are assuming equally likely outcomes—i.e., that births are equally distributed among the 12 calendar months. Although statistics on births indicate that this assumption is not completely true, it is close enough to make the results shown in our table reasonably accurate.

What is the probability that of 13 people, at least two have the same birth month?

Chapter Test 395

Exercises 16.7

1. A club with 15 members has to elect a president, a secretary, and a treasurer. In how many ways can the election turn out?

2. A restaurant offers two choices of soup, two of fish, four of the main course, and three desserts. How many different ways of choosing a four-course meal are there?

3. How many three-digit numerals can be formed using the digits 1, 2, 3, 4, and 5 if:
   a) No digit can be repeated?
   b) Digits may be repeated?
   c) The numeral must begin with 2 or 4 and digits may be repeated?
   d) The number represented is an odd number and digits may be repeated?

4. In how many different ways can a coach assign Albert, Bob, Charles, and Don to positions on the infield (first base, second base, third base, shortstop) of a baseball team?

5. A three-digit number is selected at random (i.e., it is equally likely that any of the numbers $100, 101, \ldots, 999$ is selected). Find the probability that the number selected:
   a) Has an odd number for its first digit;
   b) Is an even number;
   c) Has an even number for its first and last digit.

6. A coin is tossed four times.
   a) How many different outcomes are there for the experiment?
   b) Find the probability that exactly one head occurs in the four tosses.

7. In the table on page 394, the probability that at least two of four persons have the same birth month is given as 41/96. Verify this value.

8. A committee of three is selected by lot from the six people A, B, C, D, E, and F. The first person selected is to be chairperson, the second vice-chairperson, and the third, recorder. What is the probability that A will be chairperson, B will be vice-chairperson, and C will be recorder?

9. From a panel of 20 seniors, 15 juniors, 10 sophomores, and 5 freshmen, a panel of 5 is selected by lot. What is the probability that all five panelists will be freshmen?

10. The 11 letters of the word "Mississippi" are scrambled and then rearranged by random choice. What is the probability that the first four letters in the rearrangement will be the four "i's"?

Chapter Test

In preparation for this test you should review the meaning of the following terms and symbols:

Sample space (383)   Fundamental counting principle (392)
Event (385) $n(S)$ (384)

396   Probability

Impossible event (386) $n(E)$ (386)

Sure event (386) $P(E)$ (387)

Probability of an event (387) E' (390)

You should also review how to:

1. Find a sample space for an experiment (383–384);
2. Write events as subsets of a sample space (385–387);
3. Find the probability of an event E from the definition of $P(E)$ (387–388);
4. Find the probability of an event E using E' (390–391);
5. Use the fundamental counting principle to find $n(S)$ and $n(e)$ (392–394).

1. A sample space for an experiment is the set of all _____ _____ for the experiment.

2. An _____ is any subspace of the sample space of an experiment.

3. An event represented by the empty set is called an _____ event.

4. An event represented by the sample space S is called a _____ event.

5. We define $P(E) = \dfrac{n(E)}{n(S)}$ if all outcomes of the experiment are _____ _____ .

6. If E is an impossible event, then $P(E) =$ ____ .

7. If E is a sure event, then $P(E) =$ ____ .

8. If E is any event, then 0 ____ $P(E)$ ____ ____ .

9. If E is any event, then $P(E') =$ _____ .

10. If one task can be done in T_1 ways, a second task can then be done in T_2 ways, . . . , the nth task can then be done in T_n ways, then the total number of ways that all n tasks can be done in that order is _____ .

11. A die is thrown and then a coin tossed. Construct a sample space for the experiment.

12. For the experiment of Problem 11, write the event: "The number on the die is even and the coin comes up heads".

13. Calculate the probability of the event in Problem 12.

14. How many four-digit phone numbers are possible if the first digit cannot be zero and the last digit must be an even number?

15. If all the phone numbers described in Problem 14 are in use, what is the probability of having a phone number whose first and second digits are even numbers?

16. In the table of Section 16.7, the probability of at least two of five persons having the same birth month is given as 89/144. Verify this value.

References for Further Reading   397

TEST ANSWERS

1. Possible, outcomes 2. Event 3. Impossible
4. Sure 5. Equally, likely 6. 0
7. 1 8. $\leq, \leq, 1$ 9. $1 - P(E)$
10. $T_1 \times T_2 \times \cdots \times T_n$
11. $S = \{(1, H), (2, H), (3, H), (4, H), (5, H), (6, H), (1, T), (2, T), (3, T), (4, T), (5, T), (6, T)\}$
12. $E = \{(2, H), (4, H), (6, H)\}$
13. $P(E) = n(E)/n(S) = 3/12 = 1/4$
14. $9 \times 10 \times 10 \times 5 = 4500$
15. $n(E) = 4 \times 5 \times 10 \times 5; \quad P(E) = \dfrac{4 \times 5 \times 10 \times 5}{9 \times 10 \times 10 \times 5} = \dfrac{2}{9}$
16. $n(E') = \dfrac{12 \times 11 \times 10 \times 9 \times 8}{12 \times 12 \times 12 \times 12 \times 12} = \dfrac{55}{144}$ and

   so $\quad P(E) = 1 - P(E') = 1 - \dfrac{55}{144} = \dfrac{89}{144}$

REFERENCES FOR FURTHER READING

1. May, Lola J., "Probability; chance for a change", *Grade Teacher*, **86** (1969), pp. 31–32.

2. Mosteller, Frederick, Rourke, Robert E. K., and Thomas, George B., *Probability: A First Course*. Addison-Wesley Publishing Co., Reading, Mass. (1961).

3. Page, David A., "Probability," Chapter 6 of *The Growth of Mathematical Ideas*. National Council of Teachers of Mathematics, Reston, Va., 1959. Describes ways of introducing probability in the primary school.

4. Wilkinson, Jack D., and Nelson, Owen, "Probability and statistics—Trial teaching in the sixth grade," *AT*, **13** (1966), pp. 100–105.

Chapter 17

Statistics

17.1 INTRODUCTION

Some of the many uses of statistics were listed in the introduction to Chapter 16. Here we will be concerned only with descriptive statistics—ways of presenting data and the use of various kinds of "averages"—because it is this type of statistics that is commonly introduced into contemporary primary-school mathematics programs.

17.2 ORGANIZING STATISTICAL DATA

Suppose that as a class experiment the heights of the children in the class are measured, and the results, to the nearest centimeter, are recorded as follows:

140	125	155	135
150	145	140	157
135	150	145	140
155	137	148	150
150	148	137	143

In this "raw" form it is hard to see any pattern. Clearly, one very simple thing we could do to make the data easier to comprehend would be to order the data to read:

157	150	145	137
155	150	143	137
155	148	140	135
150	148	140	135
150	145	140	125

399

400   Statistics

17.2

Suppose, however, that two classes are involved, so that the raw data appeared as follows:

140	125	155	135	148	150	155	137	140	148
150	145	140	157	145	148	145	155	148	135
135	150	145	140	148	157	140	137	150	148
155	137	148	150	140	145	148	148	137	150
150	148	137	145	148	137	145	140	148	150

Just arranging these 50 heights in order would still leave us with a list that is difficult to analyze. A better way, then, of organizing these data is to prepare a **frequency table,** as shown below:

Height in cm	Tally marks	Frequency
125	/	1
135	///	3
137	++++ /	6
140	++++ //	7
145	++++ //	7
148	++++ ++++ //	12
150	++++ ///	8
155	////	4
157	//	2

Total frequency = Total number of pupils = 50

From this table we see how a child who is 137 cm tall compares with other children in the class. We can say how many children are of the same height, or are taller, or shorter. It would be more difficult to extract this information from the data we had originally, even after the data is ordered.

When even more data are involved, such a frequency table can still be hard to interpret, and in such cases, we often group the data. For example, suppose that we have the problem of analyzing the record of the 100 grades listed below:

50	100	66	52	36	56	68	64	52	88
18	96	56	44	30	50	50	40	10	36
20	54	64	50	58	44	12	60	46	56
74	34	50	82	51	59	54	34	30	48
52	60	18	50	46	20	42	16	68	48
20	58	73	32	56	52	52	58	44	66
82	62	52	28	38	76	86	30	72	54
12	52	21	33	54	58	58	62	52	21
30	50	58	27	24	27	66	48	32	29
28	23	19	64	29	48	36	70	42	89

If we were to construct a frequency table for this data as we did for the heights of children, we would find that there would be nearly 50 different scores represented. In such a situation it would be best to **group** the grades, in order to

17.2   **Organizing Statistical Data**   **401**

comprehend the overall picture at a glance. We could group in tens: 1–10; 11–20; 21–30; etc., or in fives, 1–5; 6–10; 11–15; 16–20; etc. We call 1–10, 11–20, 21–30, or 1–5, 6–10, 11–15, etc., the **class intervals.**

If we use the class intervals 1–10, 11–20, etc., our results will be as follows:

Class intervals	Tally marks	Frequency
1–10	/	1
11–20	卌 ////	9
21–30	卌 卌 ////	14
31–40	卌 卌	10
41–50	卌 卌 卌 ///	18
51–60	卌 卌 卌 卌 卌 /	26
61–70	卌 卌 /	11
71–80	////	4
81–90	卌	5
91–100	//	2

Note that, although the table above is concise and gives a lot of information, some information is lost by grouping. For example, although we know that nine scores fall in the interval 11–20, we cannot tell, without looking at the original data, what those scores are. Clearly, the larger the interval, the more concise the table, but the more information lost. The choice of size of interval depends on various factors, but generally the number of classes should not be more than around 20. In any event, it is important to avoid having intervals of different sizes (e.g., to avoid using as intervals 1–5, 6–10, and then 11–20).

Exercises 17.2

1. The grades on a test were as follows:

90, 63, 66, 76, 83, 76, 66, 55
80, 89, 70, 54, 76, 78, 81, 68

a) Arrange them in order from highest to lowest.
b) What percent of the grades are above 75?
c) What is the difference between the highest and the lowest grades?

2. The weights in kilograms of children in a certain class were recorded to the nearest half-kilogram, as follows:

40.5	41.5	41.5	40.5	40.0
41.5	40.5	42.0	40.0	39.0
40.5	40.0	41.0	39.0	41.0
40.5	42.0	39.5	41.5	40.5
40.0	41.0	39.0	39.5	41.0

Make a frequency table for this data.

3. The grades of 50 pupils on a test were as follows:

70	83	94	88	91
68	86	66	98	85
100	73	77	57	92
95	80	95	92	89
62	77	80	56	95
57	86	98	83	75
81	92	95	86	89
60	98	85	61	91
88	68	100	86	83
70	54	71	83	89

Using the intervals 96–100, 91–95, ..., 51–55, make a frequency distribution table for these data.

17.3 GRAPHICAL REPRESENTATION OF STATISTICAL DATA

Tables such as those discussed in the previous section are one way of presenting statistical data. A more vivid way, however, involves the use of various forms of **statistical graphs.** Many examples of these can be found in newspapers and magazines. Most of them are **bar graphs, line graphs, pie charts,** or **pictograms.** Examples of such graphs are shown in Figs. 17.1 through 17.5.

There are no specific rules for constructing graphs other than the aim of making the data vivid to the viewer. The choice of scale is largely determined by

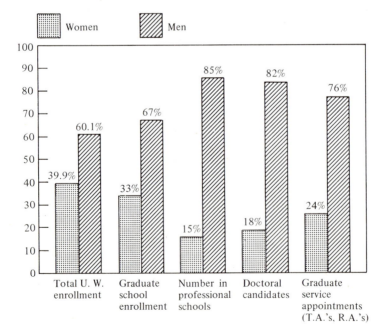

Figure 17.1 Bar graph.

the space available; the labeling of the graph should indicate clearly what information is being presented, and if possible, the source of the information.

To calculate the size of the wedges in a pie chart, some preliminary calculations need to be performed. For example, the pie chart shown above was

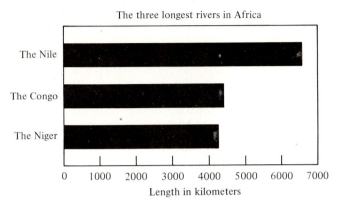

Figure 17.2 Horizontal bar graph.

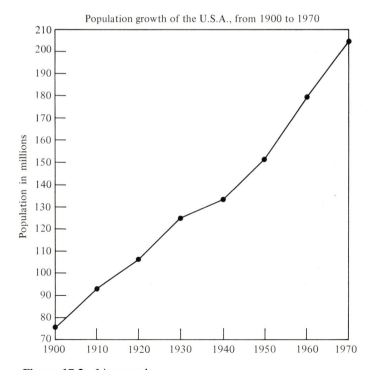

Figure 17.3 Line graph.

Share of the 1960–1970 increase in electricity demand in homes

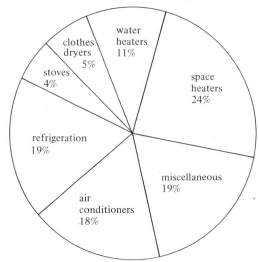

Figure 17.4 Pie chart.

Oil production of some Arab countries in 1972

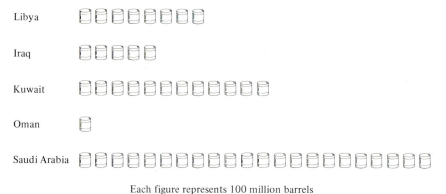

Each figure represents 100 million barrels

Figure 17.5 Pictogram.

constructed by first performing the following calculations:

0.11 × 360° = 40°   0.18 × 360° = 65°
0.05 × 360° = 18°   0.19 × 360° = 68°
0.04 × 360° = 14°   0.24 × 360° = 86°
0.19 × 360° = 68°

(The answers are given correct only to the nearest degree and so add up to 359° rather than 360°.)

Graphs can be informative but also misleading—either by accident or by intention. This is particularly true of line graphs and pictograms. For example, suppose the XYZ Company made the following profits in the years 1969 to 1975 (in millions of dollars):

1969	1970	1971	1972	1973	1974	1975
1.1	1.2	1.4	1.7	2.0	2.1	2.2

By choosing the scale for a line graph showing these profits, the company could give the impression of either a fairly rapid growth in profits or a fairly slow growth in profits, as illustrated in Figs. 17.6 and 17.7, respectively.

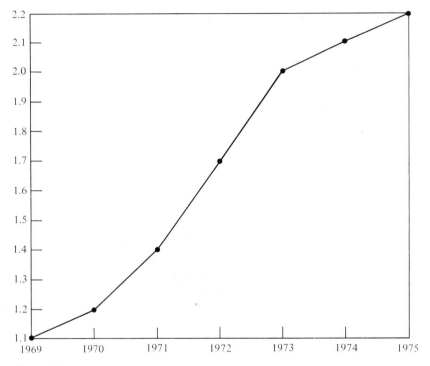

Figure 17.6

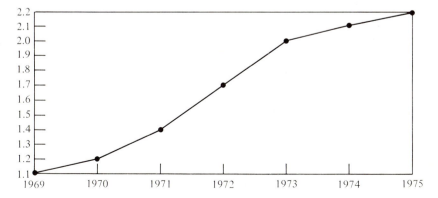

Figure 17.7

Consider now the pictogram shown in Fig. 17.8, representing production from an oil field where the production in the second year was twice the production in the first year. Here the second barrel has been drawn with a radius twice that of the first barrel. This means, however, that the volume of the second barrel is four times the volume of the first barrel, so that the picture could easily be misleading. For this reason, when we use pictograms for comparisons, we should use pictures all of the same size and, in our example, draw *two* barrels, of the same size as that for the first year, to indicate production in the second year.

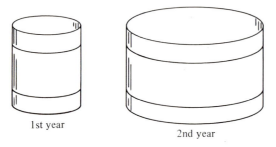

1st year   2nd year

Figure 17.8 Misleading pictogram.

Work on statistics in the primary school should include the collection and presentation of such data as school population, presentation of data obtained from such sources as almanacs, and analysis of statistical tables and graphs found in newspapers and magazines. (The article at the end of this chapter gives some suggestions for graphing activities for children.)

Exercises 17.3

1. Construct a vertical bar graph to picture the following data: The approximate federal debt in billions of dollars was 353.7 in 1969, 370.9 in 1970, 398.1 in 1971, 427.3 in 1972, 458.1 in 1973, and 475.1 in 1974. (Data from the 1975 *Information Please Almanac.*)

2. Construct a horizontal bar graph to picture the following data: The heights, in feet, of the falls in Yosemite National Park are Bridal Veil, 620; Illilouette, 370; Nevada, 594; Ribbon, 1,612; Silver Strand, 1,170; Vernal, 317; Yosemite, 2,425. (Data from the 1975 *World Almanac.*)

3. Construct a line graph to show the data in Exercise 1.

4. Draw a pie chart to show the following data: In 1972, according to the 1975 *World Almanac*, the number of new housing units begun in the U.S.A. in the Northeast was 329,500, in the North Central region 442,800, in the South 1,057,000, and in the West 527,400. (First find the per cent each region was of the total.)

5. Use Fig. 17.2 to find the approximate length of the Nile, the Congo, and the Niger rivers.

6. Use Fig. 17.5 to find the number of barrels of oil produced by Kuwait and by Saudi Arabia in 1972.

7. Use Fig. 17.6 or Fig. 17.7 to find the years in which there was the least increase in profits and the years in which there was the greatest increase for the XYZ Company. How much were these increases?

8. Do the following exercises taken from a fifth-grade text. (The information is taken from the *World Book Encyclopedia* for 1967, and so is considerably outdated!)

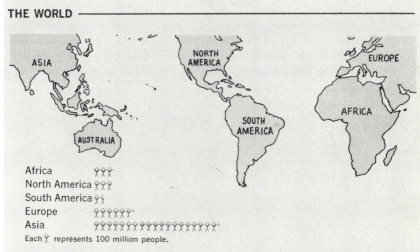

This **pictograph** shows the population* of five of the seven large blocks of land (continents) on the earth's surface. Antarctica is the only continent that is not populated. Australia's population of about 11½ million is too small to show on the pictograph.

1. List the continents given in the pictograph. Beside each continent give the approximate population.
2. Use the pictograph to tell how many times as many people live in Asia as in North America.
3. Use the pictograph to tell how many more people live in Asia than on all the other continents combined.

9. Do the following exercises taken from a sixth-grade text.

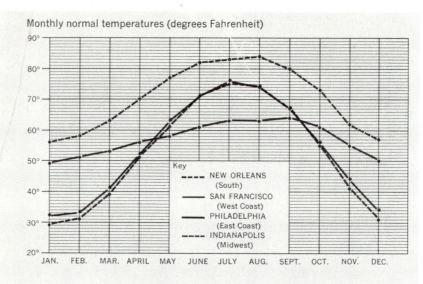

1. Give the normal July temperature for each city.
2. How much higher is the July temperature in New Orleans than in:
   [A] Indianapolis   [B] San Francisco

17.4 MEASURES OF CENTRAL TENDENCY

Suppose that the grades, in order, in an examination given to 13 persons were as follows:

96	89	80
95	87	80
90	87	79
89	86	
89	84	

The most common grade is 89, and this is called the **mode** of the data; the grade in the middle (seventh from top or bottom grade) is 87, and this is called the **median** of the data.

More than one mode is possible. For the following data:

100	73
100	70
95	70
90	65
76	64

17.4   **Measures of Central Tendency**   **409**

there are two modes: 100 and 70. (In fact, if each item in a set of data were different, we would have as many modes as items!) Since there is an even number (10) of items, the median in this case is taken to be halfway between the fifth and sixth numbers, i.e., as

$$\frac{76 + 73}{2} = 74.5$$

Both the mode and the median are called **measures of central tendency,** and sometimes, loosely, "averages". However, the word "average" most commonly refers to a third measure of central tendency, the **arithmetic mean** (or simply, the **mean**). This is the sum of all the numbers in a set of data divided by the number of items. Thus the mean for the first set of data is:

$$\frac{96 + 95 + 90 + 89 + 89 + 89 + 87 + 87 + 86 + 84 + 80 + 80 + 79}{13}$$

$$= \frac{1131}{13} = 87$$

(A shorter method for calculating such means will be given later.)

Measures of central tendency are called so because they are numbers about which the set of numbers appears to cluster. That is, in some sense, they are "typical" values. Which measure of central tendency to use depends on the data and the desire to present a true (or false!) picture. For example, suppose that a small factory owned by Mr. Jones had the following payroll:

Mr. Jones	$26,000	Mr. Hood	$7,000
Mr. Smith	9,000	Mr. Davis	6,000
Mr. Exley	9,000	Mr. Shapiro	5,000
Mr. Ramirez	8,000		

The mode for this data is clearly $9,000, the median is clearly $8,000, and the mean is easily calculated to be $10,000. To convince stockholders that he is paying high wages, Mr. Jones might announce that the average wage paid is $10,000 (the mean); if he is a little less deceitful, he might say that the most common wage paid is $9,000 (the mode). On the other hand, the *union* might use the median, and say that half the salaries paid are less than $8,000.

Here the mean, the median, and the mode all differ from each other. Any two, or even all three, may, however, coincide. Thus, for the data below, all three are equal to 33.

30
32
33
33
37

410   Statistics   **17.4**

Since it is frequently useful for a teacher to compute means in determining grades, we end this chapter with a shortcut method for calculating them—called the method of **assumed mean.**

Suppose that we have the data $X_1, X_2, \ldots, X_n$, for which we wish to calculate the mean, M. Suppose, further, we guess that the mean is A. Consider now the expression

$$S = A - \frac{(A - X_1) + (A - X_2) + \cdots + (A - X_n)}{n}$$

where

$$A - X_1, \qquad A - X_2, \qquad \ldots, \qquad A - X_n$$

are called the **deviations** from the assumed mean, A. Algebraic calculations give us

$$S = A - \frac{(A + A + \cdots + A) - (X_1 + X_2 + \cdots + X_n)}{n}$$

$$= A - \frac{nA - (X_1 + X_2 + \cdots + X_n)}{n}$$

$$= A - \frac{nA}{n} + \frac{X_1 + X_2 + \cdots + X_n}{n}$$

$$= A - A + \frac{X_1 + X_2 + \cdots + X_n}{n}$$

$$= \frac{X_1 + X_2 + \cdots + X_n}{n}$$

$$= M$$

Thus $S = M$, and so

$$M = A - \frac{(A - X_1) + (A - X_2) + \cdots + (A - X_n)}{n}$$

$$= A - \frac{\text{Sum of deviations}}{n}$$

That is, the mean is equal to the assumed mean minus a correction factor. (Note that if the assumed mean is less than the true mean, the correction factor will be negative, and we will have, for example, $A - (-2) = A + 2$.)

To illustrate the use of this formula let us return to the problem of calculating the mean for the set of data given at the beginning of this section. It doesn't really matter what we use for our assumed mean, but the arithmetic is minimized if our assumed mean is taken reasonably close to the actual mean. Let us, then, take $A = 89$. Then we have:

17.4 Measures of Central Tendency

X	$A - X$
96	−7
95	−6
90	−1
89	0
89	0
89	0
87	2
87	2
86	3
84	5
80	9
80	9
79	10

Sum of deviations = 26

Thus

$$M = A - \frac{\text{Sum of deviations}}{13} = 89 - 2 = 87$$

Exercises 17.4

1. Calculate the mean for the above data using 86 as an assumed mean. Also find the median and the mode (or modes). What would happen if you took 87 as the assumed mean?

2. Use the method of assumed mean to calculate the mean of the data of Exercise 1 of Exercises 17.2. Also find the median and the mode (or modes).

3. See Exercise 9 at the end of Section 17.3, where we showed two exercises from a sixth-grade text; the next exercise in that text asks for the average summer (June, July, and August) temperatures for San Francisco, Philadelphia, Indianapolis, and New Orleans. Find these average temperatures.

4. Do the following exercises taken from a fifth-grade text.

   Finding special averages

   1. The bar graph shows the heights in inches of 6 fifth-grade children. Find the average height of these children.

   2. Jan is 3834 days old. Fran is 4015 days old. Nan is 3923 days old. Find the average age (in days) of the 3 girls.

   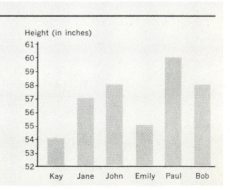

412   Statistics

3. The table gives the weights of 8 fifth-grade children. Find the average weight of the children.

Jay	76	Ann	79
Nancy	64	Ted	85
Susan	68	Bill	98
Steve	96	Mary	82

Chapter Test

In preparation for this test you should review the meaning of the following terms:

Frequency table (400)

Class interval (401)

Bar graph (402–403)

Line graph (403)

Pie chart (404)

Pictogram (404)

Measures of central tendency (408–409)

Mode (408)

Median (408)

Arithmetic mean (409)

You should also review how to:

1. Construct frequency tables (400–401);
2. Construct bar, line, and pie graphs and pictograms (402–404);
3. Calculate modes and medians (408);
4. Calculate arithmetic means both directly from the definition (409) and also by the method of assumed mean (410–411).

(Data in Problems 1 through 4 are taken from the 1975 *Information Please Almanac.*)

1. The career earnings of the top six U.S. golfers (through October 15, 1974) were as follows:

Jack Nicklaus	$2,224,931	Lee Trevino	$1,265,076
Arnold Palmer	$1,666,544	Bruce Crampton	$1,193,674
Billy Casper	$1,483,633	Gary Player	$1,091,596

   Construct a vertical bar graph to show this information.

2. The Indian population in the U.S.A. from 1920 to 1970 was as follows:

1920	244,437	1950	343,410
1930	332,397	1960	523,591
1940	333,969	1970	792,730

   Construct a line graph to show this data.

3. The U.S. domestic freight traffic in 1973 by major carriers, was apportioned among these carriers, in millions of ton-miles, as follows: railroads, 860,000; inland waterways, 351,000; trucks, 510,000; and oil pipelines, 495,000. Construct a pie chart to show this data.

4. The industrial production indices for some western European countries and the U.S.S.R. in 1973 were (with 1963 = 100) as follows:

Austria	184	Italy	171
Belgium	157	Luxembourg	147
Denmark	166	Netherlands	218
France	187	Norway	161
West Germany	174	Sweden	167
Greece	284	United Kingdom	135
Ireland	184	U.S.S.R.	216

a) Calculate the mean of these indices, correct to 2 decimal places, by using an assumed mean of 170.
b) Find the median index.
c) Find the mode or modes of the indices.

5. The grades on a test involving 50 students were as follows:

```
64  92  96  73  61  86  76  88  82  77
77  65  78  98  98  79  80  61  86  94
80  72  61  75  79  68  90  69  61  73
82  86  85  63  86  99  100 70  66  67
84  62  86  84  91  62  62  84  69  77
```

Using the intervals 96–100, 91–95, ..., 61–65, make a frequency chart for this data.

TEST ANSWERS

(Your answers to the first two problems may, of course, look somewhat different from the graphs given here.)

1.

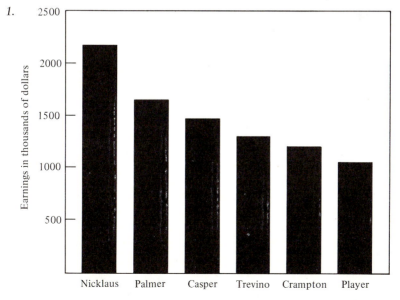

Career earnings of the top 6 golfers through Oct. 15, 1974

2.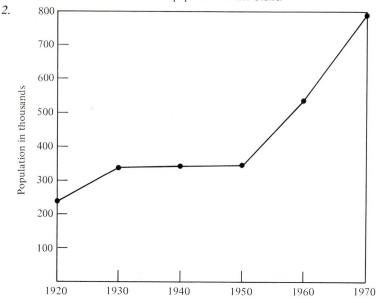
Indian population of the U.S.A.

3.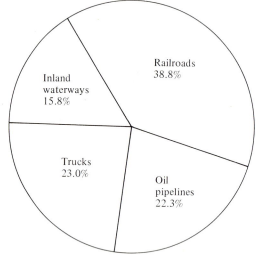
Domestic freight traffic in 1973

4. a) Sum of deviations from the mean = −14 + 13 + 4 − 17 − 4 − 114 − 14 − 1 + 23 − 48 + 9 + 3 + 35 − 46 = −171;

$$M = 170 - \frac{-171}{14} = 170 + 12.21 = 182.21$$

b) Median = $\dfrac{171 + 174}{2}$ = 172.5   c) Mode = 184

5. 96–100   卌   5
    91–95   ///   3
    86–90   卌 //   7
    81–85   卌 /   6
    76–80   卌 ////   9
    71–75   ////   4
    66–70   卌 /   6
    61–65   卌 卌   10

Graphs in the primary grades

by Morris Pincus and Frances Morgenstern

Reprinted from *The Arithmetic Teacher*, Vol. **17** (1970), pp. 499–501.

Many topics that were formerly introduced in the upper grades are now being introduced successfully in the lower grades, some as early as kindergarten. The construction and reading of graphs is one of these topics.

Information in graphic form is interesting to young children because it is presented in a concrete or pictorial way. Graphs encourage discovery inasmuch as relationships, patterns, trends, and significant changes in the data can be readily perceived. Ideas and information that might otherwise seem complicated or obscure can be grasped in this more concise and simplified form. The use of words is reduced to a minimum or eliminated. Because graphs make possible the presentation and analysis of data in an interesting and clear form, data for their construction can be drawn from other curricular areas, especially science and social studies.

Not only is information more easily seen and relationships more readily understood, but often new questions are raised stimulating further study. For example, an inquiry into the reason for a sharp increase in absence might lead to the discovery that there was a snowstorm on a particular day (see Fig. 1).

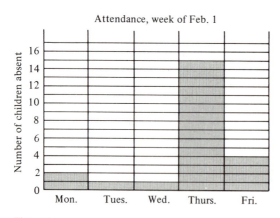

Figure 1

It is desirable to develop the ability to read and construct graphs through a sequence of activities that proceed from concrete objects to symbolic representation.

For example, in a kindergarten class, the teacher forms lines to show the number of children who walk to school, ride in cars, or ride in buses (see Fig. 2).

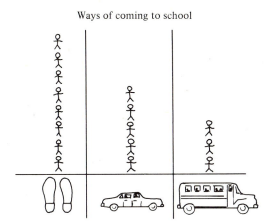

Figure 2

It is possible for the children, by just looking at the length of the lines, to answer such questions as: How do most children come? Do more children come by car or by bus?

From this experience, it is easy for children to realize that there are several difficulties in using the actual children. It is hard to look at your own line and compare it with others, children in one line might be crowded together, a child might have to leave the line, and so forth. They can consider alternatives, such as using an object (doll, block, stick figures, name cards) for each child, thus making the transition from the actual thing to a representative object.

1. **Constructing Graphs from Materials**

    Objects are used to represent in a one-to-one correspondence the actual things being discussed.
    a) *Blocks*

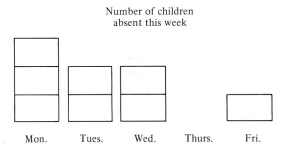

418 Statistics

b) *Empty spools or beads on dowels or wire*

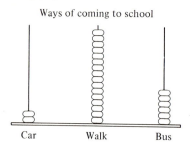

c) *Gummed squares*

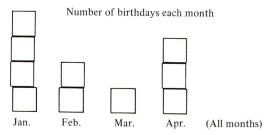

d) *Strips of paper* (oaktag) representing actual height of plants

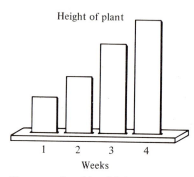

e) *Name cards*   (As children begin to learn to read their names, name cards may be substituted for actual children.)

James
Sally
Ann
Tony
Judy
David
Bob
Jeffrey
Jill
Debbie

Eat in school

Tommy
Ellen
Joseph
Donna

Eat at home

Graphs in the Primary Grades 419

2. **Converting a Record or Tally to a Graph.**

   a) *Keep a tally.* b) *Rotate the tally.*

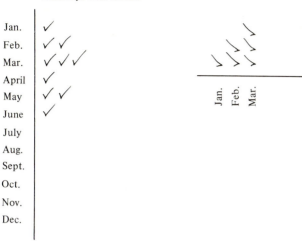

   c) *Lead children to see a need for regular spacing* in order to "read" information quickly and accurately (achieved by putting checks in adjacent boxes of same size, e.g., graph paper).
   d) *Same graph as in* (b)—each check in a box.

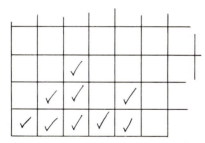

   e) *Color or shade each column* of checks.
   f) *To make it easier to obtain exact information from the graph,* we develop the idea of using a vertical axis and labeling it.

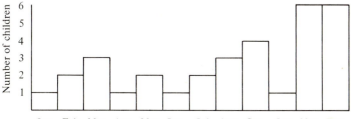

420 Statistics

3. Some Suggested Topics That Can Lead to Graph Construction.

a) *Number of times* each child has painted
b) *Number of holidays* each month
c) *Favorite TV programs*
d) *Number of children* leaving room each hour
e) *Makes of cars* on school block
f) *Kinds of sandwiches* eaten for lunch
g) *Size of families*
h) *Temperature* each day (high, low, or average)
i) *Number of cars* passing school each 5 minutes
j) *Types of clocks* in homes (electric, mechanical, type numerals, etc.)
k) *Time spent* on different activities in a day (school day, weekend, or holiday)

REFERENCES FOR FURTHER READING

1. Girard, Ruth A., "Development of critical interpretation of statistics and graphs," *AT*, **14** (1967), pp. 272–277.

2. Grass, Benjamin A., "Statistics Made Simple," *AT*, **12** (1968), pp. 196–198. Describes work done with a fourth-grade class.

3. Huff, Darrell, *How to Lie with Statistics*, W. W. Norton and Co., N.Y., 1959. Many examples of how statistical data can be presented in a misleading way.

4. Junge, Charlotte W., "Dots, plots, and profiles," *AT*, **16** (1969), pp. 371–378. Discusses the role of graphing in the primary school.

5. Nuffield Mathematics Project, *Pictorial Representation*, John Wiley and Sons, N.Y. (no date). Another publication in this fine series—amply illustrated with children's work.

6. Pieters, Richard S., and Kinsella, John J., "Statistics," Chapter 7 of *The Growth of Mathematical Ideas*, National Council of Teachers of Mathematics, Reston, Va., 1959. Describes ways of introducing statistics into the primary-school curriculum.

7. Schell, Leo M., "Horizontal enrichment with graphs," *AT*, **14** (1967), pp. 654–656.

8. Wilkinson, Jack D., and Nelson, Owen, "Probability and statistics—Trial teaching in the sixth grade," *AT*, **13** (1966), pp. 100–105.

Chapter 18

Number Theory

18.1 INTRODUCTION

Number theory, as thought of by mathematicians, deals with the properties of whole numbers. Many of these properties lend themselves very nicely to the discovery approach (which is such a prominent feature of contemporary mathematics programs); and it is indeed possible for children to make discoveries of results not previously known, even though the literature of number theory is a vast one, and new results are being obtained every year by professional mathematicians.

We have already considered several properties of whole numbers—some of them in the form of discovery exercises; among them were:

1. The existence of Kapreker's constant (Section 2.9);

2. The pattern obtained in repeating the process of adding the squares of the digits of a whole number (Exercise 7 of Exercises 2.9);

3. Patterns in the table of basic addition facts (Section 2.10);

4. The pattern obtained in repeating the process of squaring the sum of the digits of a number (Exercise 1 of Exercises 2.10);

5. Properties of the G.C.D. and L.C.M. (Exercises 6.5).

In this concluding chapter we will look at some additional results concerning whole numbers, including some in the exercises involving discovery-type experiences. All of the topics discussed may be introduced with profit at various levels in the primary school.

422   Number Theory   **18.2**

18.2 TESTS FOR DIVISIBILITY

It is easy to determine whether or not a whole number N is divisible by 2, 5, or 10 simply by looking at the last digit of the number. (Actually, of course, by looking at the last digit of the *numeral* for the number. Numbers, being abstractions, don't have digits!)

   i) A number is divisible by 2 if and only if its last digit is 0, 2, 4, 6, or 8.

  ii) A number is divisible by 5 if and only if its last digit is 0 or 5.

 iii) A number is divisible by 10 if and only if its last digit is 0.

      For a number $N \geq 100$, we have the following test for divisibility by 4:

 iv) A number is divisible by 4 if and only if the number formed by the last two digits is divisible by 4.

For example, 143,216 is divisible by 4 because 16 is divisible by 4 whereas 143,226 is not divisible by 4 because 26 is not divisible by 4.

      We will prove the validity of this test for five-digit numbers. (The proof that the result holds for any whole number $N \geq 100$ is similar but involves a somewhat complicated notation that tends to obscure the basic idea of the proof.) Now any five-digit number can be written as

$$N = 10{,}000a + 1000b + 100c + 10d + e$$

where a, b, c, d, and e are in the set $\{0, 1, 2, \ldots, 9\}$ (but $a \neq 0$). Th is, for example

$$51{,}102 = 10{,}000 \times 5 + 1000 \times 1 + 100 \times 1 + 10 \times 0 + 2$$

Then, applying the distributive property,

$$N = 4 \times (2500a + 250b + 25c) + 10d + e$$

Thus

$$N \div 4 = \frac{N}{4} = (2500a + 250b + 25c) + \frac{10d + e}{4}$$

Now $2500a + 250b + 25c$ is certainly a whole number. Hence $N/4$ is a whole number, and hence N is divisible by 4, if and only if $\frac{10d+e}{4}$ is a whole number, i.e., if and only if $10d + e$ is divisible by 4. But $10d + e$ is simply the expanded form of the two-digit number de. (For example, if $d = 2$, and $e = 4$, $24 = 10 \times 2 + 4$.)

      Note that if a number is not divisible by 2, there is no need to test further whether or not it is divisible by 4, and that, for two-digit numbers, the test for divisibility by 4 would be simply to try the division. (This is why we stipulated for the test that $N \geq 100$.)

18.2   **Tests for Divisibility**   **423**

For divisibility by 3 or 9 we have the following test:

v) A number is divisible by 3 [9] if and only if the sum of its digits is divisible by 3 [9].

For example, 141 is divisible by 3 because $1 + 4 + 1 = 6$ is divisible by 3 but is not divisible by 9 because 6 is not divisible by 9. On the other hand, 2457 is divisible by both 3 and 9 because $2 + 4 + 5 + 7 = 18$ and 18 is divisible by 9 (and hence certainly is divisible by 3). (Of course, a number not divisible by 3 is also not divisible by 9.)

We prove this result for the "9" case and, again, only for five-digit numbers, leaving the proof of the "3" case as an exercise. So we again have

$$N = 10{,}000a + 1000b + 100c + 10d + e$$

with a, b, c, d, and e as before. But now we rewrite N as

$$
\begin{aligned}
N &= (9999 + 1)a + (999 + 1)b + (99 + 1)c + (9 + 1)d + e \\
&= (9999a + a) + (999b + b) + (99c + c) + (9d + d) + e \\
&= (9999a + 999b + 99c + 9d) + (a + b + c + d + e) \\
&= 9 \times (1111a + 111b + 11c + d) + (a + b + c + d + e)
\end{aligned}
$$

Thus

$$N \div 9 = \frac{N}{9} = (1111a + 111b + 11c + d) + \frac{a + b + c + d + e}{9}$$

and, as in the proof of the test for divisibility by 4, we see that N is divisible by 9 if and only if $a + b + c + d + e$, the sum of the digits of N, is divisible by 9.

Exercises 18.2

1. Use the tests given in the text to determine whether or not the following numbers are divisible by 2, 3, 4, 5, 9, or 10.

   a) 367   b) 729   c) 12,986   d) 37,264
   e) 34,710   f) 2592   g) 647   h) 295
   i) 72,460   j) 82,396   k) 2591   l) 149,278

2. Prove the rule for divisibility by 3 for a five-digit number.

3. If a number is divisible by both 2 and 3, must it also be divisible by 6? If a number is divisible by 6, must it also be divisible by both 2 and 3?

4. If a number is divisible by both 2 and 5, must it also be divisible by 10? If a number is divisible by 10, must it also be divisible by 2 and 5?

5. If a number is divisible by both 2 and 6, must it also be divisible by 12? If a number is divisible by 12, must it also be divisible by 2 and 6?

424   Number Theory   **18.3**

6. Devise a test for divisibility by 8, for numbers greater than or equal to 1000, that is similar to the test in the text for divisibility by 4.

18.3 CASTING OUT NINES

"Casting out nines" refers to a method of checking computations with whole numbers. It is based on the fact that when any whole number is divided by 9, the remainder is the same as the remainder obtained when the sum of the digits of the number is divided by 9. Again, our proof of this fact will be for five-digit numbers, with a general proof possible along these lines.

For a five-digit number, then, we saw in the previous section that

$$N \div 9 = \frac{N}{9} = (1111a + 111b + 11c + d) + \frac{a + b + c + d + e}{9}$$

Now if $a + b + c + d + e$ divided by 9 has a quotient q and a remainder r, we have

$$N \div 9 = (1111a + 111b + 11c + d) + q + \frac{r}{9}$$

$$= (1111a + 111b + 11c + d + q) + \frac{r}{9}$$

so that N divided by 9 has the quotient $(1111a + 111b + 11c + d + q)$ and, like $a + b + c + d$, the remainder r.

For example, take the numbers 53, 76, and 135 and divide each by 9. We have:

$53 \div 9$ has remainder 8	$5 + 3 = 8$ and $8 \div 9$ also has remainder 8;
$76 \div 9$ has remainder 4	$7 + 6 = 13$ and $13 \div 9$ also has remainder 4;
$135 \div 9$ has remainder 0	$1 + 3 + 5 = 9$ and $9 \div 9$ also has remainder 0.

Testing for the correctness of the answer to an addition problem by casting out nines is illustrated by the following example:

Addition	**Sum of Digits**	**Sum of Digits**	
156	$1 + 5 + 6 = 12$	$1 + 2 = 3$	(Continue these
94	$9 + 4 = 13$	$1 + 3 = 4$	columns if
290	$2 + 9 + 0 = 11$	$1 + 1 = 2$	necessary until
+44	$4 + 4 = 8$	$+8$	one-digit
Sum = 584		Sum = 17	numbers are
$5 + 8 + 4 = 17$		$1 + 7 = 8$	obtained.)
$1 + 7 = 8$			

The remainders when the sum is divided by 9 have been calculated in two different ways to be 8, and this serves as a check on the addition.

18.3 **Casting Out Nines** **425**

Various shortcuts are possible. For example, in considering $5 + 8 + 4$ we can observe that $5 + 4 = 9$ and then "cast out" the 9 to leave 8. Similarly, in considering $9 + 4$ we can "cast out" the 9 right away to get 4.

Next, we consider an example illustrating the checking of a subtraction by this method:

$$
\begin{array}{r}
4967 \\
-2224 \\
\hline
2743
\end{array}
\qquad
\begin{array}{r}
8 \\
-1 \\
\hline
7
\end{array}
$$

$$2743 \to 7$$

Using the shortcut method, we have $4967 \to 4 + 6 + 7$ by "casting out a 9". Then $4 + 6 \to 10 \to 1$ by "casting out a 9 out of 10" and so $4967 \to 1 + 7 = 8$. Then $2224 \to 2 + 2 + 2 + 4 = 10 \to 1$ and $8 - 1 = 7$. Finally, $2743 \to 2 + 7 + 4 + 3 \to 7$ since $2 + 7 = 9$ which we can "cast out". Thus our two remainders are both 7 and, again, this serves as a check on the calculation.

It may happen that the subtraction of the sums of digits will yield a negative number, as in the following example:

$$
\begin{array}{r}
95 \\
-69 \\
\hline
26
\end{array}
\qquad
\begin{array}{r}
9 + 5 \to 5 \\
(-)6 + 9 \to 6 \\
\hline
-1
\end{array}
$$

$$2 + 6 = 8$$

Then we simply add 9 to -1 to get 8, or, alternatively, subtract 9 from 8 to get -1.

Now let us consider the checking of a multiplication problem:

$$
\begin{array}{r}
3673 \\
\times 75 \\
\hline
18365 \\
267110 \\
\hline
285475
\end{array}
\qquad
\begin{array}{l}
3 + 6 + 7 + 3 \to 10 \to 1 \\
(\times) \qquad 7 + 5 = 12 \to 3 \\
\hline
\qquad\qquad\qquad\qquad 3
\end{array}
$$

$2 + 8 + 5 + 4 + 7 + 5$ (Underlining indicates combinations
$\to \; 2 + 8 + 7 + 5$ giving a 9 to be "cast out.")
$\to 8 + 5 \to 4$

Since $4 \neq 3$, some mistake has been made in our multiplication (or possibly in our check!). Find the error.

To check a division problem, $N \div d$, by casting out nines we must rewrite the answer in the form $N = (q \times d) + r$. For example, consider the division

$$
\begin{array}{r}
426 \\
17 \overline{\smash{)}7243} \\
6800 \\
\hline
443 \\
340 \\
\hline
103 \\
102 \\
\hline
1
\end{array}
$$

426   Number Theory   **18.4**

Rewriting the solution, we have

$$7243 = (426 \times 17) + 1$$

Now $7243 \to 4 + 3 = 7$; $426 \to 12 \to 3$; $17 \to 8$. On the left, then, we have 7 and on the right,

$$(3 \times 8) + 1 = 24 + 1 = 25 \to 7$$

Thus the result of casting out nines is 7 in both cases.

It should be noted that checking by casting out nines does not enable us to detect all possible mistakes but it does enable us to catch most *likely* mistakes. For example, in our addition example, the wrong answers of 548 and 683 would "check". But these are certainly not "likely" mistakes! (For a discussion of the relationship between casting out nines and clock arithmetic, see the article at the end of this chapter.)

Exercises 18.3

1. Use casting out nines to check the following calculations and correct any errors found by this check.

a)
```
     29
     36
     72
   +75
    212
```

b)
```
   61,847
   23,973
   18,421
  +20,068
  124,309
```

c)
```
    6839
  −3471
    3468
```

d)
```
   27,212
  −13,449
   13,763
```

e)
```
      237
    ×456
     1422
    11850
    94800
   108072
```

f)
```
      512
     ×75
     2550
    35840
    38390
```

g)
```
           817
   18 ⟌ 14721
        14400
          321
          180
          141
          126
           15
```

h)
```
          1284
   39 ⟌ 50000
        39000
        11000
         7800
         3200
         3020
          180
          156
           24
```

2. What do you think would correspond to a check by casting out nines in base ten, if computations were in base five? In base seven? In base n for $n \geq 2$?

18.4 SOME UNSOLVED PROBLEMS IN NUMBER THEORY

There are many unsolved problems in mathematics, and the amount of new mathematics being developed increases every year. (Indeed, it is undoubtedly true that the amount of mathematics developed in the first half of this century is greater than in all the previous centuries put together!) Most of these unsolved problems, however, concern highly technical material. In number theory, on the other hand, there are several unsolved problems that are easy to understand. We will discuss five of them here.

The first two we will present are the oldest and deal with what are called perfect numbers. Suppose that we consider the sum of all the **proper divisors** of a number (i.e., the sum of all the divisors of a number except the number itself). Consider the following table where we do not list 1 (which has no proper divisors) or primes (which have only 1 as a proper divisor).

Number	Sum of proper divisors
4	$1 + 2 = 3$
6	$1 + 2 + 3 = 6$
8	$1 + 2 + 4 = 7$
10	$1 + 2 + 5 = 8$
12	$1 + 2 + 3 + 4 + 6 = 16$
14	$1 + 2 + 7 = 10$
15	$1 + 3 + 5 = 9$
16	$1 + 2 + 4 + 8 = 15$
18	$1 + 2 + 3 + 6 + 9 = 21$

Now for some of the numbers in our list (4, 8, 10, 14, 15, and 16), this sum is less than the number; such numbers are called **deficient** numbers. In other numbers in our list (12 and 18) the sum is greater than the number; such numbers are called **abundant** numbers. For just one number in our list (6) is the sum equal to the number; such numbers are called **perfect** numbers. (This nomenclature dates back to the ancient Greeks.)

Perfect numbers are quite rare. The next one after 6 is 28 $(1 + 2 + 4 + 7 + 14 = 28)$, and the next one after that is 496 $(1 + 2 + 4 + 8 + 16 + 31 + 62 + 124 + 248 = 496)$.

Since the time of Euclid it has been known that every *even* perfect number N is of the form

$$N = 2^{p-1}(2^p - 1)$$

where *both* p and $2^p - 1$ are prime numbers. (Primes of the form $2^p - 1$ are known as **Mersenne** primes, after the seventeenth century French mathematician Marin Mersenne.) Thus the first four perfect numbers can be obtained as follows:

p	$p - 1$	2^{p-1}	$2^p - 1$	$N = 2^{p-1}(2^p - 1)$
2	1	2	3	6
3	2	4	7	28
5	4	16	31	496
7	6	64	127	8128

At this point it may seem as if the requirement that $2^p - 1$ be a prime is unnecessary since, for all the primes we have used so far, $2^p - 1$ is a prime. For $p = 11$, however, $2^p - 1 = 2^{11} - 1 = 2047 = 23 \times 89$, so that $2^{11} - 1$ is not a prime and hence $2^{10}(2^{11} - 1) = 2,096,128$ is not a perfect number.

428   Number Theory   **18.4**

We see that the finding of *even perfect numbers* depends upon finding Mersenne primes. At the present time only 24 Mersenne primes are known and thus only 24 perfect numbers are known. The largest perfect number known in 1975 was $2^{19,936}(2^{19,937} - 1)$, corresponding to the Mersenne prime $2^{19,937} - 1$, which has 12,003 digits!

Although 24 even perfect numbers are known, it is not known whether there is an infinite number of even perfect numbers, nor is it known whether there are any odd perfect numbers. (It is known that if any odd perfect number exists, it must be greater than 10^{36}!) Here, then, are the first two of our five unsolved problems.

In historical sequence, our next unsolved problem involves what is known as **Fermat's last "theorem"** (after Pierre de Fermat, another French mathematician of the seventeenth century). As background for this "theorem" recall the Pythagorean theorem, which states that if a and b are the lengths of the sides of a right triangle whose hypotenuse is of length c, then

$$c^2 = a^2 + b^2$$

Now there are many triples of natural numbers (c, a, b) that satisfy this equation. Two of them are $(5, 3, 4)$ and $(13, 5, 12)$ since

$$5^2 = 3^2 + 4^2 \qquad \text{and} \qquad 13^2 = 5^2 + 12^2$$

Since the time of Pythagoras, mathematicians have wondered whether or not there are also triples of natural numbers (c, a, b) such that

$$c^3 = a^3 + b^3$$
$$c^4 = a^4 + b^4$$
$$c^5 = a^5 + b^5$$

etc.

In reading a book on number theory written by the Greek mathematician Diophantus (c. A.D. 250), in which ways of obtaining all triples of natural numbers (c, a, b) satisfying $c^2 = a^2 + b^2$ are described, Fermat wrote the following famous comment in the margin of the book:

> However, it is impossible to write a cube as the sum of two cubes, a fourth power as the sum of two fourth powers, and, in general, any power beyond the second, as the sum of two similar powers. For this I have discovered a truly wonderful proof, but the margin is too small to contain it.

That is, Fermat asserted that no triples (c, a, b) of natural numbers exist that satisfy the equation

$$c^n = a^n + b^n$$

for n a natural number greater than 2.

To date no one has been able either to prove Fermat's last "theorem" (that is why we have used the quotation marks) or, on the other hand, to find a

18.4   **Some Unsolved Problems in Number Theory**   **429**

counterexample (i.e., a triple of natural numbers (c, a, b) such that $c^n = a^n + b^n$ for some natural number $n > 2$). Fermat did give a proof of his "theorem" for the case $n = 4$ and it seems likely that he also had a proof for the case $n = 3$. The general consensus among mathematicians today seems to be that Fermat's conjecture is indeed correct but that what Fermat thought was a proof contained some error. (You may wonder why Fermat did not publish his proof, as a mathematician of today would. The fact is that mathematical journals did not exist in Fermat's time. Results were communicated among mathematicians by letters; and either Fermat found an error in what he thought was a proof, did not get around to writing up his proof, or his letters regarding it have been lost. Personally I'd bet on the first alternative!)

Our final examples of unsolved problems date back to around 1742, when a German mathematician, Christian Goldbach, observed that it seemed that every even number greater than 4 could be written as the sum of two odd prime numbers (not necessarily uniquely). For example,

$$6 = 3 + 3 \qquad\qquad\qquad 14 = 3 + 11 \, (= 7 + 7)$$
$$8 = 3 + 5 \qquad\qquad\qquad 16 = 5 + 11 \, (= 3 + 13)$$
$$10 = 3 + 7 \, (= 5 + 5) \qquad\quad 18 = 7 + 11 \, (= 5 + 13)$$
$$12 = 5 + 7 \qquad\qquad\qquad 20 = 3 + 17 \, (= 7 + 13)$$

where the equality in parentheses indicates an alternative way of providing the desired sum.

Goldbach also observed that it seemed that every odd number greater than 7 is the sum of three odd prime numbers (again, not necessarily uniquely). For example:

$$9 = 3 + 3 + 3 \qquad\qquad\qquad 17 = 3 + 7 + 7$$
$$(= 5 + 5 + 7 \quad = 3 + 3 + 11)$$
$$11 = 3 + 3 + 5 \qquad\qquad\qquad 19 = 3 + 5 + 11$$
$$(= 5 + 7 + 7 \quad = 3 + 3 + 13)$$
$$13 = 3 + 3 + 7 \, (= 3 + 5 + 5) \qquad 21 = 3 + 5 + 13$$
$$(= 3 + 7 + 11 = 5 + 5 + 11$$
$$= 7 + 7 + 7)$$
$$15 = 3 + 5 + 7 \, (= 5 + 5 + 5) \qquad 23 = 3 + 3 + 17$$
$$(= 3 + 7 + 13 = 5 + 7 + 11$$
$$= 5 + 5 + 13)$$

The conjectures that every even number greater than 4 can be written as the sum of two odd prime numbers and that every odd number greater than 7 can be written as the sum of three odd prime numbers are known as **Goldbach's conjectures.** They have been verified for all numbers up to 100,000; and steps toward a proof have been made—especially by Russian mathematicians. Again, however, these conjectures have as yet been neither proved nor disproved.

430   **Number Theory**   **18.4**

Programed Lesson 18.4

1. A proper divisor of a natural number N is any divisor of N other than _____.

     N

2. N is called a deficient number if the sum of the proper divisors of N is _____ than N.

     less

3. N is called an abundant number if the sum of the proper divisors of N is _____ than N.

     greater

4. N is called a perfect number if the sum of the proper divisors of N is _____ to N.

     equal

5. Fermat's last "theorem" states that the equation $c^n = a^n + b^n$ has no solution in natural numbers a, b, and c, if n ___ ___ .

     $>, 2$

6. Goldbach's conjecture concerning even numbers states that if n is an even number greater than ____, then n is the sum of _____ _____ _____ numbers.

     4, two, odd, prime

7. Goldbach's conjecture concerning odd numbers states that if n is an odd number greater than ____, then n is the sum of _____ _____ _____ numbers.

     7, three, odd, prime

Exercises 18.4

1. Classify the natural numbers from 19 to 40 as deficient, abundant, or perfect.

2. Do you think that there are any odd abundant numbers?

3. Verify Goldbach's conjectures for the natural numbers from 25 to 50.

4. If N is a natural number such that the sum of its proper divisors is $k \times N$, where k is a natural number, N is called **multiply perfect of class k.** (Thus perfect numbers are the multiply perfect numbers of class 1.) Show that 120 is a multiply perfect number of class 2.

5. Two numbers M and N are said to be **amicable** if the sum of the proper divisors of M is equal to N and the sum of the proper divisors of N is equal to M. Show that 220 and 284 are amicable numbers.

6. **Fermat's theorem** (in contrast to Fermat's *last* "theorem") has been proved. It states that $a^{p-1} - 1$ is divisible by p if p is a prime and a is not divisible by p. Verify Fermat's theorem for the cases when $p = 5$ and $a = 3$, and when $p = 3$ and $a = 8$.

7. Some prime numbers can be written as the sum of the squares of two natural numbers and others cannot. (For example, $5 = 1^2 + 2^2$ and $13 = 2^2 + 3^2$, but 3 and 7 cannot be so written.) By experimentation, with, say, the natural numbers from 1 to 50, try to find a rule for determining which prime numbers can be so written.

8. Consider the figure below, which can be extended indefinitely. In the first row there is 1 dot, in the second 2, in the third 3, in the fourth 4, etc. The sum of the dots in the first n rows is given for $n = 1, 2, 3,$ and 4 by the following table:

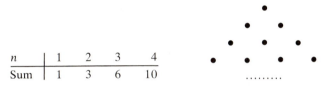

n	1	2	3	4
Sum	1	3	6	10

By extending the table if necessary, try to find the formula for the sum of the dots in n rows. (These sums are called, naturally enough, **triangular** numbers.)

9. Some natural numbers can be written as the sum of three consecutive natural numbers and others cannot. (For example, $6 = 1 + 2 + 3$ and $12 = 3 + 4 + 5$, but 3 and 10 cannot be so written.) Try to find a rule for determining which numbers can be so written.

10. Show that if the conjecture of Goldbach concerning even numbers is true, then his conjecture concerning odd numbers must also be true. (Because of this, when *the* Goldbach conjecture is referred to, the one concerning even numbers is meant.)

11. Choose any natural number. If it is even, divide it by 2; if it is odd, multiply it by 3 and add 1. Repeat this process with the number you obtain, and continue until you find something interesting about this process. (For example, beginning with 17, we get $3 \times 17 + 1 = 52$, then $52 \div 2 = 26$, $26 \div 2 = 13$, $13 \times 3 + 1 = 40$, etc.) The conjecture that you will undoubtedly make concerning this process is, at present (1976), unproved.

18.5 FIBONACCI NUMBERS

A very interesting sequence of natural numbers, which has connections with such diverse fields as art and biology, originally arose from a problem posed by the thirteenth-century mathematician Leonardo of Pisa (commonly called Fibonacci, since he was the son (*figlio*) of Bonaccio). In his book, *Liber Abaci*, he proposed the following problem: How many pairs of rabbits can be produced from a single

432   Number Theory   **18.5**

pair, if it be supposed that every month each pair begets a new pair which, from the second month on, becomes productive?

Let us make a chart showing the number of rabbits in successive months. (We assume the first pair is newborn.)

Month	Pairs at beginning of month	Nonproducing pairs at beginning of month	Pairs born during month	Pairs at end of month
1	1	1	0	1
2	1	0	1	2
3	2	1	1	3
4	3	1	2	5
5	5	2	3	8
6	8	3	5	13
7	13	5	8	21
8	21	8	13	34

So at the beginning of the first and second months we have 1 pair of rabbits, 2 at the beginning of the third month, 3 at the beginning of the fourth month, etc.

Let us write u_n for the number of pairs of rabbits at the beginning of the nth month. Thus, from the table above, we see that

$$u_1 = 1 \qquad u_3 = 2 \qquad u_5 = 5 \qquad u_7 = 13$$
$$u_2 = 1 \qquad u_4 = 3 \qquad u_6 = 8 \qquad u_8 = 21$$

Now we note that

$$u_3 = u_1 + u_2 \qquad u_5 = u_3 + u_4 \qquad u_7 = u_5 + u_6$$
$$u_4 = u_2 + u_3 \qquad u_6 = u_4 + u_5 \qquad u_8 = u_6 + u_7$$

In general, for $n \geq 3$ we have

$$u_n = u_{n-2} + u_{n-1} \tag{1}$$

A formula like this, which gives successive terms of a sequence of numbers in terms of one or more preceding terms, is called a **recursion** formula. From Eq. (1), given u_1 and u_2, we can find u_n for any $n \geq 3$. For example,

$$u_9 = u_7 + u_8 \quad = 13 + 21 = 34$$
$$u_{10} = u_8 + u_9 \quad = 21 + 34 = 55$$
$$u_{11} = u_9 + u_{10} = 34 + 55 = 89$$

and so on.

Using Eq. (1) we thus obtain the **Fibonacci sequence** where each successive term is obtained by adding the two previous terms in the sequence.

(Other Fibonacci sequences can be obtained by choosing different values for u_1 and u_2. Thus if $u_1 = 2$ and $u_2 = 5$,

$$u_3 = u_1 + u_2 = 2 + 5 = 7, \qquad u_4 = u_2 + u_3 = 5 + 7 = 12, \qquad \text{etc.}$$

Here, however, we will always mean, by a Fibonacci sequence, the sequence with $u_1 = u_2 = 1$—the original one considered by Fibonacci.) The numbers of the Fibonacci sequence are called **Fibonacci numbers.**

There are many, many known properties of Fibonacci numbers, and new ones are being found every year. (Indeed, there is a journal, *The Fibonacci Quarterly*, concerned entirely with Fibonacci numbers, their generalizations, and their applications!) Many of these properties can be discovered through experimentation by children.

For example, suppose we find the sum, S_n, of the first n Fibonacci numbers, as shown in the following table:

n	$S_n = u_1 + u_2 + \cdots + u_n$
1	$S_1 = 1$
2	$S_2 = 1 + 1 = 2$
3	$S_3 = 1 + 1 + 2 = 4$
4	$S_4 = 1 + 1 + 2 + 3 = 7$
5	$S_5 = 1 + 1 + 2 + 3 + 5 = 12$
6	$S_6 = 1 + 1 + 2 + 3 + 5 + 8 = 20$

Compare, now, this sequence of sums $S_1, S_2, \ldots$ with the Fibonacci sequence $u_1, u_2, \ldots$, as shown in the following table:

n	1	2	3	4	5	6	7	8
u_n	1	1	2	3	5	8	13	21
S_n	1	2	4	7	12	20	—	—

It is not hard to see that each sum listed is one less than a term of the Fibonacci sequence. More specifically, it appears that, for all natural numbers n,

$$S_n = u_{n+2} - 1 \tag{2}$$

Thus

$$S_1 = u_3 - 1 = 2 - 1 = 1$$
$$S_2 = u_4 - 1 = 3 - 1 = 2$$
$$S_3 = u_5 - 1 = 5 - 1 = 4$$
$$S_4 = u_6 - 1 = 8 - 1 = 7$$
$$S_5 = u_7 - 1 = 13 - 1 = 12$$
$$S_6 = u_8 - 1 = 21 - 1 = 20$$

434 Number Theory

Of course, the results of our experimentation do not *prove* that Eq. (2) holds for all natural numbers n. We have only verified this result for $n = 1, 2, \ldots, 6$. A proof, however, is easily possible. We have

$$u_3 = u_1 + u_2, \quad \text{so that} \quad u_1 = u_3 - u_2;$$
$$u_4 = u_2 + u_3, \quad \text{so that} \quad u_2 = u_4 - u_3;$$

and, in general,

$$u_{n+2} = u_n + u_{n+1}, \quad \text{so that} \quad u_n = u_{n+2} - u_{n+1}.$$

Thus we have

$$u_1 = u_3 - u_2$$
$$u_2 = u_4 - u_3$$
$$u_3 = u_5 - u_4$$
$$\vdots$$
$$u_{n-1} = u_{n+1} - u_n$$
$$u_n = u_{n+2} - u_{n+1}$$

Adding up both sides of these equalities, we have

$$S_n = u_1 + u_2 + \cdots + u_{n-1} + u_n \quad \text{(on the left)}$$

and, on the right,

$$(u_3 - u_2) + (u_4 - u_3) + (u_5 - u_4) + \cdots + (u_{n+1} - u_n) + (u_{n+2} - u_{n+1}) \tag{3}$$

But

$$(u_3 - u_2) + (u_4 - u_3) = u_4 - u_2,$$
$$(u_4 - u_2) + (u_5 - u_4) = u_5 - u_2$$

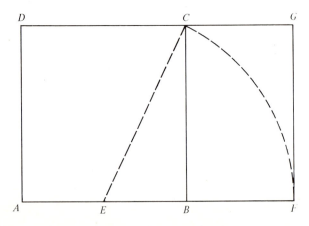

Figure 18.1

so that, as we continue along with the additions in (3), we eliminate all the u's except u_2 and u_{n+2}, and wind up with $-u_2 + u_{n+2}$. The proof is then completed by observing that $u_2 = 1$.

For example, if $n = 6$, Eq. (3) becomes:

$$(u_3 - u_2) + (u_4 - u_3) + (u_5 - u_4) + (u_6 - u_5) + (u_7 - u_6) + (u_8 - u_7)$$

$$= (u_4 - u_2) + (u_5 - u_4) + (u_6 - u_5) + (u_7 - u_6) + (u_8 - u_7)$$

$$= (u_5 - u_2) + (u_6 - u_5) + (u_7 - u_6) + (u_8 - u_7)$$

$$= (u_6 - u_2) + (u_7 - u_6) + (u_8 - u_7)$$

$$= (u_7 - u_2) + (u_8 - u_7)$$

$$= u_8 - u_2 = u_8 - 1$$

Additional discovery-type work with Fibonacci numbers is given in the exercises for this section, and we conclude our remarks concerning the Fibonacci sequence by showing an example of a relationship between the Fibonacci sequence and art.

The **golden rectangle** has found application by artists since the time of the Greeks (it is regarded by many as the rectangle of the most "pleasing" shape to the eye). We can construct a golden rectangle as follows:

Consider a square $ABCD$ with E the midpoint of side $\overline{AB}$, as shown in Fig. 18.1. With E as center and radius EC, strike an arc intersecting $\overleftrightarrow{AB}$ at F. Now erect a perpendicular to $\overleftrightarrow{AB}$ at F intersecting $\overleftrightarrow{DC}$ at G. Then the rectangle, $AFGD$ is said to be a golden rectangle. (All other golden rectangles are similar to $AFGD$; that is, the lengths of their sides are in the same ratio as the lengths of the sides of $AFGD$.)

Now what does a golden rectangle have to do with Fibonacci numbers? To answer this question, let us calculate the ratio of the length of the sides of a golden rectangle. That is, letting $x = AD$ and $y = AF$, we want to find $y:x$. Then

$$y = AF = AE + EF = \tfrac{1}{2}AB + EF = \tfrac{1}{2}AD + EF = \tfrac{1}{2}x + EC$$

(Recall that F was determined by striking an arc of radius EC from E and that $ABCD$ is a square.) To find EC we use the Pythagorean theorem on the right triangle EBC with sides x and $\tfrac{x}{2}$, to obtain

$$(EC)^2 = x^2 + \left(\frac{x}{2}\right)^2 = x^2 + \frac{x^2}{4} = \frac{5x^2}{4}$$

436 Number Theory

18.5

Hence

$$EC = \sqrt{\frac{5x^2}{4}} = \frac{\sqrt{5}\,x}{2} = \frac{\sqrt{5}}{2}\,x$$

Thus

$$y = \tfrac{1}{2}x + EC = \tfrac{1}{2}x + \frac{\sqrt{5}}{2}\,x = \left(\frac{1}{2} + \frac{\sqrt{5}}{2}\right)x = \left(\frac{1 + \sqrt{5}}{2}\right)x$$

and so

$$y : x = \frac{1 + \sqrt{5}}{2}\,x : x = (1 + \sqrt{5}) : 2$$

It is still difficult to see any connection with Fibonacci numbers! Note now, however, that, to five decimal places,

$$\frac{1 + \sqrt{5}}{2} = \frac{1 + 2.23606}{2} = \frac{3.23606}{2} = 1.61803$$

so that

$$y : x = 1.61803 : 1$$

Now let us calculate the ratio of pairs of successive terms of the Fibonacci sequence. We have, to five decimal places,

$u_2 : u_1 = 1 : 1 = 1.00000 : 1$	$u_6 : u_5 = 8 : 5 = 1.60000 : 1$	$u_{10} : u_9 = 55 : 34 = 1.61765 : 1$
$u_3 : u_2 = 2 : 1 = 2.00000 : 1$	$u_7 : u_6 = 13 : 8 = 1.62500 : 1$	$u_{11} : u_{10} = 89 : 55 = 1.61818 : 1$
$u_4 : u_3 = 3 : 2 = 1.50000 : 1$	$u_8 : u_7 = 21 : 13 = 1.61538 : 1$	$u_{12} : u_{11} = 144 : 89 = 1.61798 : 1$
$u_5 : u_4 = 5 : 3 = 1.66667 : 1$	$u_9 : u_8 = 34 : 21 = 1.61905 : 1$	$u_{13} : u_{12} = 233 : 144 = 1.61806 : 1$

Our calculations seem to indicate that the ratio of u_{n+1} to u_n gets closer and closer to $(1 + \sqrt{5}) : 2$ as n increases (alternately less than $(1 + \sqrt{5}) : 2$ and greater than $(1 + \sqrt{5}) : 2$). Indeed, it may be shown (but not easily) that this is the case.

Exercises 18.5

1. Use Eq. (2) to calculate S_9 and S_{10} and check your result by direct addition.

2. Take $u_1 = 4$ and $u_2 = 6$, and calculate the first six terms of the resulting Fibonacci sequence.

3. Experiment to find a simple way of calculating the sums

$$u_1 + u_3, \qquad u_1 + u_3 + u_5, \qquad \ldots, \qquad u_1 + u_3 + u_5 + \ldots + u_{2n-1}$$

4. Do as in Exercise 3 for the sums

$$u_2 + u_4, \qquad u_2 + u_4 + u_6, \qquad \ldots, \qquad u_2 + u_4 + u_6 + \ldots + u_{2n}$$

18.5   **Fibonacci Numbers**   **437**

5. Do as in Exercise 3 for the sums

$$u_1^2 + u_2^2, \qquad u_1^2 + u_2^2 + u_3^2, \qquad \ldots, \qquad u_1^2 + u_2^2 + u_3^2 + \ldots + u_n^2$$

6. What are the conditions on n so that u_n is divisible to 2? By 3? By 4? By 5?

Modular arithmetic

by Margaret Haines

Reprinted from *The Arithmetic Teacher*, Vol. 9 (1962), pp. 127–129.

A new type arithmetic obtained from a "clocklike" number system is called *modular arithmetic*. It may also be called *finite number systems* because these have only a specific number of numbers in the system. It is used for all machines with dials or with a finite number of numbers. "Scientists find it useful in the study of atoms which have shells of electrons." The limited number of numbers in the entire system makes it relatively simple.

Let's begin with a mod five system. In technical language the system is a "closed system" since every addition, subtraction, multiplication, or division with mod five numbers will give an answer which is contained in the original five numbers. To simplify matters, teachers should make a dial as illustrated in Fig. 1. Add numerals and a pointer to the dial. We count clockwise. Please note 5 is not shown. In any modular system the "name numeral" does not appear. To add 3 and 4 in mod five, place the pointer at 3 and then move it 4 spaces clockwise. It will stop at 2. So 3 and 4 are 2 mod five. Don't count the numeral you start with, in this case, 3. Begin with 4, 0, 1, 2, and 2 is the answer because it is four spaces removed from 3. Should the sum ever be larger than 4, divide the sum by 5 (since this is mod five), discard the quotient, and use only the remainder for the answer.

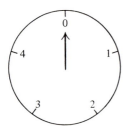

Figure 1

Example. 3 + 4 = 7. In mod five this would be 7 ÷ 5 = 1(r) 2; therefore, 3 + 4 = 2 mod five. Likewise, 2 + 3 = 0 mod five and 4 + 4 + 4 = 12, or 12 ÷ 5 = 2(r) 2, so 4 × 3 = 12, 12 ÷ 5 = 2(r) 2. So we may conclude that 3 + 4 = 2 mod five and 4 × 3 = 2 mod five.

Other systems may be built from days of the week or months of the year. These should be in this order, mod seven and mod twelve. Think of Sunday as 1, Monday as 2, Tuesday as 3, Wednesday 4, Thursday 5, Friday 6, and Saturday as 0. This finite system has no number larger than 7. The sum of 4 and 5 is 2 mod seven. Since Wednesday is 4, count 5 days beyond, beginning with Thursday. We end on Monday, which is 2. Another way to check this is 4 + 5 = 9 ÷ 7 = 1(r) 2. Hence our answer is still 4 and 5 = 2 mod seven.

In mod twelve, the system has only twelve numbers. So the sum of 8 and 9 is written 8 + 9 = 5 mod twelve. Again one may use a dial or a clockface to simplify matters. Count 9 places from 8 going clockwise, and we get 5.

Subtraction is the inverse or opposite of addition, so we begin at a given number and count counterclockwise to arrive at the answer, 8 − 9 = 11 mod twelve if we count, 7, 6, 5, 4, 3, 2, 1, 12, 11, nine places counterclockwise.

Modulo seven is used constantly when teachers teach days of the week. Modulo twelve (see Fig. 2 for example of mod twelve dial) is used both in the study of months and the study of telling time.

Figure 2

Modulo nine may be used to check addition, subtraction, multiplication, and division. Most children despise checking, so why not introduce this method? Some teachers may know it as "casting out nines." I've tried this in my fifth grade and even the slower pupils love it when they master the techniques.

The method is similar in all operations except division, but I'll demonstrate all operations for clarity. Remember, one uses only remainders after dividing by the *modular number*, in this case, 9.

Example.

```
       326        Check:
      +218        3 + 2 + 6 = 11 =  2 mod 9
       544        2 + 1 + 8 = 11 =  2 mod 9
                                   ④mod 9
                  5 + 4 + 4 = 13 = ④mod 9
```

When the encircled numerals are the same, the problem should be correct. However, this check is not infallible. As in other check systems, if the pupil uses reversals such as 454 instead of 544, he will still get the same number, 4.

The idea is to add 3, 2, and 6, divide this number by 9, and place the remainder, which is 2. Do the same for 2, 1, and 8; then since the operation is addition, one must add 2 and 2. Next one must add 5, 4, 4, again divide by 9, and if the remainder is the same, the problem is generally correct.

Example

```
      115,171     Check:
      −66,287     1 + 1 + 5 + 1 + 7 + 1 = 16 =  7 mod 9
       48,884        6 + 6 + 2 + 8 + 7 = 29 =  2 mod 9
                                               ⑤mod 9
                  4 + 8 + 8 + 8 + 4 = 32 =    ⑤mod 9
```

Subtraction is similar except one must subtract 2 from 7 instead of adding.

If one should have a problem in which the minuend would result in a remainder less than the subtrahend remainder, one can always add 9 without changing the final result.

440   Number Theory

Example

	Check:	
69	$6 + 9 = 15 = 6 \bmod 9$	$15 \bmod 9$
-43	$4 + 3 = 7 = 7 \bmod 9$	$-7 \bmod 9$
26	$2 + 6 = \circled{8} \bmod 9$	$\circled{8} \bmod 9$

Multiplication is must as simple as the others. Just remember all final numbers should be one-place numbers, such as 5, instead of 14, for in 14 there is still a 9 remaining and if divided again the remainder is 5.

Example

	Check:
473	$4 + 7 + 3 = 14 = 5 \bmod 9$
$\times 232$	$2 + 3 + 2 = 7 = 7 \bmod 9$
946	$5 \times 7 = 35 = \circled{8} \bmod 9$
14190	
94600	
109736	$1 + 9 + 7 + 3 + 6 = 26 = \circled{8} \bmod 9$

Division is a little more tricky because of the remainder.

Example

$$
\begin{array}{r}
20\ (\mathrm{r})\ 19 \\
23\overline{)479} \\
46 \\
\overline{19} \\
0 \\
\overline{19}
\end{array}
$$

Add the divisor digits $2 + 3 = 5$. Add the quotient digits, $2 + 0 = 2$. Now multiply $5 \times 2 = 10 = 1$; add the remainder $1 + 9 = 10 = 1$, and add $1 + 1 = 2$. Next add the dividend digits $4 + 7 + 9 = 20 = 2$.

Teachers may question the value of using these new methods in the intermediate grades. It is a means of introducing a fascination for numbers. This may prove to be of great value in producing the mathematicians needed throughout our nation and may also reduce the number of people that reach maturity "hating" anything connected with numbers.

REFERENCES FOR FURTHER READING

1. Hosford, Philip L., "Enrichment motivation using general Fibonacci sequences," *The Mathematics Teacher*, **68** (1975), pp. 430–437.

2. Loftus, Sonja, "Fibonacci numbers: Fun and fundamentals for the slow learner," *AT*, **17** (1970), pp. 204–208.

3. Morton, Robert L., "Fractional numbers with a sum of 1," *AT*, **13** (1966), pp. 658–661. Describes some interesting activities involving perfect, abundant, and deficient numbers.

4. Oliver, Charlene, "Gus's magic numbers: a key to the divisibility tests for primes," *AT*, **19** (1972), pp. 183–189. Describes how Gus developed some of these tests for himself.

5. Shoemaker, Richard W., *Perfect Numbers*, National Council of Teachers of Mathematics, Reston, Va. (1973).

6. Smith, Frank, "Divisibility rules for the first fifteen primes," *AT*, **18** (1971), pp. 85–87.

7. Tassone, Sister Ann Dominic, "A pair of rabbits and a mathematician," *AT*, **14** (1967), pp. 285–288.

List of Symbols
(and where first introduced)

{ }	to indicate sets	2
$\in$	to indicate set membership	3
/	drawn through a symbol to indicate negation as in $\notin$ and $\neq$	3
$\subseteq$	to indicate subsets	3
{ }	the null set	3
$\emptyset$	the null set	3
$\sim$	to indicate that two sets are matching sets	7
$\sim$	to indicate similarity of geometrical figures	180
$\sim$	to indicate equivalent fractions	238
$n(A)$	for the number of elements of a set A	12
$\overline{AB}$	for the line segment joining points A and B	14
$\overrightarrow{AB}$	for the ray from point A through the point B	14
$\overleftrightarrow{AB}$	for the line joining points A and B	15
$\cup$	to indicate the union of sets	22
$<$	for "less than"	22
$>$	for "greater than"	22
$\cap$	to indicate the intersection of sets	29
$A \backslash B$	for the relative complement of set B with respect to set A	31
$\geq$	for "greater than or equal to"	33
$A \times B$	for the Cartesian product of sets A and B	105
(a, b)	for ordered pair	105

444   List of Symbols

(a, b)	for the G.C.D. of a and b	139
$\leq$	for "less than or equal to"	109
$[a, b]$	for the L.C.M. of a and b	139
$\lvert a \rvert$	for the absolute value of a	166
$\cong$	to indicate congruence of geometrical figures	179
$\measuredangle$	for an angle	179
$\triangle$	for a triangle	179
$*$	to indicate a binary operation	317
$f(x)$	to indicate a functional value	333
$P(E)$	for the probability of an event E	387

Answers to Summary Tests and Selected Exercises

Exercises 3.7

1. 434_{ten}

5. $110{,}100_{\text{four}}$

3. 110_{ten}

7. $11{,}002_{\text{three}}$

Exercises 4.5

1. a)

$$
\begin{array}{r}
\overset{8}{\cancel{9}}{}^{1}2 \\
-7\,8 \\
\hline
1\,4
\end{array}
$$

$$
\begin{array}{r}
92 \\
-78 \\
\hline
\end{array}
\rightarrow
\begin{array}{r}
90 + 2 \\
(-)70 + 8 \\
\hline
\end{array}
\rightarrow
\begin{array}{r}
80 + 12 \\
(-)70 + 8 \\
\hline
10 + 4 = 14
\end{array}
$$

c)

$$
\begin{array}{r}
\overset{4}{\cancel{5}}{}^{1}\overset{1}{\cancel{2}}6 \\
-1\,8\,8 \\
\hline
3\,3\,8
\end{array}
$$

$$
\begin{array}{r}
526 \\
-188 \\
\hline
\end{array}
\rightarrow
\begin{array}{r}
500 + 20 + 6 \\
(-)100 + 80 + 8 \\
\hline
\end{array}
\rightarrow
\begin{array}{r}
500 + 10 + 16 \\
(-)100 + 80 + 8 \\
\hline
\end{array}
\rightarrow
$$

$$
\begin{array}{r}
400 + 110 + 16 \\
(-)100 + 80 + 8 \\
\hline
300 + 30 + 8 = 338
\end{array}
$$

2. a)

$$
\begin{array}{r}
9\overset{1}{2} \\
8 \\
-7\,8 \\
\hline
14
\end{array}
$$

$$
\begin{array}{r}
92 \\
-78 \\
\hline
\end{array}
\rightarrow
\begin{array}{r}
90 + 2 \\
(-)70 + 8 \\
\hline
\end{array}
\rightarrow
\begin{array}{r}
90 + 12 \\
(-)80 + 8 \\
\hline
10 + 4 = 14
\end{array}
$$

c)

$$
\begin{array}{r}
5\overset{1}{2}\overset{1}{6} \\
2\,9 \\
-\cancel{1}\,\cancel{8}\,8 \\
\hline
3\,3\,8
\end{array}
$$

$$
\begin{array}{r}
526 \\
-188 \\
\hline
\end{array}
\rightarrow
\begin{array}{r}
500 + 20 + 6 \\
(-)100 + 80 + 8 \\
\hline
\end{array}
\rightarrow
\begin{array}{r}
500 + 10 + 16 \\
(-)100 + 90 + 8 \\
\hline
\end{array}
\rightarrow
$$

$$
\begin{array}{r}
500 + 120 + 16 \\
(-)200 + 90 + 8 \\
\hline
300 + 30 + 8 = 338
\end{array}
$$

445

446   Answers to Summary Tests and Selected Exercises

Exercises 4.6

1. a) 1231   c) 11,102

2. a) $\overset{\overset{2}{\cancel{3}}\overset{1}{2}}{\underset{}{}}$
    $\underline{-1\,4}$
    $1\,3$

   $\begin{array}{r}32\\-14\\\hline\end{array}\rightarrow\begin{array}{r}30+2\\(-)10+4\\\hline\end{array}\rightarrow\begin{array}{r}20+12\\(-)10+4\\\hline 10+3=13\end{array}$

  c) $\overset{\overset{3}{\cancel{4}}\overset{1}{\cancel{2}}\overset{1}{2}}{}$
    $\underline{-1\,3\,4}$
    $2\,3\,3$

   $\begin{array}{r}422\\-134\\\hline\end{array}\rightarrow\begin{array}{r}400+20+2\\(-)100+30+4\\\hline\end{array}\rightarrow\begin{array}{r}400+10+12\\(-)100+30+4\\\hline\end{array}\rightarrow$

                     $\begin{array}{r}300+110+12\\(-)100+30+4\\\hline 200+30+3=233\end{array}$

3. a) $\overset{\overset{1}{3}2}{}$
    $\overset{2}{-\cancel{1}\,4}$
    $1\,3$

   $\begin{array}{r}32\\-14\\\hline\end{array}\rightarrow\begin{array}{r}30+2\\(-)10+4\\\hline\end{array}\rightarrow\begin{array}{r}30+12\\(-)20+4\\\hline 10+3=13\end{array}$

  c) $\overset{\overset{1}{2}\overset{1}{2}}{4\,2\,2}$
    $2\,4$
    $-\cancel{1}\,\cancel{3}\,4$
    $2\,3\,3$

   $\begin{array}{r}422\\-134\\\hline\end{array}\rightarrow\begin{array}{r}400+20+2\\(-)100+30+4\\\hline\end{array}\rightarrow\begin{array}{r}400+20+12\\(-)100+40+4\\\hline\end{array}\rightarrow$

                     $\begin{array}{r}400+120+12\\(-)200+40+4\\\hline 200+30+3=233\end{array}$

4. a) 1231   c) 2001

5. a) $\overset{\overset{2}{\cancel{3}}\overset{1}{\cancel{2}}\overset{1}{2}}{}$
    $\underline{-1\,3\,3}$
    $1\,2\,3$

   $\begin{array}{r}322\\-133\\\hline\end{array}\rightarrow\begin{array}{r}300+20+2\\(-)100+30+3\\\hline\end{array}\rightarrow\begin{array}{r}300+10+12\\(-)100+30+3\\\hline\end{array}\rightarrow$

                     $\begin{array}{r}200+110+12\\(-)100+30+3\\\hline 100+20+3=123\end{array}$

6. a) $\overset{\overset{1}{2}\overset{1}{2}}{3\,2\,2}$
    $\overset{2\,10}{}$
    $-\cancel{1}\,\cancel{3}\,3$
    $1\,2\,3$

   $\begin{array}{r}322\\-133\\\hline\end{array}\rightarrow\begin{array}{r}300+20+2\\(-)100+30+3\\\hline\end{array}\rightarrow\begin{array}{r}300+20+13\\(-)100+100+3\\\hline\end{array}\rightarrow$

                     $\begin{array}{r}300+120+13\\(-)200+100+3\\\hline 100+20+3=123\end{array}$

7. a) seven   c) five

Answers to Summary Tests and Selected Exercises 447

Exercises 6.1

1.
```
   232
  ×574
  1160
  1624
   928
133168
```

3.
```
   218
  ×607
  1308
  1526
132326
```

Exercises 6.4

1. 12,223   *3.* 121,314   *5.* 132 r 13   *7.* 1020 r 23

9. 21,102   *11.* 133 r 20   *13.* 35,205   *15.* 5,634 r 31

Exercises 6.5

1. G.C.D. = 5,   L.C.M. = 75   *3.* G.C.D. = 1,   L.C.M. = 120

5. G.C.D. = 1,   L.C.M. = 168   *7.* G.C.D. = 3,   L.C.M. = 72

9. G.C.D. = 4,   L.C.M. = 240   *15.* [8, 12] = 24

Answers to Summary Test for Chapters 1 through 6

1. i) b, e, f, g, h;   ii) a, b, d, f, g, h;   iii) b, f, g, h

2. a) F;   b) F;   c) F;   d) T;   e) T;   f) T

3. (a) and (d);   (b) and (c)

4. $\subseteq$

5. $x \in A$ or $x \in B$

6. $x \in A$ and $x \in B$

7. For all sets A, B, and C. $A \cup B = B \cup A$, $A \cup (B \cup C) = (A \cup B) \cup C$, $A \cap B = B \cap A$, $A \cap (B \cap C) = (A \cap B) \cap C$.

8. For all whole numbers a, b, and c, $a \times (b + c) = (a \times b) + (a \times c)$.

9. $A, \emptyset, A, A$ *10.* $\subseteq, \in, \notin$ *11.* $\emptyset, A$

12. Additive

13. Identity

448 Answers to Summary Tests and Selected Exercises

14. i)

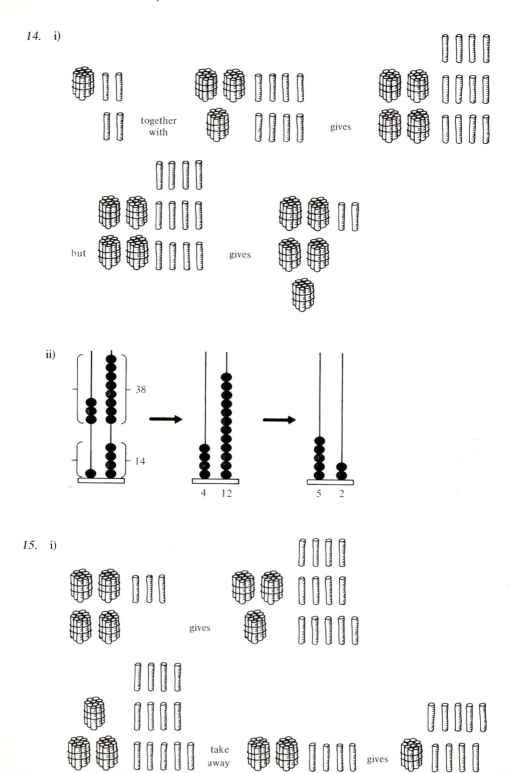

ii)

15. i)

Answers to Summary Tests and Selected Exercises   449

15. ii)

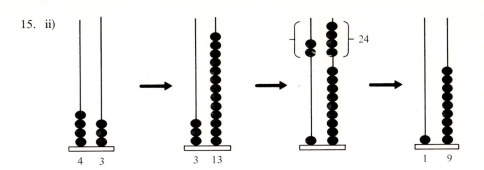

16.   234
       578
      +385
        17
       180
      1000
      1197

17. i) $\overset{\;\;1}{\underset{\;}{2}}\overset{3_1}{\underset{\;}{}}$
        $\cancel{3}\cancel{4}2$
        -156
        $\overline{186}$

    $\begin{array}{r}342\\-156\\\hline\end{array}$ →
    $\begin{array}{r}300+40+2\\(-)100+50+6\\\hline\end{array}$ →
    $\begin{array}{r}300+30+12\\(-)100+50+6\\\hline\end{array}$ →

    $\begin{array}{r}200+130+12\\(-)100+50+6\\\hline 100+80+6=186\end{array}$

  ii) $\overset{11}{3\,4\,2}$
        $2\,6$
        $-\cancel{1}\,\cancel{5}\,6$
        $\overline{186}$

    $\begin{array}{r}342\\-156\\\hline\end{array}$ →
    $\begin{array}{r}300+40+2\\(-)100+50+6\\\hline\end{array}$ →
    $\begin{array}{r}300+40+12\\(-)100+60+6\\\hline\end{array}$ →

    $\begin{array}{r}300+140+12\\(-)200+60+6\\\hline 100+80+6=186\end{array}$

  iii) $\begin{array}{r}342\\-156\\\hline\end{array}$

       $\begin{array}{r}342\\+843\\\hline \cancel{1}185\\+1\\\hline 186\end{array}$

18.   232   19.   232   20.   3142   21. 22$\overline{)342}$ 13 r 1
       144   ×43 -1343 $\overline{220}$
      +341 $\overline{11}$ $\overline{1244}$ 122
      $\overline{1322}$ 140 $\overline{121}$
                       1100 1
                        130
                       2200
                      13000
                      $\overline{22131}$

22. $\{(a, x), (a, y), (a, z), (b, x), (b, y), (b, z)\}$

23. $2, 3, 6, 3, 6, A \times B$ 24. $12 \div 6$ 25. 3

26. Commutative   27. b, c (or c, b)

28. $a \times (b + c)$, $(a \times b) + (a \times c)$ 29. $\emptyset, 0, 0$

30. $(6 \div 2) \div (2 + 1) = 1$ is one possible answer.

31. 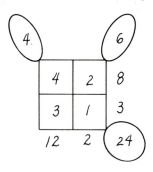   32. Associative, intersection

33. i) $D_{15} = \{1, 3, 5, 15\}$, $D_{18} = \{1, 2, 3, 6, 9, 18\}$,
    $D_{30} = \{1, 2, 3, 5, 6, 10, 15, 30\}$;
    $D_{15} \cap D_{18} \cap D_{30} = \{1, 3\}$, so 3 is the G.C.D.
    ii) $15 = 3 \times 5$, $18 = 2 \times 3^2$, $30 = 2 \times 3 \times 5$, so G.C.D. $= 3$.

34. i) $M_9 = \{9, 18, 27, 36, 45, 54, 63, 72, 81, \ldots\}$,
    $M_{18} = \{18, 36, 54, 72, 90, \ldots\}$, $M_{24} = \{24, 48, 72, 96, 120, \ldots\}$;
    $M_9 \cap M_{18} \cap M_{24} = \{72, \ldots\}$, so L.C.M. is 72.
    ii) $9 = 3^2$, $18 = 2 \times 3^2$, $24 = 2^3 \times 3$, so L.C.M. is $2^3 \times 3^2 = 72$.

Exercises 7.3

1. $-$ 3. $+$ 5. $+, +, -$ 7. $+, -, -$ 9. $-, +, -$

Exercises 9.5

1. a) 0.12   c) 0.325   e) 0.8

2. a) $0.\overline{4}$ c) $0.\overline{27}$

3. a) 43/99   c) 1222/9900 ($= 611/4950$)

Exercises 9.6

1. a) $(1 \times 10^2 + 6 \times 10 + 2 + 5 \times \frac{1}{10})_{\text{seven}}$

Answers to Summary Tests and Selected Exercises   451

2. a) $(\frac{3}{8})_{ten}$ c) $(\frac{1}{18})_{ten}$ e) $(\frac{22}{9})_{ten}$

3. a) 0.2_{four} c) $0.\overline{1}_{three}$

Exercises 9.10

1. a) 3.5   c) 4.5
2. a) 6   c) 150
3. a) 64   c) 48
5. $5000   7. 9%   9. 21,000

Exercises 11.6

1. 20 hours = 8 P.M.

3. 41°F = 5°C,   −10°F = −23$\frac{1}{3}$°C

4. 45°C = 113°F,   −15°C = 5°F

7. 3.84 kg

Answers to Summary Test for Chapters 7 through 12

1. Negative, additive inverse   2. Natural

3. 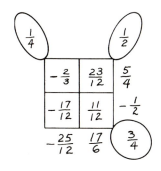   4.

5. >   6. $\frac{11}{4}$ 7. Congruent, congruent

8. Parallel   9. Sphere

10. Rectangular parallelopiped   11. Equilateral

12. Proper: $\frac{2}{9}, \frac{3}{6}$; in lowest terms: $\frac{12}{5}, \frac{2}{9}$

13. Nonterminating, nonrepeating (either order)

452   Answers to Summary Tests and Selected Exercises

14. $0.\overline{153846}$ 15. $0.023\overline{0}$ 16. $\frac{73}{55}$

17. 7%   18. $\frac{24}{36} + \frac{32}{36} + \frac{15}{36} = \frac{71}{36} = 1\frac{35}{36}$

19.

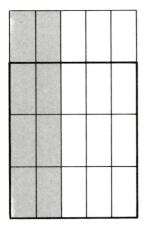

20.

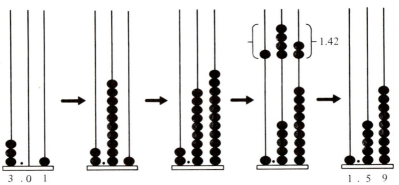

21. $\frac{4}{3} \div \frac{5}{6} = \frac{4 \times 5 \times 6}{3 \times 5 \times 6} \div \frac{5}{6} = \frac{(4 \times 5 \times 6) \div 5}{(3 \times 5 \times 6) \div 6} = \frac{4 \times 6}{3 \times 5} = \frac{24}{15} = \frac{8}{5};$

$\frac{4}{3} \div \frac{5}{6} = \frac{4}{3} \times \frac{6}{5} = \frac{24}{15} = \frac{8}{5}$

22.
```
         0.321
0̸.014.)̸004.494
        4 2
        ‾‾‾
         29
         28
        ‾‾‾
         14
         14
        ‾‾‾
          0
```

```
0̸.014.)̸004.494
        4.200   0.300
        ‾‾‾‾‾
        0.294
        0.280   0.020
        ‾‾‾‾‾
        0.014
        0.014   0.001
        ‾‾‾‾‾   ‾‾‾‾‾
            0   0.321
```

```
0.014)0.004494
      0.004200   0.300
      ‾‾‾‾‾‾‾‾
      0.000294
      0.000280   0.020
      ‾‾‾‾‾‾‾‾
      0.000014
      0.000014   0.001
      ‾‾‾‾‾‾‾‾   ‾‾‾‾‾
             0   0.321
```

23. 8   24. 100; 10,000; 1000; 1000

25. 5°; 104°   26. 132 cm, 1386 cm^2 27. 2.21 cm^2

28. 69.2 cm^3 29. $\frac{5}{9}$ 30. $\frac{44}{23}$

Answers to Summary Tests and Selected Exercises   453

Exercises 13.1

1. Only on the set of positive rational numbers
3. "×" only   5. Yes

Exercises 13.2

1. Commutative and associative
3. Commutative and associative for both ∪ and ∩.
5. Commutative and associative
7. Commutative and associative
9. Neither commutative nor associative

Exercises 14.3

1. $f(0) = 3$, $f(-1) = 1$, $f(\sqrt{2}) = 2\sqrt{2} + 3$ 3. $\{2, 5, 10\}$
5. $h(2) = h(\frac{1}{2}) = h(-2.14) = 0$; $h(\sqrt{2}) = h(\pi) = 1$; range $= \{0, 1\}$
7. The set of even whole numbers   9. -7

Exercises 14.4

1. 2; 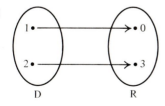 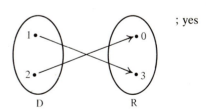 ; yes

3. No, one member of the range cannot be obtained from an element of the domain—contrary to the definition of a function.

5.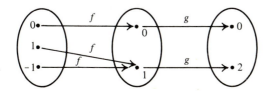

7. The lefthand one

9.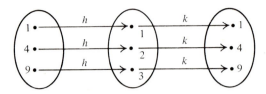

Exercises 14.5

1. (a), (c), (d), (e), (g), and (i) are functions.

2. a) $\{(4,7),(5,8),(6,9)\}$ c) $\{(x, x^3): x \text{ a real number}\}$

3. $f = \{(4, 8), (7, 14), (10, 20)\}$, $g = \{(8, 4), (14, 7), (20, 10)\}$

5. a) $\{(1, 2), (0, 3), (4, 5)\}$ c) $\{(2, 1), (4, 3), (6, 5), (7, 6)\}$

Exercises 16.3

1. $\{\text{Sun., Mon., Tues., Wed., Thurs., Fri., Sat.}\}$, $n(S) = 7$

3. $\{(x, y): x \text{ and } y \text{ any of } 1, 2, \ldots, 10\}$, $n(S) = 100$

Exercises 16.4

1. a) $E = \{\text{Tues., Thurs.}\}$, $n(E) = 2$
   b) $E = \emptyset$, $n(E) = 0$ c) $E = S$, $n(E) = 7$

3. a) $E = \{(1, 1), (1, 2), \ldots, (1, 10)\}$, $n(E) = 10$
   b) $E = \{(1, 1), (2, 1), \ldots, (10, 1)\}$, $n(E) = 10$
   c) $E = \{(1, 1)\}$, $n(E) = 1$
   d) $E = \{(2, 2), (4, 2), (6, 2), (8, 2), (10, 2)\}$, $n(E) = 5$

Exercises 16.5

1. a) $\frac{2}{7}$ b) 0   c) 1

3. a) $\frac{1}{10}$ b) $\frac{1}{10}$ c) $\frac{1}{100}$ d) $\frac{1}{20}$

5. $S = \{(x, y): x \text{ and } y \text{ any of } 1, 2, 3, \ldots, 6\}$, $n(S) = 36$
   a) $E = \{(6, 6)\}$, $n(E) = 1$, $P(E) = \frac{1}{36}$
   b) $E = \{(2, 6), (3, 5), (4, 4), (6, 2), (5, 3)\}$, $n(E) = 5$, $P(E) = \frac{5}{36}$
   c) $E = \{(1, 2), (1, 3), (1, 4), (1, 5), (1, 6), (2, 3), (2, 4), (2, 5), (2, 6),$
       $(3, 4), (3, 5), (3, 6), (4, 5), (4, 6), (5, 6)\}$, $n(E) = 15$, $P(E) = \frac{5}{12}$
   d) $E = S$, $n(E) = 36$, $P(E) = 1$
   e) $E = \emptyset$, $n(E) = 0$, $P(E) = 0$

Answers to Summary Tests and Selected Exercises   **455**

7. $S = \{BB, BG, GB, GG\}, \quad n(S) = 4$
    a) $E = \{BB\}, \quad n(E) = 1, \quad P(E) = \frac{1}{4}$
    b) $E = \{BG, GB\}, \quad n(E) = 2, \quad P(E) = \frac{1}{2}$
    c) $E = \{BG, GB, GG\}, \quad n(E) = 3, \quad P(E) = \frac{3}{4}$
    d) $E = \{BB, BG, GB\}, \quad n(E) = 3, \quad P(E) = \frac{3}{4}$

9. $S = \{(x, y, z): x, y, \text{ and } z \text{ any of Jan., Feb., } \ldots, \text{Dec.}\}, \quad n(S) = 1728$
    a) $E = \{(\text{Jan., Jan., Jan.}), (\text{Feb., Feb., Feb.}), \ldots, (\text{Dec., Dec., Dec.})\}$,
       $n(E) = 12, \quad P(E) = \frac{1}{144}$
    b) $E = \{(x, y, z): x, y, \text{ and } z \text{ any one of Jan., June, July}\}, \quad n(E) = 27$,
       $P(E) = \frac{1}{64}$

Exercises 16.6

1.   a) E': The sum of the two numbers on the dice is greater than or equal to 11.
    b) $E' = \{(5, 6), (6, 5), (6, 6)\}, \quad n(E') = 3, \quad n(S) = 36, \quad P(E') = \frac{1}{12}$
    c) $P(E) = \frac{11}{12}$

3.   a) At least one of the two slips of paper drawn has a prime number written on it.
    b) $E' = \{(2, 1), (2, 4), (2, 6), (2, 8), (2, 10), (2, 12), (1, 2), (4, 2),$
           $(6, 2), (8, 2), (10, 2), (12, 2)\}$
    c) $n(E') = 12, \quad n(S) = 42, \quad P(E') = \frac{12}{42} = \frac{2}{7}$
    d) $P(E) = \frac{5}{7}$

5. E': All of the four persons have different birth months.

7. E': At least one of the numbers shown on the dice is a prime number.

Exercises 16.7

1.   2730 (assuming no one can hold two offices!)

3.   a) 60   b) 125   c) 50   d) 75

5.   a) $\frac{5}{9}$ b) $\frac{1}{2}$ c) $\frac{2}{9}$

9.   1/2,118,760

Index

Index

Abacus, 59
Absolute value, 166, 236
Abundant number, 427
Addition, with abacus, 77–78
  of integers, 152
  by method of partial sums, 79
  of rational numbers, 249–251
  with stick bundles, 76–77
Additive identity, 25
Additive inverse, 149
Additive property of inequalities, 163, 270
Additive property of 0, 25
Altitude, 288
Amicable numbers, 431
Arithmetic mean, 409
Associative property, of addition, 25
  of multiplication, 112
  of set union, 25
Assumed mean, 410
Austrian method of subtraction, 83

Bar graphs, 402–403
Base, 36, 288
Basimal fractions, 232
Binary operation, 317
Borrowing method of subtraction, 82–83
Braces, 2

Cardinal number, 12
Cartesian plane, 240

Cartesian product, 105
Casting out nines, 424–426
Celsius, 297
Center of rotation, 358
Centimeter, 282
Chord, 200
Circumference, 284
Class interval, 401
Clock arithmetic, 320
Closed, 157
Closed curve, 15
Common divisor, 137
Common fraction, 223
Common multiple, 137
Commutative property, of addition, 25
  of multiplication, 112
  of set union, 25
Complement of a set, 31
Composition of functions, 337
Cone, right circular, 189
Congruent figures, 178
Coordinate plane, 340
Counting numbers, 12
Cube, 189
Cubic centimeter, 295
Curly brackets, 2
Curve, 15
  closed, 15
  open, 15
  simple, 15
Cylinder, right circular, 189

460   Index

Decimal fraction, 223
  repeating, 226
  terminating, 226
Deficient numbers, 427
Degree, 297
Denominator, 222
Deviation from the mean, 410
Diameter, 285
Digits, 58
Diophantus, 428
Disjoint sets, 24
Distrbutive property, 113
Divisibility tests, 422–423
Division, of integers, 161
  as inverse of multiplication, 110
  of rational numbers, 260–264
  of whole numbers, 132–133
Domain of a function, 333

Edge of a polyhedron, 191
Equal additions method of subtraction,
  83–84
Equivalence classes of fractions, 238
Equivalent fractions, 218
Equivalent sets, 7
Ethiopian numerals, 55, 58
Even numbers, 9
Events, 385–386
  impossible, 386
  sure, 386
Expanded notation, 63
Experimental probability, 381–382

Face of a polyhedron, 191
Factorization, 138
Fahrenheit, 297
Fermat's last theorem, 428
Fermat's therorem, 431
Fibonacci numbers, 431–436
Fibonacci sequence, 432
Field, 231, 324
Field properties, 270
Finite sets, 9, 11
Fractions, 215
  basimal, 232
  common, 223
  decimal, 223
  equivalence classes of, 238
  equivalent, 218
  fundamental principle for, 218

  improper, 222
  in lowest terms, 222
  mixed, 222
  proper, 222
  vulgar, 223
Frames, 37
Frequency table, 400
Folding symmetries, 368
Functional notation, 334
Functions, 329–345
  composition of, 337
  definition of, 333
  domain of, 333
  graphs of, 340–345
  inverses of, 337
  many-to-one, 337
  as mappings, 336–337
  one-to-one, 337
  range of, 333
  as sets of ordered pairs, 338–339
Fundamental counting principle, 392
Fundamental principle for fractions, 218

Geoboard, 185
Geometric motions, 355
Geometric transformations, 355
Goldbach conjectures, 429
Golden rectangle, 435
Gram, 298
Graphs, bar, 402–403
  of functions, 340–344
  line, 403
  statistical, 402–406
Greatest common divisor, 137
Greatest common factor, 139
Grouping property, 26

Hectare, 290
Highest common factor, 139
Hindu-Arabic numerals, 58

Image, 357
Impossible event, 386
Improper fraction, 222
Inequalities, 35–36, 149, 162–165, 270
Infinite sets, 9, 11
Integers, 150–166
  addition of, 152–155
  division of, 161–162
  inequality of, 163–165

multiplication of, 159–160
subtraction of, 155–156
Intersection of sets, 29
Inverse, additive, 149
of a function, 337
multiplicative, 258
Irrational number, 217

Japanese number names, 56

Kapreker's constant, 41
Kilogram, 298
Kilometer, 282
Kuanyan number language, 57

Least common denominator, 250
Least common multiple, 137
Lefthand method of multiplication, 127
Length, unit of, 281
Line, 15
graphs, 403
number, 9
segment, 14
Lines, parallel, 191
Liter, 295
Logic, language of, 24
Long method of multiplication, 130
Long ton, 298
Lowest terms, of a fraction, 222
Lowest terms, of a ratio, 267

Many-to-one function, 337
Mapping, 336
Matching sets, 47
Mean, 409
arithmetic, 409
assumed, 410
Measures of central tendency, 409
Median, 408
Mersenne primes, 427
Meter, 282
Milligram, 298
Milliliter, 295
Millimeter, 282
Minute, 297
Mixed fraction, 222
Mode, 408
Modular arithmetic, 320
Modulo, 320
Monte Carlo methods, 382
Motion geometry, 355

Motion, rigid, 366
Multiple, 106
Multiplication, associative property of, 112
of integers, 159–160
lefthand method of, 127
by the long method, 130
of rational numbers, 255–258
righthand method of, 127
by the Russian peasant method, 128
in terms of Cartesian product of sets, 104–105
Multiplication–addition principle, 114
Multiplication properties of inequalities, 163, 170
Multiplication properties of 0 and 1, 113
Multiplicative inverse, 256

Natural number, 12
Negative integer, 150
Negative rational number, 215, 233
Negative whole number, 150
Nonnegative integer, 156
Nonterminating decimal, 226
Number, abundant, 427
amicable, 431
counting, 12
deficient, 427
even, 9
irrational, 217
line, 9
natural, 12
negative, 149
perfect, 427
positive whole, 150
prime, 138
rational, 215
real, 217
triangular, 431
whole, 9
Numerals, 13
Numerals, Hindu-Arabic, 58
Numerator, 222

One, multiplicative property of, 113
One-to-one correspondence, 7
One-to-one function, 337
Open curve, 15
Operation, binary, 317
Order properties, 26
Ordered pair, 105

462 **Index**

Parallel lines, 191
Parallel movement, 356
Parallelogram, 288
Parallelopiped, rectangular, 189
Partial sums, 79
Partition of a set, 108
Percent, 239
Perfect number, 427
Perimeter, 283
Piaget, 1
Pictogram, 404
Pie chart, 404
Place value, 58
Plane, 189
   Cartesian, 340
   coordinate, 340
Point, 189
   turning, 358
Polygons, similar, 369
Polyhedron, 191
Positive integer, 150
Positive rational number, 215
Positive whole number, 150
Pre-image, 357
Prime, factorization, 138
   Mersenne, 427
   number, 138
Probability, definition of, 387–390
Probability, experimental, 381–382
Proper divisor, 427
Proper fraction, 222
Proper subset, 10
Proportion, 268
Protractor, 297
Pythagorean theorem, 428

Rabbits, addition, 39–40
   breeding of, 432
   multiplication, 117–118
Radius, 284
Range of a function, 333
Ratio, 267
Ratio of similitude, 269
Rational numbers, 215–238, 249–265, 270
   addition of, 249–251
   division of, 260–264
   multiplication of, 255–258
   negative, 215, 233
   positive, 215
   subtraction of, 252–254

Ray, 14
Real number, 217
Reciprocal, 258
Rectangle, golden, 435
Rectangular parallelopiped, 189
Recursion formulas, 432
Reflections, 362–364
Reflection symmetry, 368
Reflexive property of matching, 7
Relation, 339
Relative complement, 31
Relatively prime, 139
Remainder, 109
Repeating decimal, 226
Right circular cone, 189
Righthand method of multiplication, 127
Rigid motion, 366
Rotational symmetry, 369
Rotation, center of, 358
Rotations, 358–360
Russian peasant method of multiplication, 128

Sample space, 383–384
Second, 297
Sets, Cartesian product of, 105
   concept of, 2
   disjoint, 24
   equal, 3
   equivalent, 7
   finite, 9, 11
   infinite, 9, 11
   intersection of, 29
   matching, 7
   partition of, 108
   solution, 12
   union of, 22
Similar figures, 179
Similar polygons, 269
Similitude, ratio of, 269
Simple curve, 19
Solution set, 12
Sphere, 189
Square centimeter, meter, kilometer, 290
Statistical graphs, 402–406
Stick bundles, 59
Subset, 3
Subtraction, with abacus, 81–82
   by Austrian method, 83–84
   by borrowing method, 82–83

by equal additions method, 83–84
of integers, 155–156
by method of complements. 73–74
of rational numbers, 252–254
as related to addition, 33
with stick bundles, 80–81
Sure event, 386
Symmetric property of matching, 7
Symmetries, folding, 368
  reflection, 368
  rotational, 369

Terminating decimals, 226
Tessellations, 182–185
Transformation geometry, 355–369
Transitive property of inequalities, 163,
  234
Transitive property of matching, 356–357
Translations, 356–357
Transversal, 191
Trapezoid, 288

Tree-diagram, 392
Triangular numbers, 431
Trichotomy property of inequalities, 163,
  234
Tsimchian number words, 12
Turning movements, 358
Turning point, 358

Union of sets, 22
Unit of length, 281

Venn diagram, 4
Vertex of a polyhedron, 191
Vertex of a tessellation, 183
Vulgar fraction, 223

Whole number, 9

Zero, additive property of, 25
Zero, multiplicative property of, 113